Welcome to the 70th International Symposium on Molecular Spectroscopy June 22-26, 2015 Urbana-Champaign, IL

On behalf of the Executive Committee, I extend a heartfelt welcome to all the attendees of the 70th Symposium and welcome you to the University of Illinois at Urbana-Champaign.

The Symposium presents research in fundamental molecular spectroscopy and a wide variety of related fields and applications. The continued vitality and significance of spectroscopy is annually re-affirmed by the number of talks, their variety, and the fact that many are given by students. These presentations are the heart of the meeting and are documented by this Abstract Book. Equally important is the information flowing from informal exchanges and discussions. As organizers, we strive to provide an environment that facilitates both kinds of interactions.

The essence of the meeting lies in the scientific discussions and your personal experiences this week independent of the number of times that you have attended this meeting. It is our sincere hope that you will find this meeting informative and enjoyable both scientifically and personally, whether it is your first or 50th meeting. If we can help to enhance your experience, please do not hesitate to ask the Symposium staff or the Executive Committee.

Ben McCall
Symposium Chair

SCHEDULE OF TALKS

Monday (M) 1
Tuesday (T) 7
Wednesday (W) 17
Thursday (R) 23
Friday (F) 33

ABSTRACTS

Monday (M) 38
Tuesday (T) 72
Wednesday (W) 126
Thursday (R) 160
Friday (F) 215

AUTHOR INDEX 238

VENUE AND SPONSOR INFORMATION FOLLOWS AUTHOR INDEX

70th INTERNATIONAL SYMPOSIUM ON MOLECULAR SPECTROSCOPY

Please send correspondence to
Ben McCall
International Symposium on Molecular Spectroscopy
Department of Chemistry
600 S. Mathews Avenue
Champaign IL 61801 USA
e-mail: chair@isms.illinois.edu
http://isms.illinois.edu

Mini-Symposia

ACCELERATOR-BASED SPECTROSCOPY

Organized by **Jos Oomens** (Radboud University, Nijmegen) and **Jennifer van Wijngaarden** (University of Manitoba). This mini-symposium will highlight spectroscopic techniques and the latest results from research conducted at synchrotron and free electron laser facilities. Invited Speakers: **Robert Georges** (Université de Rennes), **Terry McMahon** (University of Waterloo), **Gaël Mouret** (Université du Littoral Côte d'Opale), and **Gert von Helden** (Fritz-Haber-Institut, Berlin, Germany)

SPECTROSCOPY IN THE CLASSROOM

Organized by **Stephen Cooke** (Purchase College, SUNY) and **Andrea Minei** (College of Mount St. Vincent). This mini-symposium will focus on pedagogic innovation in molecular spectroscopy including affordable experiments, teacher resources, misused scientific terms, and perspectives on education in our field. Invited Speakers: **Geoffrey A. Blake** (California Institute of Technology) and **Colin Western** (University of Bristol)

HIGH-PRECISION SPECTROSCOPY

Organized by **Mike Heaven** (Emory University) and **Trevor Sears** (Brookhaven National Lab). This mini-symposium will cover all aspects of frequency comb generation, metrology and precision spectroscopy. Invited Speakers: **Thomas K. Allison** (Stony Brook University), **Ian Coddington** (National Institute of Standards and Technology), **David Long** (National Institute of Standards and Technology), **Hiroyuki Sasada** (Keio University), and **Wim Ubachs** (VU University Amsterdam)

Picnic

The Symposium picnic will be held on Wednesday evening at Ikenberry Commons. The cost of the picnic is included in your registration (at below cost to students), so that all may attend the event. The **Coblentz Society** is the host for refreshments for one hour starting at 6:15 PM. Food will be served starting at 7:00 PM.

Sponsorship

We are pleased to acknowledge the many organizations that support the 70th Symposium. Principal funding comes from the **Army Office of Research** (ARO). We are most grateful to ARO for their long-standing support. We also acknowledge the many efforts and contributions of **The University of Illinois** in hosting the meeting, including financial contributions from the Office of the Vice Chancellor for Research and the Departments of Chemistry, Electrical and Computer Engineering, Astronomy, Physics, and Mechanical Science and Engineering.

Our Corporate Sponsors are **Bristol Instruments, Elsevier/JMS, Ideal Vacuum Products, Journal of Physical Chemistry A, Newport/Spectra-Physics, Quantel, and Virginia Diodes**. Please see the back of this book for their advertisements.

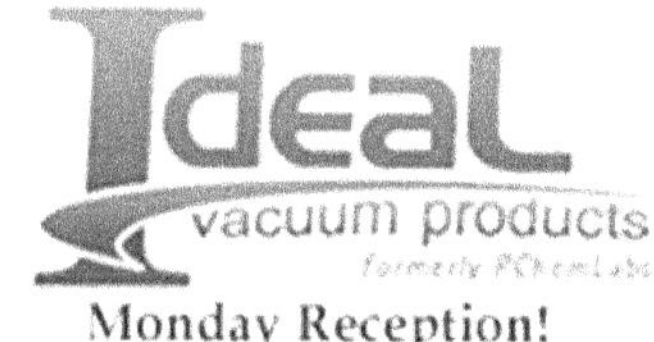

Newport

We are also pleased to acknowledge **Agilent Technologies, Andor Technology, JASCO, and Menlo Systems** as Contributing Sponsors.

IOS Press has a special inserts in your conferee packet. Our sponsors will have exhibits at the Symposium and we encourage you to visit their displays.

Rao Prize

The three Rao Prizes for the most outstanding student talks at the 2014 meeting will be presented. The winners are **Grant Buckingham**, University of Colorado; **Kathryn Chew**, Yale University; and **Yu-Hsuan Huang**, National Chiao Tung University. The Rao Prize was created by a group of spectroscopists who, as graduate students, benefited from the emphasis on graduate student participation, which has been a unique characteristic of the Symposium. This year three more Rao Prize winners will be selected.

The award is administered by a Prize Committee chaired by Gary Douberly, University of Georgia, and comprised of David Anderson, University of Wyoming; Brooks Pate, University of Virginia; Rebecca Peebles, Eastern Illinois University; Jennifer van Wijngaarden, University of Manitoba; and Tim Zwier, Purdue University. Any questions or suggestions about the Prize should be addressed to the Committee. Anyone (especially post-docs) willing to serve on a panel of judges should contact Gary Douberly (douberly@uga.edu).

Miller Prize

The Miller Prize was created in honor of Professor Terry A. Miller, who served as chair of the International Symposium on Molecular Spectroscopy from 1992 to 2013.

The Miller Prize for the best presentation given by a recent PhD at the 2014 meeting will be presented. The winner, **Christopher Leavitt** (University of Georgia), will give a lecture on Wednesday.

The Miller Prize winner and his or her co-authors will be invited to submit an article to the Journal of Molecular Spectroscopy based on the research in the prize-winning talk. After passing the normal review process, the article will appear in the Journal with a caption identifying the paper with the talk that received the Miller Prize.

The award is administered by a Prize Committee chaired by Michael Heaven, Emory University and comprised of Frank De Lucia, The Ohio State University; Jinjun Liu, University of Louisville; David Perry, The University of Akron; Cristina Puzzarini, University of Bologna; Scott Reid, Marquette University; Trevor Sears, Brookhaven National Laboratory; Jaime Stearns, Air Force Research Laboratory; Tim Steimle, Arizona State University; and Susanna Widicus Weaver, Emory University. Any questions or suggestions about the Prize should be addressed to the Committee. Anyone willing to serve on a panel of judges should contact Michael Heaven (mheaven@emory.edu).

Information

ACCOMMODATIONS

The check-in for dormitory accommodations is located in Bousfield Hall, 1214 South First Street, opens at noon on Sunday, June 21st, and remains open 24 hours a day through the Symposium. Hotel information is listed on the ISMS website.

PARKING

Parking permits are for lot E14 (see map at end of book). Please purchase parking as part of your check-in process at the dorm. If you need to purchase meter hang-tags for parking near the meeting rooms, you can do so at the registration desk.

REGISTRATION

The registration desk is located in Room 165 Noyes Lab, and is open on Sunday from 4:00-6:00 PM, and Monday through Friday from 8:00 AM-4:30 PM. Refreshments will be available from 8:00 AM-4:30 PM.

CHEMISTRY LIBRARY

The Chemistry Library will be open and available for your use during Symposium hours. The library has a number of computers, desks and tables to work at, and comfy chairs (and books!).

READY ROOM/STATION

We have set up Noyes Lab 164 as a "Ready Room" with computers that you can use to test your powerpoint presentation. If you have any problems, we will also have a staffed "Ready Station" in Noyes Lab 165 (right next to registration) where you can come for assistance.

COMPUTER LAB (VizLab)

Noyes Lab 151 is a small computer lab with Apple computers that is available for your use during the meeting. Please look in your packet for an access code to enter the room.

INTERNET ACCESS/Wi-Fi

Each attendee will receive a login and password to access campus WiFi (SSID: IllinoisNet or UIUCnet) as a guest. This access should work in most locations through campus. Please read the Internet Acceptable Use Policy below.

AUDIO/VIDEO INFORMATION

Each session room is equipped with a computer, onto which presentation files will be pre-loaded by Symposium staff. To submit your presentation file, you must go to the **Manage Presentations** link on our web site and follow the instructions. All files must be submitted by **11:59 PM CDT THE DAY BEFORE** your presentation session. All submitted files will be loaded onto the presentation computer one half-hour prior to the beginning of the session.

ACKNOWLEDGMENTS

The Symposium Chair wishes to acknowledge the hard work of numerous people who made this meeting possible. First and foremost is the Symposium Coordinator Birgit McCall, who has smoothly and single-handedly taken care of almost all of the electronic and logistical aspects of the meeting. Second are our student assistants, Brad Gibson, Charlie Markus, and Scott Dubowsky, who have handled innumerable important details to ensure the sessions and exhibitions go well. The other students in my group also play vital roles in monitoring the audiovisual systems and other aspects of the meeting. I wish to acknowledge the hospitality of the Chemistry Department and the School of Chemical Sciences (as well as the School of Molecular and Cell Biology) in tolerating our takeover of their buildings.

LIABILITY

The Symposium fees DO NOT include provisions for the insurance of participants against personal injuries, sickness, theft, or property damage. Participants and companions are advised to obtain whatever insurance they consider necessary. The Symposium organizing committee, its sponsors, and individual committee members DO NOT assume any responsibility for loss, injury, sickness, or damages to persons or belongings, however caused. The statements and opinions stated during oral presentations or in written abstracts are solely the author's responsibilities and do not necessarily reflect the opinions of the organizers.

DISCLAIMER

The views, opinions, and/or findings contained in this report are those of the authors and should not be construed as an official Department of the Army position, policy, or decision, unless so designated by other documentation.

INTERNET ACCEPTABLE USE POLICY

Each attendee will receive a login and password to access campus WiFi (SSID: IllinoisNet or UIUCnet) as a guest. Guest accounts are intended to support a broad range of communications. Professional and appropriate etiquette is required. Anonymous access and posting through guest accounts is forbidden. All users must accept that their identity may be associated with any content they provide while using the service. By accessing the campus WiFi network, you expressly acknowledge and agree to the following:

Use of the guest account service is at your sole risk and the entire risk as to satisfactory quality and performance is with you. You agree not to use the guest account intentionally or unintentionally to violate any applicable local, state, national or international law, including, but not limited to, any regulations having the force of law. To the extent not prohibited by law, in no event shall the university be liable for personal injury, or any incidental, special, indirect or consequential damages whatsoever, including, without limitation, damages for loss of profits, loss of data, business interruption or any other commercial damages or losses, arising out of or related to your use or inability to use the guest account, however caused, regardless of the theory of liability (contract, tort or otherwise) and even if the university has been advised of the possibility of such damages. The use of the guest account is subject, but not limited to, all University policies and regulations detailed at the Campus Administrative Manual (http://www.cam.illinois.edu). See the University's Web Privacy Notice (http://www.vpaa.uillinois.edu/policies/web_privacy.cfm) for all applicable laws and policies.

MA. Plenary

Monday, June 22, 2015 – 8:30 AM

Room: Foellinger Auditorium

Chair: Gregory S. Girolami, University of Illinois at Urbana-Champaign, Urbana, IL, USA

Welcome **8:30**

Phyllis M. Wise, Chancellor

University of Illinois at Urbana-Champaign

MA01 **8:40 – 9:20**
BREATHING EASIER THROUGH SPECTROSCOPY: STUDYING FREE RADICAL REACTIONS IN AIR POLLUTION CHEMISTRY, Mitchio Okumura

MA02 **9:25 – 10:05**
MOLECULAR ROTATION SIGNALS: MOLECULE CHEMISTRY AND PARTICLE PHYSICS, Jens-Uwe Grabow

Intermission

MA03 **10:35 – 11:15**
IT IS WATER WHAT MATTERS: THz SPECTROSCOPY AS A TOOL TO STUDY HYDRATION DYNAMICS, Martina Havenith

MA04 **11:20 – 12:00**
CPUF: CHIRPED-PULSE MICROWAVE SPECTROSCOPY IN PULSED UNIFORM SUPERSONIC FLOWS, Arthur Suits, Chamara Abeysekera, Lindsay N. Zack, Baptiste Joalland, Nuwandi M Ariyasingha, Barratt Park, Robert W Field, Ian Sims

MF. Mini-symposium: High-Precision Spectroscopy

Monday, June 22, 2015 – 1:30 PM

Room: 116 Roger Adams Lab

Chair: Michael Heaven, Emory University, Atlanta, GA, USA

MF01 *Journal of Molecular Spectroscopy Review Lecture* **1:30 – 2:00**
PHYSICS BEYOND THE STANDARD MODEL FROM MOLECULAR HYDROGEN SPECTROSCOPY, Wim Ubachs, Edcel John Salumbides, Julija Bagdonaite

MF02 **2:05 – 2:20**
PRECISION SPECTROSCOPY ON HIGHLY-EXCITED VIBRATIONAL LEVELS OF H_2, Ming Li Niu, Edcel John Salumbides, Wim Ubachs

MF03 **2:22 – 2:37**
BOUNDS ON THE NUMBER AND SIZE OF EXTRA DIMENSIONS FROM MOLECULAR SPECTROSCOPY, Edcel John Salumbides, Bert Schellekens, Beatriz Gato-Rivera, Wim Ubachs

MF04 **2:39 – 2:54**
CONTINUOUS SUPERSONIC EXPANSION DISCHARGE SOURCE FOR HIGH-PRECISION MID-INFRARED SPECTROSCOPY OF COLD MOLECULAR IONS, Courtney Talicska, Michael Porambo, Benjamin J. McCall

MF05 **2:56 – 3:11**
PROGRESS TOWARDS A HIGH-PRECISION INFRARED SPECTROSCOPIC SURVEY OF THE H_3^+ ION, Adam J. Perry, James N. Hodges, Charles R. Markus, G. Stephen Kocheril, Paul A. Jenkins II, Benjamin J. McCall

Intermission

MF06 **3:30 – 3:45**
HIGH PRECISION INFRARED SPECTROSCOPY OF OH^+, Charles R. Markus, Adam J. Perry, James N. Hodges, G. Stephen Kocheril, Paul A. Jenkins II, Benjamin J. McCall

MF07 **3:47 – 4:02**
TOWARD TWO-COLOR SUB-DOPPLER SATURATION RECOVERY KINETICS IN CN (X, v = 0, J), Hong Xu, Damien Forthomme, Trevor Sears, Gregory Hall, Paul Dagdigian

MF08 **4:04 – 4:19**
AN EMPIRICAL DIPOLE POLARIZABILITY FOR He FROM A FIT TO SPECTROSCOPIC DATA YIELDING ANALYTIC EMPIRICAL POTENTIALS FOR ALL ISOTOPOLOGUES OF HeH^+ , Young-Sang Cho, Robert J. Le Roy, Nikesh S. Dattani

MF09 **4:21 – 4:36**
ANALYTIC EMPIRICAL POTENTIALS FOR BeH^+, BeD^+, AND BeT^+ INCLUDING UP TO 4TH ORDER QED IN THE LONG-RANGE, AND PREDICTIONS FOR THE HALO NUCLEONIC MOLECULES $^{11}BeH^+$ and $^{14}BeH^+$., Lena C. M. Li Chun Fong, Grzegorz Lach, Robert J. Le Roy, Nikesh S. Dattani

MF10 **4:38 – 4:53**
ANALYTIC EMPIRICAL POTENTIAL AND ITS COMPARISON TO STATE OF THE ART *ab initio* CALCULATIONS FOR THE $6e^-$ EXCITED $b(1^3\Pi_u)$-STATE OF Li_2., Nikesh S. Dattani, Robert J. Le Roy

MF11 **4:55 – 5:10**
PRECISION SPECTROSCOPY OF TRAPPED HfF^+ WITH A COHERENCE TIME OF 1 SECOND, Kevin Cossel, William Cairncross, Matt Grau, Dan Gresh, Yan Zhou, Jun Ye, Eric Cornell

MF12 **5:12 – 5:27**
BROADBAND FREQUENCY COMB AND CW-LASER VELOCITY MODULATION SPECTROSCOPY OF ThF^+, Dan Gresh, Kevin Cossel, Jun Ye, Eric Cornell

MF13 **5:29 – 5:44**
PURE MW DATA FOR $v = 0 - 6$ OF PbI GIVE VIBRATIONAL SPACINGS AND A FULL ANALYTIC POTENTIAL ENERGY FUNCTION, Ji Ho (Chris) Yoo, Corey Evans, Nick Walker, Robert J. Le Roy

MG. Structure determination

Monday, June 22, 2015 – 1:30 PM

Room: 100 Noyes Laboratory

Chair: Ha Vinh Lam Nguyen, Université Paris-Est Créteil, Créteil, France

MG01 **1:30 – 1:45**
DETECTION OF HSNO, A CRUCIAL INTERMEDIATE LINKING NO AND H_2S CHEMISTRIES, Marie-Aline Martin-Drumel, Carrie Womack, Kyle N Crabtree, Sven Thorwirth, Michael C McCarthy

MG02 **1:47 – 2:02**
DETECTION AND STRUCTURAL CHARACTERIZATION OF NITROSAMIDE H_2NNO: A CENTRAL INTERMEDIATE IN deNO$_x$ PROCESSES, Michael C McCarthy, Kelvin Lee, John F. Stanton

MG03 **2:04 – 2:19**
MICROWAVE SPECTRA OF 1- AND 2-BROMOBUTANE, Soohyun Ka, Jihyun Kim, Heesu Jang, JUNG JIN Oh

MG04 **2:21 – 2:36**
ACCURATE EQUILIBRIUM STRUCTURES FOR *trans*-HEXATRIENE BY THE MIXED ESTIMATION METHOD AND FOR THE THREE ISOMERS OF OCTATETRAENE FROM THEORY; STRUCTURAL CONSEQUENCES OF ELECTRON DELOCALIZATION, Norman C. Craig, Jean Demaison, Peter Groner, Heinz Dieter Rudolph, Natalja Vogt

MG05 **2:38 – 2:53**
RING PUCKERING POTENTIALS OF THREE FLUORINATED CYCLOPENTENES: C_5F_8, C_5HF_7, and $C_5H_2F_6$, E. A. Arsenault, B. E. Long, Wallace C. Pringle, Yoon Jeong Choi, S. A. Cooke, Esther J Ocola, Jaan Laane

MG06 **2:55 – 3:10**
CONFORMATIONAL TRANSFORMATION OF FIVE-MEMBERED RINGS: THE GAS PHASE STRUCTURE OF 2-METHYLTETRAHYDROFURAN, Vinh Van, Ha Vinh Lam Nguyen, Wolfgang Stahl

MG07 **3:12 – 3:27**
ASSIGNMENT OF THE MICROWAVE SPECTRUM OF 1,2-DIFLUOROBENZENE$\cdots$HCCH: LESSONS LEARNED FROM ANALYSIS OF A DENSE BROADBAND SPECTRUM, Anuradha Akmeemana, Rebecca D. Nelson, Mikayla L. Grant, Rebecca A. Peebles, Sean A. Peebles, Justin M. Kang, Nathan A Seifert, Brooks Pate

MG08 **3:29 – 3:39**
STRUCTURE DETERMINATION AND CH$\cdots$F INTERACTIONS IN $H_2C{=}CHF\cdots H_2C{=}CF_2$ BY FOURIER-TRANSFORM MICROWAVE SPECTROSCOPY, Rachel E. Dorris, Rebecca A. Peebles, Sean A. Peebles

Intermission

MG09 **3:58 – 4:13**
MILLIMETER WAVE SPECTROSCOPY AND EQUILIBRIUM STRUCTURE DETERMINATION OF PYRIMIDINE (*m*-$C_4H_4N_2$), Zachary N. Heim, Brent K. Amberger, Brian J. Esselman, R. Claude Woods, Robert J. McMahon

MG10 **4:15 – 4:30**
MILLIMETER-WAVE SPECTROSCOPY OF PHENYL ISOCYANATE, Cara E. Schwarz, Brent K. Amberger, Benjamin C. Haenni, Brian J. Esselman, R. Claude Woods, Robert J. McMahon

MG11 **4:32 – 4:47**
BROADBAND MICROWAVE SPECTROSCOPY AS A TOOL TO STUDY THE STRUCTURES OF ODORANT MOLECULES AND WEAKLY BOUND COMPLEXES IN THE GAS PHASE , Sabrina Zinn, Thomas Betz, Chris Medcraft, Melanie Schnell

MG12 **4:49 – 5:04**
MICROWAVE SPECTRA OF 9-FLUORENONE AND BENZOPHENONE, Channing West, Galen Sedo, Jennifer van Wijngaarden

MG13 **5:06 – 5:21**
ASSESSING THE IMPACT OF BACKBONE LENGTH AND CAPPING AGENT ON THE CONFORMATIONAL PREFERENCES OF A MODEL PEPTIDE: CONFORMATION SPECIFIC IR AND UV SPECTROSCOPY OF 2-AMINOISOBUTYRIC ACID, Joseph R. Gord, Daniel M. Hewett, Matthew A. Kubasik, Timothy S. Zwier

MG14 **5:23 – 5:38**
COMPARISON OF INTRAMOLECULAR FORCES IN DIPEPTIDES WITH TWO AROMATIC RINGS: DOES DISPERSION DOMINATE?, Jessica A. Thomas

MH. Linelists

Monday, June 22, 2015 – 1:30 PM

Room: B102 Chemical and Life Sciences

Chair: Shanshan Yu, California Institute of Technology, Pasadena, CA, USA

MH01 **1:30 – 1:45**
HITRAN IN THE XXIst CENTURY: BEYOND VOIGT AND BEYOND EARTH, Laurence S. Rothman, Iouli E Gordon, Christian Hill, Roman V Kochanov, Piotr Wcislo, Jonas Wilzewski

MH02 **1:47 – 2:02**
HITRANonline: A NEW STRUCTURE AND INTERFACE FOR HITRAN LINE LISTS AND CROSS SECTIONS , Christian Hill, Laurence S. Rothman, Iouli E Gordon, Roman V Kochanov, Piotr Wcislo, Jonas Wilzewski

MH03 **2:04 – 2:19**
WORKING WITH HITRAN DATABASE USING HAPI: HITRAN APPLICATION PROGRAMMING INTERFACE, Roman V Kochanov, Christian Hill, Piotr Wcislo, Iouli E Gordon, Laurence S. Rothman, Jonas Wilzewski

MH04 **2:21 – 2:31**
GPU ACCELERATED INTENSITIES: A NEW METHOD OF COMPUTING EINSTEIN-A COEFFICIENTS, Ahmed Faris Al-Refaie, Sergei N. Yurchenko, Jonathan Tennyson

MH05 **2:33 – 2:48**
LINE SHAPE PARAMETERS FOR NEAR INFRARED CO_2 BANDS IN THE 1.61 AND 2.06 MICRON SPECTRAL REGIONS, V. Malathy Devi, D. Chris Benner, Keeyoon Sung, Linda Brown, Timothy J Crawford, Mary Ann H. Smith, Arlan Mantz

MH06 **2:50 – 3:05**
RELIABLE IR LINE LISTS FOR SO_2 AND CO_2 ISOTOPOLOGUES COMPUTED FOR ATMOSPHERIC MODELING ON VENUS AND EXOPLANETS, Xinchuan Huang, David Schwenke, Timothy Lee, Robert R. Gamache

MH07 **3:07 – 3:22**
LASER SPECTROSCOPIC STUDY OF CaH IN THE $B^2\Sigma^+$ AND $D^2\Sigma^+$ STATES, Kyohei Watanabe, Kanako Uchida, Kaori Kobayashi, Fusakazu Matsushima, Yoshiki Moriwaki

Intermission

MH08 **3:41 – 3:56**
ADDITIONAL MEASUREMENTS AND ANALYSES OF $H_2^{17}O$ AND $H_2^{18}O$, John Pearson, Shanshan Yu, Adam Walters

MH09 **3:58 – 4:13**
EXPERIMENTAL LINE LISTS OF HOT METHANE, Robert J. Hargreaves, Peter F. Bernath, Jeremy Bailey, Michael Dulick

MH10 **4:15 – 4:30**
EXPERIMENTAL TRANSMISSION SPECTRA OF HOT AMMONIA IN THE INFRARED, Christopher A. Beale, Robert J. Hargreaves, Michael Dulick, Peter F. Bernath

MH11 **4:32 – 4:47**
HYPERSONIC POST-SHOCK CAVITY RING-DOWN SPECTROSCOPY, Nicolas Suas-David, Samir Kassi, Abdessamad Benidar, Robert Georges

MH12 **4:49 – 5:04**
CH_3D NEAR INFRARED CAVITY RING-DOWN SPECTRUM REANALYSIS AND IR-IR DOUBLE RESONANCE, Shaoyue Yang, George Schwartz, Kevin Lehmann

MH13 **5:06 – 5:21**
AYTY: A NEW LINE-LIST FOR HOT FORMALDEHYDE, Ahmed Faris Al-Refaie, Sergei N. Yurchenko, Jonathan Tennyson, Andrey Yachmenev

MH14 **5:23 – 5:38**
THE MICROWAVE SPECTROSCOPY OF AMINOACETONITRILE IN THE VIBRATIONAL EXCITED STATE, Chiho Fujita, Hiroyuki Ozeki, Kaori Kobayashi

MI. Ions

Monday, June 22, 2015 – 1:30 PM

Room: 274 Medical Sciences Building

Chair: Mark Johnson, Yale University, New Haven, CT, USA

MI01 **1:30 – 1:45**
ROTATIONAL ACTION SPECTROSCOPY VIA STATE-SELECTIVE HELIUM ATTACHMENT, Lars Kluge, Alexander Stoffels, Sandra Brünken, Oskar Asvany, Stephan Schlemmer

MI02 **1:47 – 2:02**
SYMMETRY BEYOND PERTURBATION THEORY: FLOPPY MOLECULES AND ROTATION-VIBRATION STATES, Hanno Schmiedt, Stephan Schlemmer, Per Jensen

MI03 **2:04 – 2:19**
STUDYING ROTATION/TORSION COUPLING IN H_5^+ USING DIFFUSION MONTE CARLO, Melanie L. Marlett, Zhou Lin, Anne B McCoy

MI04 **2:21 – 2:36**
HIGH-J ROTATIONAL LINES OF ^{13}C ISOTOPOLOGUES OF HCO^+ MEASURED BY USING EVENSON-TYPE TUNABLE FIR SPECTROMETER, Mari Suzuki, Ryo Oishi, Yoshiki Moriwaki, Fusakazu Matsushima, Takayoshi Amano

MI05 **2:38 – 2:48**
UV-UV HOLE-BURNING SPECTROSCOPY OF A PROTONATED ADENINE DIMER IN A COLD QUADRUPOLE ION TRAP, Hyuk Kang

MI06 **2:50 – 3:05**
SPECTROSCOPIC INVESTIGATION OF PROTON-COUPLED ELECTRON TRANSFER IN WATER OXIDATION CATALYZED BY A RUTHENIUM COMPLEX, $[Ru(tpy)(bpy)(H_2O)]^{2+}$, Erin M. Duffy, Brett Marsh, Jonathan Voss, Etienne Garand

MI07 **3:07 – 3:22**
PROBING SOLVATAION SHELLS OF $Ni(H_2O)_m{}^{2+}$ (m=4-10) AND $NiOH(H_2O)_n{}^+$ (n=2-5) WITH CRYOGENIC ION VIBRATIONAL SPECTROSCOPY. , Jonathan Voss, Brett Marsh, Jia Zhou, Etienne Garand

MI08 **3:24 – 3:39**
MICROSOLVATION OF THE $Mg_2SO_4{}^{2+}$ CATION: CRYOGENIC VIBRATIONAL SPECTROSCOPY OF $(Mg^{2+})_2SO_4{}^{2-}(H_2O)_{n=4-11}$, Patrick J Kelleher, Joseph W DePalma, Christopher J Johnson, Joseph Fournier, Mark Johnson

Intermission

MI09 **3:58 – 4:13**
CAPTURE AND STRUCTURAL DETERMINATION OF ACTIVATED INTERMEDIATES IN NICKEL CATALYZED CO_2 REDUCTION, Stephanie Craig, Fabian Menges, Arron Wolk, Joseph Fournier, Niklas Tötsch, Mark Johnson

MI10 **4:15 – 4:30**
THRESHOLD IONIZATION SPECTROSCOPIC CHARACTERIZATION OF La ATOM REACTION WITH ISOPRENE, Wenjin Cao, Dong-Sheng Yang

MI11 **4:32 – 4:47**
Ce-PROMOTED BOND ACTIVATION OF ETHYLENE PROBED BY MASS-ANALYZED THRESHOLD IONIZATION SPECTROSCOPY, Yuchen Zhang, Sudesh Kumari, Wenjin Cao, Dong-Sheng Yang

MI12 **4:49 – 5:04**
STRUCTURE DETERMINATION OF CISPLATIN-AMINO ACID ANALOGUES BY INFRARED MULTIPLE PHOTON DISSOCIATION ACTION SPECTROSCOPY, Chenchen He, Xun Bao, Yanlong Zhu, Stephen Strobehn, Bett Kimutai, Y-W Nei, C S Chow, M T Rodgers, Juehan Gao, J. Oomens

MI13 **5:06 – 5:21**
STRUCTUAL EFFECTS OF CYTIDINE 2′ RIBOSE MODIFICATIONS AS DETERMINED BY IRMPD ACTION SPECTROSCOPY, Lucas Hamlow, Chenchen He, Lin Fan, Ranran Wu, Bo Yang, M T Rodgers, Giel Berden, J. Oomens

MI14 **5:23 – 5:38**
GAS-PHASE CONFORMATIONS AND ENERGETICS OF SODIUM CATIONIZED 2′-DEOXYGUANOSINE AND GUANOSINE: IRMPD ACTION SPECTROSCOPY AND THEORETICAL STUDIES, Yanlong Zhu, Lucas Hamlow, Chenchen He, Xun Bao, M T Rodgers, Juehan Gao, J. Oomens

MI15 **5:40 – 5:55**
UNRAVELING PROTON TRANSFER IN STEPWISE HYDRATED N-HETEROCYCLIC ANIONS, John T. Kelly, Nathan I Hammer, Kit Bowen, Gregory S. Tschumper

MJ. Small molecules

Monday, June 22, 2015 – 1:30 PM
Room: 217 Noyes Laboratory

Chair: Leah C O'Brien, Southern Illinois University, Edwardsville, IL, USA

MJ01 **1:30 – 1:45**
DEPERTURBATION ANALYSIS FOR THE $a^3\Pi$ AND $c^3\Sigma^-$ STATES OF C_2, Jian Tang, Wang Chen, Kentarou Kawaguchi

MJ02 **1:47 – 2:02**
HIGH – RESOLUTION LASER SPECTROSCOPY OF THE $A\,^3\Pi_1 \leftarrow X\,^1\Sigma^+$ SYSTEM OF ICl IN 0.7 μm REGION. , Nobuo Nishimiya, Tokio Yukiya, Masao Suzuki, Robert J. Le Roy

MJ03 **2:04 – 2:19**
HIGH RESOLUTION LASER SPECTROSCOPY FOR ABSORPTION TO LEVELS LYING NEAR THE DISSOCIATION LIMIT OF THE $A\,^3\Pi_1$ STATE OF IBr, Tokio Yukiya, Nobuo Nishimiya, Masao Suzuki, Robert J. Le Roy

MJ04 **2:21 – 2:36**
THE NEAR-INFRARED SPECTRUM OF NiCl: ANALYSES OF THE (0,1), (1,0), & (2,1) BANDS OF SYSTEM G AND THE (1,0) BAND OF SYSTEM H, Jack C Harms, Courtney N Gipson, Ethan M Grames, James J O'Brien, Leah C O'Brien

MJ05 **2:38 – 2:53**
ANALYSIS OF EMISSION SPECTRA OF YTTRIUM MONOIODIDE PRODUCED BY THE PHOTODISSOCIATION OF YI_3, Wenting Wendy Chen, Thomas C. Galvin, Thomas J. Houlahan, Jr., J. Gary Eden

MJ06 **2:55 – 3:05**
GENERATION OF VIBRATIONALLY EXCITED HCP FROM A STABLE SYNTHETIC PRECURSOR, Alexander W. Hull, Jun Jiang, Trevor J. Erickson, Carrie Womack, Matthew Nava, Christopher Cummins, Robert W Field

Intermission

MJ07 **3:24 – 3:39**
DPF ANALYSES YIELD FULLY ANALYTIC POTENTIALS FOR THE $B\,^1\Pi_u$ "BARRIER" STATES OF Rb_2 and Li_2 AND AN IMPROVED GROUND-STATE WELL DEPTH FOR Rb_2, Kai Slaughter, Nikesh S. Dattani, Claude S. Amiot, Amanda J. Ross, Robert J. Le Roy

MJ08 **3:41 – 3:56**
LASER SPECTROSCOPY OF THE PHOTOASSOCIATION OF Rb–Ar AND Rb–Kr THERMAL PAIRS: STRUCTURE OF THE Rb–RARE GAS $A^2\Pi_{1/2}$ STATE NEAR THE CLASSICAL LIMIT, Andrey E. Mironov, William Goldshlag, Kyle T Raymond, J. Gary Eden

MJ09 **3:58 – 4:13**
COLLISION-INDUCED ABSORPTION WITH EXCHANGE EFFECTS AND ANISOTROPIC INTERACTIONS: THEORY AND APPLICATION TO $H_2 - H_2$ and $N_2 - N_2$., Tijs Karman, Evangelos Miliordos, Katharine Hunt, Ad van der Avoird, Gerrit Groenenboom

MJ10 ***Post-Deadline Abstract*** **4:15 – 4:30**
PHOTO-DISSOCIATION RESONANCES OF JET-COOLED NO_2 AT THE DISSOCIATION THRESHOLD BY CW-CRDS, CHALLENGING RRKM THEORIES, Patrick Dupré

MJ11 **4:32 – 4:47**
SELF- AND CO_2-BROADENED LINE SHAPE PARAMETERS FOR THE ν_2 AND ν_3 BANDS OF HDO, V. Malathy Devi, D. Chris Benner, Keeyoon Sung, Linda Brown, Arlan Mantz, Mary Ann H. Smith, Robert R. Gamache, Geronimo L. Villanueva

MJ12 **4:49 – 5:04**
DISPERSED FLUORESCENCE SPECTRA OF JET COOLED SiCN, Masaru Fukushima, Takashi Ishiwata

MJ13 **5:06 – 5:21**
INTERNAL FORCE FIELD DETERMINATION OF $\tilde{C}^1\,B_2$ STATE of SO_2, Jun Jiang, Barratt Park, Carrie Womack, Robert W Field

MJ14 **5:23 – 5:38**
MEASUREMENT AND MODELING OF COLD $^{13}CH_4$ SPECTRA FROM 2.1 TO 2.7 μm, Linda Brown, Keeyoon Sung, Timothy J Crawford, Andrei V. Nikitin, Sergey Tashkun, Michael Rey, Vladimir Tyuterev, Mary Ann H. Smith, Arlan Mantz

TA. Metal containing

Tuesday, June 23, 2015 – 8:30 AM

Room: 116 Roger Adams Lab

Chair: Jacob Stewart, Emory University, Atlanta, GA, USA

TA01 **8:30 – 8:45**
BONDING AT THE EXTREME. DETECTION AND CHARACTERIZATION OF THORIUM DIMER, Th_2, Timothy Steimle, Seth Muscarella, Damian L Kokkin

TA02 **8:47 – 9:02**
THE QUINTESSENTIAL BOND OF MODERN SCIENCE. THE DETECTION AND CHARACTERIZATION OF DIATOMIC GOLD SULFIDE, AuS., Damian L Kokkin, Ruohan Zhang, Timothy Steimle, Bradley W Pearlman, Ian A Wyse, Thomas D. Varberg

TA03 **9:04 – 9:19**
LASER SPECTROSCOPY OF RUTHENIUM CONTAINING DIATOMIC MOLECULES: RuH/D AND RuP., Allan G. Adam, Ricarda M. Konder, Nicole M. Nickerson, Colan Linton, D. W. Tokaryk

TA04 **9:21 – 9:36**
OPTICAL ZEEMAN SPECTROSCOPY OF CALCIUM FLUORIDE, CaF., Timothy Steimle, Damian L Kokkin, Jack Delvin, Michael Tarbutt

TA05 **9:38 – 9:53**
ELECTRONIC TRANSITIONS OF YTTRIUM MONOPHOSPHIDE, Allan S.C. Cheung, Biu Wa Li, MAN-CHOR Chan

TA06 **9:55 – 10:10**
ROTATIONALLY RESOLVED SPECTROSCOPY OF THE $B^1\Pi \leftarrow X^1\Sigma^+$ AND $C^1\Sigma^+ \leftarrow X^1\Sigma^+$ ELECTRONIC BANDS OF CaO, Michael Sullivan, Jacob Stewart, Michael Heaven

Intermission

TA07 **10:29 – 10:44**
HIGH RESOLUTION LASER SPECTROSCOPY OF NICKEL MONOBORIDE, NiB, E. S. Goudreau, Colan Linton, D. W. Tokaryk, Allan G. Adam

TA08 **10:46 – 11:01**
MOLECULAR LINE LISTS FOR SCANDIUM AND TITANIUM HYDRIDE USING THE DUO PROGRAM, Lorenzo Lodi, Sergei N. Yurchenko, Jonathan Tennyson

TA09 **11:03 – 11:18**
UV SPECTROSCOPY ON GAS PHASE Cu(I)-BIPYRIDYL COMPLEXES , Shuang Xu, Casey Christopher, J. Mathias Weber

TA10 **11:20 – 11:35**
ANION PHOTOELECTRON SPECTROSCOPY OF NbW^- and $W_2{}^-$, D. Alex Schnepper, Melissa A. Baudhuin, Doreen Leopold, Sean M. Casey

TB. Mini-symposium: Accelerator-Based Spectroscopy

Tuesday, June 23, 2015 – 8:30 AM

Room: 100 Noyes Laboratory

Chair: Jennifer van Wijngaarden, University of Manitoba, Winnipeg, MB, Canada

TB01 *INVITED TALK* **8:30 – 9:00**
JET-COOLED SPECTROSCOPY ON THE AILES INFRARED BEAMLINE OF THE SYNCHROTRON RADIATION FACILITY SOLEIL, Robert Georges

TB02 **9:05 – 9:20**
LOWEST VIBRATIONAL STATES OF ACRYLONITRILE, Zbigniew Kisiel, Marie-Aline Martin-Drumel, Olivier Pirali

TB03 **9:22 – 9:37**
FIR SYNCHROTRON SPECTROSCOPY OF HIGH TORSIONAL LEVELS OF CD_3OH: THE TAU OF METHANOL, Ronald M. Lees, Li-Hong Xu, Brant E Billinghurst

TB04 **9:39 – 9:54**
FAR-INFRARED SYNCHROTRON-BASED SPECTROSCOPY OF PROTON TUNNELLING IN MALONALDEHYDE, E. S. Goudreau, D. W. Tokaryk, Stephen Cary Ross

Intermission

TB05 *INVITED TALK* **10:13 – 10:43**
THE DISCRETE NATURE OF THE COHERENT SYNCHROTRON RADIATION, Stefano Tammaro, Olivier Pirali, P. Roy, Jean François LAMPIN, Gaël DUCOURNEAU, Arnaud Cuisset, Francis Hindle, Gaël Mouret

TB06 **10:48 – 11:03**
LOW-TEMPERATURE COLLISIONAL BROADENING IN THE FAR-INFRARED CENTRIFUGAL DISTORTION SPECTRUM OF CH_4, Vincent Boudon, Jean Vander Auwera, Laurent Manceron, F. Kwabia Tchana, Tony GABARD, Badr AMYAY, Mbaye Faye

TB07 **11:05 – 11:20**
HYDROGEN AND NITROGEN BROADENED ETHANE AND PROPANE ABSORPTION CROSS SECTIONS, Robert J. Hargreaves, Dominique Appadoo, Brant E Billinghurst, Peter F. Bernath

TC. Mini-symposium: Spectroscopy in the Classroom

Tuesday, June 23, 2015 – 8:30 AM

Room: B102 Chemical and Life Sciences

Chair: S. A. Cooke, Purchase College SUNY, Purchase, NY, USA

TC01 *INVITED TALK* **8:30 – 9:00**
PGOPHER IN THE CLASSROOM AND THE LABORATORY, Colin Western

TC02 **9:05 – 9:20**
SPECTROSCOPY FOR THE MASSES, Robert J. Le Roy, Scott Hopkins, William P. Power, Tong Leung, John Hepburn

TC03 **9:22 – 9:37**
RESEARCH AT A LIBERAL ARTS COLLEGE: MAKE SURE YOU HAVE A NET FOR YOUR HIGH WIRE ACT, Mark D. Marshall, Helen O. Leung

TC04 **9:39 – 9:54**
A SPECTROSCOPY BASED P-CHEM LAB, INCLUDING A DETAILED TEXT AND LAB MANUAL, John Muenter

Intermission

TC05 **10:13 – 10:28**
HOW WE KNOW: SPECTROSCOPY IN THE FIRST YEAR AND BEYOND, Kristopher J Ooms

TC06 **10:30 – 10:40**
EXPANDED CHOICES FOR VIBRATION-ROTATION SPECTROSCOPY IN THE PHYSICAL CHEMISTRY TEACHING LABORATORY, Joel R Schmitz, David A Dolson

TC07 **10:42 – 10:57**
SPECTROSCOPIC CASE-BASED STUDIES IN A FLIPPED QUANTUM MECHANICS COURSE, Steven Shipman

TC08 **10:59 – 11:09**
THE H-ATOM SPECTRUM: NOT A CLASSROOM DEMONSTRATION . . . , Wolfgang Jäger

TC09 **11:11 – 11:26**
RAMAN INVESTIGATION OF TEMPERATURE PROFILES OF PHOSPHOLIPID DISPERSIONS IN THE BIOCHEMISTRY LABORATORY, Norman C. Craig

TC10 ***Post-Deadline Abstract*** **11:28 – 11:43**
ONLINE AND CERTIFIABLE SPECTROSCOPY COURSES USING INFORMATION AND COMMUNICATION TOOLS. A MODEL FOR CLASSROOMS AND BEYOND, Mangala Sunder Krishnan

TD. Conformers, isomers, chirality, stereochemistry

Tuesday, June 23, 2015 – 8:30 AM

Room: 274 Medical Sciences Building

Chair: Emilio J. Cocinero, Universidad del País Vasco (UPV-EHU), Leioa, Spain

TD01 **8:30 – 8:45**
A JOINT THEORETICAL AND EXPERIMENTAL STUDY OF THE SiH_2OO ISOMERIC SYSTEM, Michael C McCarthy, Jürgen Gauss

TD02 **8:47 – 9:02**
A MINTY MICROWAVE MENAGERIE: THE ROTATIONAL SPECTRA OF MENTHONE, MENTHOL, CARVACROL, AND THYMOL, David Schmitz, V. Alvin Shubert, Thomas Betz, Barbara Michela Giuliano, Melanie Schnell

TD03 **9:04 – 9:19**
THE ROTATIONAL SPECTRUM AND CONFORMATIONAL STRUCTURES OF METHYL VALERATE, Ha Vinh Lam Nguyen, Wolfgang Stahl

TD04 **9:21 – 9:36**
ROTATIONAL SPECTRUM OF THE METHYL SALICYLATE-WATER COMPLEX: THE MISSING CONFORMER AND THE TUNNELING MOTIONS, Supriya Ghosh, Javix Thomas, Yunjie Xu, Wolfgang Jäger

TD05 **9:38 – 9:53**
UNRAVELLING THE CONFORMATIONAL LANDSCAPE OF NICOTINOIDS: THE STRUCTURE OF COTININE BY BROADBAND ROTATIONAL SPECTROSCOPY, Iciar Uriarte, Patricia Ecija, Emilio J. Cocinero, Cristobal Perez, Elena Caballero-Mancebo, Alberto Lesarri

TD06 **9:55 – 10:10**
CONFORMATIONALLY RESOLVED STRUCTURES OF JET-COOLED PHENACETIN AND ITS HYDRATED CLUSTERS, Cheol Joo Moon, Ahreum Min, Ahreum Ahn, Myong Yong Choi

TD07 **10:12 – 10:27**
CONFORMATIONAL STRUCTURES OF JET-COOLED ACETAMINOPHEN-WATER CLUSTERS BY IR-DIP SPECTROSCOPY AND COMPUTATIONAL CALCULATIONS, Ahreum Min, Ahreum Ahn, Cheol Joo Moon, Myong Yong Choi

Intermission

TD08 **10:46 – 11:01**
THE INHERENT CONFORMATIONAL PREFERENCES OF GLUTAMINE-CONTAINING PEPTIDES: THE ROLE FOR SIDE-CHAIN BACKBONE HYDROGEN BONDS, Patrick S. Walsh, Carl McBurney, Samuel H. Gellman, Timothy S. Zwier

TD09 **11:03 – 11:18**
APPLICATIONS OF STRUCTURAL MASS SPECTROMETRY TO METABOLOMICS: CLARIFYING BOND SPECIFIC SPECTRAL SIGNATURES WITH ISOTOPE EDITED SPECTROSCOPY, Olga Gorlova, Conrad T. Wolke, Joseph Fournier, Sean Colvin, Mark Johnson, Scott Miller

TD10 **11:20 – 11:35**
ALKALI METAL-GLUCOSE INTERACTION PROBED WITH INFRARED PRE-DISSOCIATION SPECTROSCOPY, Steven J. Kregel, Brett Marsh, Jia Zhou, Etienne Garand

TD11 **11:37 – 11:52**
PROBING THE CONFORMATIONAL LANDSCAPE OF A POLYETHER BUILDING BLOCK BY RAMAN JET SPECTROSCOPY, Sebastian Bocklitz, Martin A. Suhm

TE. Instrument/Technique Demonstration

Tuesday, June 23, 2015 – 8:30 AM

Room: 217 Noyes Laboratory

Chair: Ken Leopold, University of Minnesota, Minneapolis, MN, USA

TE01 **8:30 – 8:45**
ELIMINATION OF THE VACUUM PUMP REQUIREMENT FOR HIGH-RESOLUTION ROTATIONAL SPECTROSCOPY., Jennifer Holt, Ryan W Daly, Christopher F. Neese, Frank C. De Lucia

TE02 **8:47 – 8:57**
3-D PRINTED SLIT NOZZLES FOR FOURIER TRANSFORM MICROWAVE SPECTROSCOPY, Chris Dewberry, Becca Mackenzie, Susan Green, Ken Leopold

TE03 **8:59 – 9:14**
IMPLEMENTATION OF CMOS MILLIMETER-WAVE DEVICES FOR ROTATIONAL SPECTROSCOPY, Brian Drouin, Adrian Tang, Erich T Schlecht, Adam M Daly, Emily Brageot, Qun Jane Gu, Yu Ye, Ran Shu, M.-C. Frank Chang, Rod M. Kim

TE04 **9:16 – 9:31**
FAST SWEEPING DIRECT ABSORPTION (SUB)MILLIMETER SPECTROSCOPY BASED ON CHIRPED-PULSE TECHNOLOGY, Brian Hays, Steven Shipman, Susanna L. Widicus Weaver

TE05 **9:33 – 9:48**
FAST SWEEPING DOUBLE RESONANCE MICROWAVE-(SUB)MILLIMETER SPECTROSCOPY BASED ON CHIRPED PULSE TECHNOLOGY, Brian Hays, Susanna L. Widicus Weaver, Steven Shipman

Intermission

TE06 **10:07 – 10:22**
ON THE PHASE DEPENDENCE OF DOUBLE-RESONANCE EXPERIMENTS IN ROTATIONAL SPECTROSCOPY, David Schmitz, V. Alvin Shubert, Anna Krin, David Patterson, Melanie Schnell

TE07 **10:24 – 10:39**
MICROWAVE THREE-WAVE MIXING EXPERIMENTS FOR CHIRALITY DETERMINATION: CURRENT STATUS, Cristobal Perez, V. Alvin Shubert, David Schmitz, Chris Medcraft, Anna Krin, Melanie Schnell

TE08 **10:41 – 10:56**
A SEMI-AUTOMATED COMBINATION OF CHIRPED-PULSE AND CAVITY FOURIER TRANSFORM MICROWAVE SPECTROSCOPY, Kyle N Crabtree, Marie-Aline Martin-Drumel, Michael C McCarthy, Sydney A Gaster, Taylor M Hall, Deondre L Parks, Gordon G Brown

TE09 **10:58 – 11:13**
SUBMILLIMETER ABSORPTION SPECTROSCOPY IN SEMICONDUCTOR MANUFACTURING PLASMAS AND COMPARISON TO THEORETICAL MODELS, Yaser H. Helal, Christopher F. Neese, Frank C. De Lucia, Paul R. Ewing, Ankur Agarwal, Barry Craver, Phillip J. Stout, Michael D. Armacost

TE10 **11:15 – 11:30**
CLOUD COMPUTING FOR THE AUTOMATED ASSIGNMENT OF BROADBAND ROTATIONAL SPECTRA: PORTING AUTOFIT TO AMAZON EC2, Aaron C Olinger, Steven Shipman

TF. Mini-symposium: High-Precision Spectroscopy

Tuesday, June 23, 2015 – 1:30 PM

Room: 116 Roger Adams Lab

Chair: Trevor Sears, Brookhaven National Laboratory, Upton, NY, USA

TF01 *INVITED TALK* **1:30 – 2:00**
COMB-REFERENCED SUB-DOPPLER RESOLUTION INFRARED SPECTROMETER, Hiroyuki Sasada

TF02 **2:05 – 2:20**
SUB-DOPPLER RESOLUTION SPECTROSCOPY OF THE FUNDAMENTAL VIBRATION BAND OF HCl WITH A COMB-REFERENCED SPECTROMETER, Kana Iwakuni, Hideyuki Sera, Masashi Abe, Hiroyuki Sasada

TF03 **2:22 – 2:37**
OBSERVATION AND ANALYSIS OF THE A_1-A_2 SPLITTING OF CH_3D, Masashi Abe, Hideyuki Sera, Hiroyuki Sasada

TF04 **2:39 – 2:54**
HIGH RESOLUTION SPECTROSCOPY OF NAPHTHALENE CALIBRATED BY AN OPTICAL FREQUENCY COMB, Akiko Nishiyama, Kazuki Nakashima, Ayumi Matsuba, Masatoshi Misono

TF05 **2:56 – 3:11**
OPTICAL FREQUENCY COMB FOURIER TRANSFORM SPECTROSCOPY WITH RESOLUTION EXCEEDING THE LIMIT SET BY THE OPTICAL PATH DIFFERENCE, Aleksandra Foltynowicz, Lucile Rutkowski, Alexandra C Johanssson, Amir Khodabakhsh, Piotr Maslowski, Grzegorz Kowzan, Kevin Lee, Martin Fermann

TF06 ***Post-Deadline Abstract*** **3:13 – 3:23**
METROLOGY WITH AN OPTICAL FEEDBACK FREQUENCY STABILIZED CRDS, Samir Kassi, Johannes Burkart

Intermission

TF07 *INVITED TALK* **3:42 – 4:12**
CAVITY ENHANCED ULTRAFAST TRANSIENT ABSORPTION SPECTROSCOPY, Thomas K Allison, Melanie Roberts Reber, Yuning Chen

TF08 **4:17 – 4:32**
NOISE-IMMUNE CAVITY-ENHANCED OPTICAL FREQUENCY COMB SPECTROSCOPY, Lucile Rutkowski, Amir Khodabakhsh, Alexandra C Johanssson, Aleksandra Foltynowicz

TF09 **4:34 – 4:49**
A NEW BROADBAND CAVITY ENHANCED FREQUENCY COMB SPECTROSCOPY TECHNIQUE USING GHz VERNIER FILTERING., Jérôme Morville, Lucile Rutkowski, Georgi Dobrev, Patrick Crozet

TF10 **4:51 – 5:06**
A DECADE-SPANNING HIGH-RESOLUTION ASYNCHRONOUS OPTICAL SAMPLING BASED TERAHERTZ TIME-DOMAIN SPECTROMETER, Jacob T Good, Daniel Holland, Ian A Finneran, Brandon Carroll, Marco A. Allodi, Geoffrey Blake

TF11 **5:08 – 5:23**
DOPPLER-LIMITED SPECTROSCOPY WITH A DECADE-SPANNING TERAHERTZ FREQUENCY COMB, Ian A Finneran, Jacob T Good, Daniel Holland, Brandon Carroll, Marco A. Allodi, Geoffrey Blake

TF12 **5:25 – 5:40**
DUAL COMB RAMAN SPECTROSCOPY ON CESIUM HYPERFINE TRANSITIONS-TOWARD A STIMULATE RAMAN SPECTRUM ON CF_4 MOLECULE, Tze-Wei Liu, Yen-Chu Hsu, Wang-Yau Cheng

TG. Large amplitude motions, internal rotation

Tuesday, June 23, 2015 – 1:30 PM

Room: 100 Noyes Laboratory

Chair: Kaori Kobayashi, University of Toyama, Toyama, Japan

TG01 **1:30 – 1:45**
THE BAND OF CH_3CH_2D FROM 770-880 cm^{-1}, Adam M Daly, Brian Drouin, John Pearson, Peter Groner, Keeyoon Sung, Linda Brown, Arlan Mantz, Mary Ann H. Smith

TG02 **1:47 – 2:02**
LOW-TEMPERATURE HIGH-RESOLUTION INFRARED SPECTRUM OF ETHANE-1D, C_2H_5D: ROTATIONAL ANALYSIS OF THE ν_{17} BAND NEAR 805 cm^{-1}using ERHAM., Peter Groner, Adam M Daly, Brian Drouin, John Pearson, Keeyoon Sung, Linda Brown, Arlan Mantz, Mary Ann H. Smith

TG03 **2:04 – 2:19**
MICROWAVE SPECTROSCOPY OF THE EXCITED VIBRATIONAL STATES OF METHANOL, John Pearson, Adam M Daly

TG04 **2:21 – 2:36**
FIRST HIGH RESOLUTION ANALYSIS OF THE ν_{21} BAND OF PROPANE AT 921.4 cm^{-1}: EVIDENCE OF LARGE-AMPLITUDE-MOTION TUNNELLING EFFECTS, Agnes Perrin, F. Kwabia Tchana, Jean-Marie Flaud, Laurent Manceron, Jean Demaison, Natalja Vogt, Peter Groner, Walter Lafferty

TG05 **2:38 – 2:53**
TORSIONAL STRUCTURE IN THE $\tilde{A} - \tilde{X}$ SPECTRUM OF THE CH_3O_2 AND CH_2XO_2 RADICALS, Meng Huang, Anne B McCoy, Terry A. Miller

TG06 **2:55 – 3:10**
UPDATE OF THE ANALYSIS OF THE PURE ROTATIONAL SPECTRUM OF EXCITED VIBRATIONS OF CH_3CH_2CN , Adam M Daly, John Pearson, Shanshan Yu, Brian Drouin, Celina Bermúdez, José L. Alonso

Intermission

TG07 **3:29 – 3:44**
UNUSUAL INTERNAL ROTATION COUPLING IN THE MICROWAVE SPECTRUM OF PINACOLONE, YueYue Zhao, Ha Vinh Lam Nguyen, Wolfgang Stahl, Jon T. Hougen

TG08 **3:46 – 4:01**
THE COMPLETE ROTATIONAL SPECTRUM OF CH_3NCO UP TO 376 GHz, Zbigniew Kisiel, Lucie Kolesniková, José L. Alonso, Ivan Medvedev, Sarah Fortman, Manfred Winnewisser, Frank C. De Lucia

TG09 **4:03 – 4:18**
GAS PHASE CONFORMATIONS AND METHYL INTERNAL ROTATION FOR 2-PHENYLETHYL METHYL ETHER AND ITS ARGON VAN DER WAALS COMPLEX FROM FOURIER TRANSFORM MICROWAVE SPECTROSCOPY, Ranil M. Gurusinghe, Michael Tubergen

TG10 **4:20 – 4:30**
A COMPARISON OF BARRIER TO METHYL INTERNAL ROTATION OF METHYLSTYRENES: MICROWAVE SPECTROSCOPIC STUDY , Ranil M. Gurusinghe, Michael Tubergen

TG11 **4:32 – 4:47**
MICROWAVE SPECTRA AND AB INITIO STUDIES OF THE NE-ACETONE COMPLEX, Jiao Gao, Javix Thomas, Yunjie Xu, Wolfgang Jäger

TG12 **4:49 – 5:04**
THE EFFECTS OF INTERNAL ROTATION AND ^{14}N QUADRUPOLE COUPLING IN N-METHYLDIACETAMIDE, Raphaela Kannengießer, Konrad Eibl, Ha Vinh Lam Nguyen, Wolfgang Stahl

TG13 **5:06 – 5:21**
A NEW HYBRID PROGRAM FOR FITTING ROTATIONALLY RESOLVED SPECTRA OF METHYLAMINE-LIKE MOLECULES: APPLICATION TO 2-METHYLMALONALDEHYDE, Isabelle Kleiner, Jon T. Hougen

TG14 **5:23 – 5:38**
DETERMINATION OF TORSIONAL BARRIERS OF ITACONIC ACID AND N-ACETYLETHANOLAMINE USING CHIRPED-PULSED FTMW SPECTROSCOPY, Josiah R Bailey, Timothy J McMahon, Ryan G Bird, David Pratt

TH. Radicals

Tuesday, June 23, 2015 – 1:30 PM
Room: B102 Chemical and Life Sciences

Chair: Bernadette M. Broderick, University of Georgia, Athens, GA, USA

TH01 **1:30 – 1:45**
AB INITIO SIMULATION OF THE PHOTOELECTRON SPECTRUM FOR METHOXY RADICAL, Lan Cheng, Marissa L. Weichman, Jongjin B. Kim, Takatoshi Ichino, Daniel Neumark, John F. Stanton

TH02 **1:47 – 2:02**
JAHN-TELLER COUPLING IN THE METHOXY RADICAL: INSIGHTS INTO THE INFRARED SPECTRUM OF MOLECULES WITH VIBRONIC COUPLING, Britta Johnson, Edwin Sibert

TH03 **2:04 – 2:19**
RE-EVALUATION OF HO_3 STRUCTURE USING MILLIMETER-SUBMILLIMETER SPECTROSCOPY, Luyao Zou, Brian Hays, Susanna L. Widicus Weaver

TH04 **2:21 – 2:36**
ON THE STARK EFFECT IN OPEN SHELL COMPLEXES EXHIBITING PARTIALLY QUENCHED ELECTRONIC ANGULAR MOMENTUM, Gary E. Douberly, Christopher P. Moradi

TH05 **2:38 – 2:53**
INFRARED LASER SPECTROSCOPY AND AB INITIO COMPUTATIONS OF $OH\cdots(D_2O)_N$ COMPLEXES IN HELIUM NANODROPLETS, Joseph T. Brice, Christopher M. Leavitt, Christopher P. Moradi, Gary E. Douberly, Federico J Hernandez, Gustavo A Pino

TH06 **2:55 – 3:10**
VIBRATIONAL-TORSIONAL COUPLING REVEALED IN THE INFRARED SPECTRUM OF HE-SOLVATED *n*-PROPYL RADICAL, Christopher P. Moradi, Bernadette M. Broderick, Jay Agarwal, Henry F. Schaefer III., Gary E. Douberly

TH07 **3:12 – 3:27**
VIBRONIC SPECTROSCOPY OF HETERO DIHALO-BENZYL RADICALS GENERATED BY CORONA DISCHARGE : JET-COOLED CHLOROFLUOROBENZYL RADICALS, Young Yoon, Sang Lee

TH08 **3:29 – 3:44**
GROWING UP RADICAL: INVESTIGATION OF BENZYL-LIKE RADICALS WITH INCREASING CHAIN LENGTHS, Joseph A. Korn, Khadija M. Jawad, Daniel M. Hewett, Timothy S. Zwier

Intermission

TH09 **4:03 – 4:18**
ANALYSIS OF ROTATIONALLY RESOLVED SPECTRA TO NON-DEGENERATE (a_1'') UPPER-STATE VIBRONIC LEVELS IN THE $\tilde{A}^2E'' - \tilde{X}^2A_2'$ ELECTRONIC TRANSITION OF NO_3, Mourad Roudjane, Terrance Joseph Codd, Ming-Wei Chen, Henry Tran, Dmitry G. Melnik, Terry A. Miller, John F. Stanton

TH10 **4:20 – 4:35**
ANALYSIS OF ROTATIONALLY RESOLVED SPECTRA TO DEGENERATE (e′) UPPER-STATE VIBRONIC LEVELS IN THE $\tilde{A}^2E'' - \tilde{X}^2A_2'$ ELECTRONIC TRANSITION OF NO_3, Henry Tran, Terry A. Miller

TH11 **4:37 – 4:52**
ROVIBRONIC VARIATIONAL CALCULATIONS OF THE NITRATE RADICAL, Bryan Changala, Joshua H Baraban, John F. Stanton

TH12 **4:54 – 5:09**
VIBRONIC STRUCTURE OF THE $\tilde{X}\ ^2A_2'$ STATE OF NO_3 , Masaru Fukushima

TH13 **5:11 – 5:26**
HIGH-RESOLUTION LASER SPECTROSCOPY OF $^{14}NO_3$ RADICAL: VIBRATIONALLY EXCITED STATES OF THE B^2E' STATE, Kohei Tada, Shunji Kasahara, Takashi Ishiwata, Eizi Hirota

TH14 **5:28 – 5:43**
STRUCTURAL CHARACTERIZATION OF HYDROXYL RADICAL ADDUCTS IN AQUEOUS MEDIA, Ireneusz Janik, G. N. R. Tripathi

TI. Dynamics/Kinetics/Ultrafast

Tuesday, June 23, 2015 – 1:30 PM

Room: 274 Medical Sciences Building

Chair: Patrick Vaccaro, Yale University, New Haven, CT, USA

TI01 **1:30 – 1:45**
MULTISCALE SPECTROSCOPY OF DIFFUSING MOLECULES IN CROWDED ENVIRONMENTS, Ahmed A Heikal

TI02 **1:47 – 1:57**
INVESTIGATING THE ROLE OF HUMAN SERUM ALBUMIN ON THE EXCITED STATE DYNAMICS OF INDOCYANINE GREEN USING SHAPED FEMTOSECOND LASER PULSES, Muath Nairat, Arkaprabha Konar, Marie Kaniecki, Vadim V. Lozovoy, Marcos Dantus

TI03 **1:59 – 2:14**
ULTRAFAST SPECTROSCOPIC AND *AB INITIO* COMPUTATIONAL INVESTIGATIONS ON SOLVATION DYNAMICS OF NEUTRAL AND DEPROTONATED TYROSINE., Takashige Fujiwara, Marek Z. Zgierski

TI04 **2:16 – 2:31**
WHICH ELECTRONIC AND STRUCTURAL FACTORS CONTROL THE PHOTOSTABILITY OF DNA AND RNA PURINE NUCLEOBASES?, Marvin Pollum, Christian Reichardt, Carlos E. Crespo-Hernández, Lara Martínez-Fernández, Inés Corral, Clemens Rauer, Sebastian Mai, Philipp Marquetand, Leticia González

TI05 **2:33 – 2:48**
ULTRAFAST DYNAMICS IN DNA AND RNA DERIVATIVES MONITORED BY BROADBAND TRANSIENT ABSORPTION SPECTRSCOPY, Matthew M Brister, Carlos E. Crespo-Hernández

TI06 **2:50 – 3:05**
CAN FEMTOSECOND TRANSIENT ABSORPTION SPECTROSCOPY PREDICT THE POTENTIAL OF SMALL MOLECULES AS PERSPECTIVE DONORS FOR ORGANIC PHOTOVOLTAICS?, Regina DiScipio, Genevieve Sauve, Carlos E. Crespo-Hernández

TI07 **3:07 – 3:22**
MOLECULE-LIKE CdSe NANOCLUSTERS PASSIVATED WITH STRONGLY INTERACTING LIGANDS: ENERGY LEVEL ALIGNMENT AND PHOTOINDUCED ULTRAFAST CHARGE TRANSFER PROCESSES, Yizhou Xie, Meghan B Teunis, Bill Pandit, Rajesh Sardar, Jinjun Liu

TI08 **3:24 – 3:39**
TOWARD THE ACCURATE SIMULATION OF TWO-DIMENSIONAL ELECTRONIC SPECTRA, Angelo Giussani, Artur Nenov, Javier Segarra-Martí, Vishal K. Jaiswal, Ivan Rivalta, Elise Dumont, Shaul Mukamel, Marco Garavelli

Intermission

TI09 **3:58 – 4:13**
ULTRAFAST TERAHERTZ KERR EFFECT SPECTROSCOPY OF LIQUIDS AND BINARY MIXTURES, Marco A. Allodi, Ian A Finneran, Geoffrey Blake

TI10 **4:15 – 4:30**
ULTRAFAST TERAHERTZ KERR EFFECT SPECTROSCOPY OF AROMATIC LIQUIDS, Ian A Finneran, Marco A. Allodi, Geoffrey Blake

TI11 **4:32 – 4:47**
VIBRATIONALLY-RESOLVED KINETIC ISOTOPE EFFECTS IN THE PROTON-TRANSFER DYNAMICS OF GROUND-STATE TROPOLONE, Kathryn Chew, Zachary Vealey, Patrick Vaccaro

TI12 **4:49 – 5:04**
CHARACTERIZATION OF $CHBrCl_2$ PHOTOLYSIS BY VELOCITY MAP IMAGING, W G Merrill, Amanda Case, Benjamin C. Haenni, Robert J. McMahon, Fleming Crim

TI13 **5:06 – 5:21**
REVERSIBILITY OF INTERSYSTEM CROSSING IN THE $\tilde{a}^1A_1(000)$ and $\tilde{a}^1A_1(010)$ STATES OF METHYLENE, CH_2, Anh T. Le, Trevor Sears, Gregory Hall

TI14 **5:23 – 5:38**
EFFICIENT SUPER ENERGY TRANSFER COLLISIONS THROUGH REACTIVE-COMPLEX FORMATION: H + SO_2 , Jonathan M. Smith, Michael J. Wilhelm, Jianqiang Ma, HAI-LUNG Dai

TI15 **5:40 – 5:55**
FOURTH-ORDER VIBRATIONAL TRANSITION STATE THEORY AND CHEMICAL KINETICS, John F. Stanton, Devin A. Matthews, Justin Z Gong

TJ. Rydberg Atoms and Molecules

Tuesday, June 23, 2015 – 1:30 PM

Room: 217 Noyes Laboratory

Chair: Brian DeMarco, University of Illinois, Urbana, IL, USA

TJ01 **1:30 – 1:45**
PRECISION SPECTROSCOPY IN COLD MOLECULES: THE FIRST ROTATIONAL INTERVALS OF He_2^+ BY HIGH-RESOLUTION SPECTROSCOPY AND RYDBERG-SERIES EXTRAPOLATION, Paul Jansen, Luca Semeria, Simon Scheidegger, Frederic Merkt

TJ02 **1:47 – 2:02**
MICROWAVE SPECTROSCOPY OF THE CALCIUM $4snf \rightarrow 4s(n+1)d$, $4sng$, $4snh$, $4sni$, AND $4snk$ TRANSITIONS, Jirakan Nunkaew, Tom Gallagher

TJ03 **2:04 – 2:14**
PHASE DEPENDENCE IN ABOVE THRESHOLD IONIZATION IN THE PRESENCE OF A MICROWAVE FIELD, Vincent Carrat, Eric Magnuson, Tom Gallagher

TJ04 **2:16 – 2:26**
MICROWAVE TRANSITIONS BETWEEN PAIR STATES COMPOSED OF TWO Rb RYDBERG ATOMS, Jeonghun Lee, Tom Gallagher

TJ05 **2:28 – 2:43**
HIGH-RESOLUTION SPECTROSCOPY OF LONG-RANGE MOLECULAR STATES OF $^{85}Rb_2$, Ryan Carollo, Edward E. Eyler, Yoann Bruneau, Phillip Gould, W.C. Stwalley

TJ06 **2:45 – 3:00**
DOUBLE RESONANCE SPECTROSCOPY OF BaF AUTOIONIZING RYDBERG STATES, Timothy J Barnum, David Grimes, Yan Zhou, Robert W Field

TJ07 **3:02 – 3:17**
MILLIMETER WAVE SPECTROSCOPY OF RYDBERG STATES OF MOLECULES IN THE REGION OF 260-295 GHz, David Grimes, Yan Zhou, Timothy J Barnum, Robert W Field

Intermission

TJ08 **3:36 – 3:51**
EFFECTIVE ION-IN-MOLECULE POTENTIALS FOR NON-PENETRATING RYDBERG STATES OF POLAR MOLECULES, Stephen Coy, David Grimes, Yan Zhou, Robert W Field, Bryan M. Wong

TJ09 **3:53 – 4:08**
ELECTRONIC STRUCTURE OF THE X $^1\Sigma^+$ ION CORE OF CaF RYDBERG STATES, Stephen Coy, Joshua H Baraban, David Grimes, Timothy J Barnum, Robert W Field, Bryan M. Wong

TJ10 **4:10 – 4:25**
SYSTEMATICS OF RYDBERG SERIES OF DIATOMIC MOLECULES AND CORRELATION DIAGRAMS, Chun-Woo Lee

TJ11 *INVITED TALK* **4:27 – 4:42**
OBSERVATION OF CS TRILOBITE MOLECULES WITH KILO-DEBYE MOLECULAR FRAME PERMANENT ELECTRIC DIPOLE MOMENTS, James P Shaffer

TJ12 **4:44 – 4:59**
MOLECULE FORMATION AND STATE-CHANGING COLLISIONS OF SINGLE RYDBERG ATOMS IN A BEC, Kathrin Sophie Kleinbach, Michael Schlagmüller, Tara Cubel Liebisch, Karl Magnus Westphal, Fabian Böttcher, Robert Löw, Sebastian Hofferberth, Tilman Pfau, Jesús Pérez-Ríos, C. H. Greene

TJ13 *Post-Deadline Abstract* **5:01 – 5:16**
RYDBERG, VALENCE AND ION-PAIR QUINTET STATES OF O_2, Gabriel J. Vazquez, Hans P. Liebermann, H. Lefebvre-Brion

TJ14 *Post-Deadline Abstract* **5:18 – 5:33**
AB INITIO STUDY OF THE H, J, I, I′ AND I″ $^3\Pi_u$ SUPEREXCITED STATES OF O_2, Gabriel J. Vazquez, Hans P. Liebermann, H. Lefebvre-Brion

WA. Plenary

Wednesday, June 24, 2015 – 8:30 AM

Room: Foellinger Auditorium

Chair: Leslie Looney, University of Illinois, Urbana, IL, USA

RAO AWARDS **8:30**
Presentation of Awards by Yunjie Xu, University of Alberta

2014 Rao Award Winners
Grant Buckingham, University of Colorado
Kathryn Chew, Yale University
Yu-Hsuan Huang, National Chiao Tung University

MILLER PRIZE **8:40**
Introduction by Mike Heaven, Emory University

WA01 **8:45 – 9:00**
INFRARED LASER STARK SPECTROSCOPY OF THE OH···CH_3OH COMPLEX ISOLATED IN SUPERFLUID HELIUM DROPLETS, Christopher M. Leavitt, Joseph T. Brice, Gary E. Douberly, Federico J Hernandez, Gustavo A Pino

FLYGARE AWARDS **9:05**
Introduction by Trevor Sears, Brookhaven National Laboratory

WA02 **9:10 – 9:25**
WHAT CAN WE EXPECT OF HIGH-RESOLUTION SPECTROSCOPIES ON CARBOHYDRATES?, Emilio J. Cocinero, Patricia Ecija, Iciar Uriarte, Imanol Usabiaga, José A. Fernández, Francisco J. Basterretxea, Alberto Lesarri, Benjamin G. Davis

WA03 **9:30 – 9:45**
CONSTRUCTION OF POTENTIAL ENERGY SURFACES FOR THEORETICAL STUDIES OF SPECTROSCOPY AND DYNAMICS, Richard Dawes

Intermission

WA04 **10:35 – 10:50**
MILLIMETER AND SUBMILLIMETER STUDIES OF $O(^1D)$ INSERTION REACTIONS TO FORM MOLECULES OF ASTROPHYSICAL INTEREST, Brian Hays, Nadine Wehres, Bridget Alligood Deprince, Althea A. M. Roy, Jacob Laas, Susanna L. Widicus Weaver

WA05 **10:55 – 11:10**
TERAHERTZ AND INFRARED LABORATORY SPECTROSCOPY IN SUPPORT OF NASA MISSIONS, Shanshan Yu

COBLENTZ AWARD **11:15**
Presentation of Award by Mark Druy, Coblentz Society

WA06 *Coblentz Society Award Lecture* **11:20 – 12:00**
LASER SPECTROSCOPY OF RADICALS, CARBENES, AND IONS IN SUPERFLUID HELIUM DROPLETS, Gary E. Douberly

WF. Mini-symposium: High-Precision Spectroscopy

Wednesday, June 24, 2015 – 1:30 PM

Room: 116 Roger Adams Lab

Chair: Kevin Cossel, JILA - University of Colorado, Boulder, CO, USA

WF01 *INVITED TALK* **1:30 – 2:00**
ULTRASENSITIVE, HIGH ACCURACY MEASUREMENTS OF TRACE GAS SPECIES, David A. Long, Adam J. Fleisher, David F. Plusquellic, Joseph Hodges

WF02 **2:05 – 2:20**
PROBING BUFFER-GAS COOLED MOLECULES WITH DIRECT FREQUENCY COMB SPECTROSCOPY IN THE MID-INFRRARED, Ben Spaun, Bryan Changala, Bryce J Bjork, Oliver H Heckl, David Patterson, John M. Doyle, Jun Ye

WF03 **2:22 – 2:37**
FREQUENCY-AGILE DIFFERENTIAL CAVITY RING-DOWN SPECTROSCOPY, Zachary Reed, Joseph Hodges

WF04 **2:39 – 2:54**
QUANTUM-NOISE-LIMITED CAVITY RING-DOWN SPECTROSCOPY IN THE MID-INFRARED, Adam J. Fleisher, David A. Long, Qingnan Liu, Joseph Hodges

WF05 **2:56 – 3:11**
MOLECULAR LINE PARAMETERS PRECISELY DETERMINED BY A CAVITY RING-DOWN SPECTROMETER, Shui-Ming Hu, Yan Tan, Jin Wang, Yan Lu, Cunfeng Cheng, Yu Robert Sun, An-Wen Liu

WF06 **3:13 – 3:23**
BROADBAND COMB-RESOLVED CAVITY ENHANCED SPECTROMETER WITH GRAPHENE MODULATOR, Kevin Lee, Christian Mohr, Jie Jiang, Martin Fermann, Chien-Chung Lee, Thomas R Schibli, Grzegorz Kowzan, Piotr Maslowski

Intermission

WF07 *INVITED TALK* **3:42 – 4:12**
DUAL-COMB SPECTROSCOPY IN THE OPEN AIR, Greg B Rieker, Andrew Klose, Scott Diddams, Ian Coddington, Fabrizio Giorgetta, Laura Sinclair, Esther Baumann, Gar-Wing Truong, Gabriel Ycas, William C Swann, Nathan R. Newbury

WF08 **4:17 – 4:32**
FREQUENCY-COMB REFERENCED SPECTROSCOPY OF ν_4 AND ν_5 HOT BANDS IN THE $\nu_1+\nu_3$ COMBINATION BAND OF C_2H_2, Sylvestre Twagirayezu, Matthew Cich, Trevor Sears, C. McRaven, Gregory Hall

WF09 **4:34 – 4:49**
LOCAL PERTURBATIONS IN THE (10110) AND (10101) LEVELS OF C_2H_2 FROM FREQUENCY COMB-REFERENCED SPECTROSCOPY, Trevor Sears, Sylvestre Twagirayezu, Damien Forthomme, Gregory Hall, Matthew Cich

WF10 *Post-Deadline Abstract* **4:51 – 5:06**
NOISE-IMMUNE CAVITY-ENHANCED OPTICAL HETERODYNE MOLECULAR SPECTROMETRY MODELLING UNDER SATURATED ABSORPTION, Patrick Dupré

WF11 **5:08 – 5:23**
MAGNETIC SPIN-TORSION COUPLING IN METHANOL, L. H. Coudert, C. Gutle, T. R. Huet, Jens-Uwe Grabow

WF12 **5:25 – 5:40**
SPIN-ROTATION HYPERFINE SPLITTINGS AT MODERATE TO HIGH J VALUES IN METHANOL, Li-Hong Xu, Jon T. Hougen, Sergey Belov, G Yu GOLUBIATNIKOV, Alexander Lapinov, V. Ilyushin, E. A. Alekseev, A. A. Mescheryakov

WG. Mini-symposium: Accelerator-Based Spectroscopy

Wednesday, June 24, 2015 – 1:30 PM

Room: 100 Noyes Laboratory

Chair: J. Oomens, Radboud University, Nijmegen, The Netherlands

WG01 *INVITED TALK* **1:30 – 2:00**
IR SPECTROSCOPY ON PEPTIDES AND PROTEINS AFTER ION MOBILITY SELECTION AND IN LIQUID HELIUM DROPLETS , Gert von Helden

WG02 **2:05 – 2:20**
COMBINING THE POWER OF IRMPD WITH ION-MOLECULE REACTIONS: THE STRUCTURE AND REACTIVITY OF RADICAL IONS OF CYSTEINE AND ITS DERIVATIVES, Michael Lesslie, Sandra Osburn, Giel Berden, J. Oomens, Victor Ryzhov

WG03 **2:22 – 2:37**
FAR-IR ACTION SPECTROSCOPY OF AMINOPHENOL AND ETHYLVANILLIN: EXPERIMENT AND THEORY, Vasyl Yatsyna, Raimund Feifel, Vitali Zhaunerchyk, Daniël Bakker, Anouk Rijs

WG04 **2:39 – 2:54**
OPPORTUNITIES FOR GAS-PHAS MOLECULAR SPECTROSCOPY ON THE VLS-PGM BEAMLINE AT THE CANADIAN LIGHT SOURCE, Michael A MacDonald

WG05 **2:56 – 3:11**
THERMAL DECOMPOSITION OF C_7H_7 RADICALS; BENZYL, TROPYL, AND NORBORNADIENYL, Grant Buckingham, Barney Ellison, John W Daily, Musahid Ahmed

Intermission

WG06 **3:30 – 3:40**
NONDIPOLE EFFECTS IN CHIRAL SYSTEMS MEASURED WITH LINEARLY POLARIZED LIGHT, K P Bowen, O Hemmers, R Guillemin, W C Stolte, M N Piancastelli, D W Lindle

WG07 **3:42 – 3:57**
APPLICATIONS OF THE VUV FOURIER TRANSFORM SPECTROMETER AT SYNCHROTRON SOLEIL , Nelson de Oliveira, Denis Joyeux, Kenji Ito, Berenger Gans, Laurent Nahon

WG08 **3:59 – 4:14**
FORBIDDEN TRANSITIONS IN THE VUV SPECTRUM OF N_2, Alan Heays, Ming Li Niu, Nelson de Oliveira, Edcel John Salumbides, Brenton R Lewis, Wim Ubachs, Ewine van Dishoeck

WG09 **4:16 – 4:31**
SYNCHROTRON INFRARED SPECTROSCOPY OF ν_4, ν_{10}, ν_{11} AND ν_{14} STATES OF THIIRANE, Corey Evans, Jason P Carter, Don McNaughton, Andy Wong, Dominique Appadoo

WG10 **4:33 – 4:48**
FINGERPRINTS OF INTRAMOLECULAR HYDROGEN BONDS: SYNCHROTRON-BASED FAR IR STUDY OF THE CIS AND TRANS CONFORMERS OF 2-FLUOROPHENOL, Aimee Bell, James Singer, Jennifer van Wijngaarden

WG11 **4:50 – 5:05**
INFRARED CROSS-SECTIONS OF NITRO-DERIVATIVE VAPORS: NEW SPECTROSCOPIC SIGNATURES OF EXPLOSIVE TAGGANTS AND DEGRADATION PRODUCTS , Arnaud Cuisset, Gaël Mouret, Olivier Pirali, Sébastien Gruet, Gérard Pascal Piau, Gilles Fournier

WG12 ***Post-Deadline Abstract*** **5:07 – 5:22**
CHARACTERIZATION OF REACTION PATHWAYS IN LOW TEMPERATURE OXIDATION OF TETRAHYDROFURAN WITH MULTIPLEXED PHOTOIONIZATION MASS SPECTROMETRY TECHNIQUE, Ivan Antonov, Leonid Sheps

WH. Clusters/Complexes

Wednesday, June 24, 2015 – 1:30 PM

Room: B102 Chemical and Life Sciences

Chair: Elangannan Arunan, Indian Institute of Science, Bangalore, India

WH01 **1:30 – 1:45**
A STRANGE COMBINATION BAND OF THE CROSS-SHAPED COMPLEX CO_2-CS_2, Nasser Moazzen-Ahmadi, Bob McKellar

WH02 **1:47 – 2:02**
RE-ANALYSIS OF THE DISPERSED FLUORESCENCE SPECTRA OF THE C_3-RARE GAS ATOM COMPLEXES, Yi-Jen Wang, Anthony Merer, Yen-Chu Hsu

WH03 **2:04 – 2:19**
INFARED SPECTROSCOPY OF $Mn(CO_2)_n^-$ CLUSTER ANIONS, Michael C Thompson, J. Mathias Weber

WH04 **2:21 – 2:36**
INFRARED SPECTROSCOPY OF $(N_2O)_n^-$ AND $(N_2O)_mO^-$ CLUSTER ANIONS, Michael C Thompson, J. Mathias Weber

WH05 **2:38 – 2:53**
INFRARED SPECTROSCOPY OF PHENOL$^+$-TRIETHYLSILANE DIHYDROGEN-BONDED CLUSTER: INTRINSIC STRENGTH OF THE Si-H$\cdots$H-O DYHYDROGEN BOND, Haruki Ishikawa, Takayuki Kawasaki, Risa Inomata

WH06 **2:55 – 3:10**
INFRARED SPECTROSCOPY OF HYDROGEN-BONDED CLUSTERS OF PROTONATED HISTIDINE, Makoto Kondo, Yasutoshi Kasahara, Haruki Ishikawa

WH07 **3:12 – 3:27**
THEORETICAL INVESTIGATION OF THE UV/VIS PHOTODISSOCIATION DYNAMICS OF $ICN^-(Ar)_n$ and $BrCN^-(Ar)_n$, Bernice Opoku-Agyeman, Anne B McCoy

WH08 **3:29 – 3:44**
DISPERSION-DOMINATED π-STACKED COMPLEXES CONSTRUCTED ON A DYNAMIC SCAFFOLD, Deacon Nemchick, Michael Cohen, Patrick Vaccaro

Intermission

WH09 **4:03 – 4:18**
THE COMPETITION BETWEEN INSERTION AND SURFACE BINDING OF BENZENE TO THE WATER HEPTAMER, Patrick S. Walsh, Daniel P. Tabor, Edwin Sibert, Timothy S. Zwier

WH10 **4:20 – 4:35**
VIBRATIONAL SPECTROSCOPY OF BENZENE-(WATER)$_N$ CLUSTERS WITH $N = 6, 7$, Daniel P. Tabor, Edwin Sibert, Ryoji Kusaka, Patrick S. Walsh, Timothy S. Zwier

WH11 **4:37 – 4:52**
THEORETICAL STUDY OF THE IR SPECTROSCOPY OF BENZENE-(WATER)$_N$ CLUSTERS, Daniel P. Tabor, Edwin Sibert, Ryoji Kusaka, Patrick S. Walsh, Timothy S. Zwier

WH12 **4:54 – 5:09**
SPECTROSCOPIC INVESTIGATION OF TEMPERATURE EFFECTS ON THE HYDRATION STRUCTURE OF THE PHENOL CLUSTER CATION, Reona Yagi, Yasutoshi Kasahara, Haruki Ishikawa

WH13 **5:11 – 5:26**
ULTRAVIORET AND INFRARED PHOTODISSOCIATION SPECTROSCOPY OF HYDRATED ANILINIUM ION, Itaru Kurusu, Reona Yagi, Yasutoshi Kasahara, Haruki Ishikawa

WH14 **5:28 – 5:43**
WATER-NETWORK MEDIATED, ELECTRON INDUCED PROTON TRANSFER IN ANIONIC $[C_5H_5N{\cdot}(H_2O)_n]^-$ CLUSTERS: SIZE DEPENDENT FORMATION OF THE PYRIDINIUM RADICAL FOR $n \geq 3$, Andrew F DeBlase, Gary H Weddle, Kaye A Archer, Kenneth D. Jordan, Mark Johnson

WI. Astronomy

Wednesday, June 24, 2015 – 1:30 PM
Room: 274 Medical Sciences Building

Chair: Holger S. P. Müller, Universität zu Köln, Köln, NRW, Germany

WI01 **1:30 – 1:45**
THE NEW ALMA PROTOTYPE 12 M TELSCOPE OF THE ARIZONA RADIO OBSERVATORY: TRANSPORT, RECOMMISSIONING, AND FIRST LIGHT, Lucy Ziurys, N J Emerson, T W Folkers, R W Freund, D Forbes, G P Reiland, M McColl, S C Keel, S H Warner, J Kingsley, DeWayne T Halfen

WI02 **1:47 – 2:02**
FIRST SCIENTIFIC OBSERVATIONS WITH THE NEW ALMA PROTOTYPE ANTENNA OF THE ARIZONA RADIO OBSERVATORY: HCN AND CCH IN THE HELIX NEBULA, Lucy Ziurys, Deborah Schmidt

WI03 **2:04 – 2:19**
CCH AND HNC IN PLANETARY NEBULAE, Deborah Schmidt, Lucy Ziurys

WI04 **2:21 – 2:36**
MAPPING THE SPATIAL DISTRIBUTION OF METAL-BEARING OXIDES IN VY CANIS MAJORIS, Andrew Burkhardt, S. Tom Booth, Anthony Remijan, Brandon Carroll, Lucy Ziurys

WI05 **2:38 – 2:53**
C^+ AND THE CONNECTION BETWEEN DIFFERENT TRACERS OF THE DIFFUSE INTERSTELLAR MEDIUM, Steven Federman, Johnathan S Rice, Jorge L Pineda, William D Langer, Paul F Goldsmith, Nicolas Flagey

WI06 **2:55 – 3:10**
INFERRING THE TEMPERATURE AND DENSITY OF DIFFUSE INTERSTELLAR CLOUDS FROM C_3 OBSERVATIONS, Nicole Koeppen, Benjamin J. McCall

WI07 **3:12 – 3:27**
NEW BACKGROUND INFRARED SOURCES FOR STUDYING THE GALACTIC CENTER'S INTERSTELLAR GAS , Thomas R. Geballe, Takeshi Oka, E. Lambrides, S. C. C. Yeh, B. Schlegelmilch, Miwa Goto, C W Westrick

WI08 **3:29 – 3:44**
CO SPECTRAL LINE ENERGY DISTRIBUTIONS IN ORION SOURCES: TEMPLATES FOR EXTRAGALACTIC OBSERVATIONS, Nick Indriolo, Edwin Bergin

Intermission

WI09 **4:03 – 4:18**
THE DISTRIBUTION, EXCITATION, AND ABUNDANCE OF C^+, CH^+, AND CH IN ORION KL, Harshal Gupta, Patrick Morris, Zsofia Nagy, John Pearson, Volker Ossenkopf

WI10 **4:20 – 4:35**
THE DISTRIBUTION OF SH^+ AROUND ORION KL , Harshal Gupta, Karl M. Menten, Zsofia Nagy, Patrick Morris, Volker Ossenkopf, Nathan Crockett, John Pearson

WI11 **4:37 – 4:52**
CARMA 1 CM LINE SURVEY OF ORION-KL, Douglas Friedel, Leslie Looney, Joanna F. Corby, Anthony Remijan

WI12 **4:54 – 5:09**
CHEMICAL COMPLEXITY IN THE SHOCKED OUTFLOW L1157-B REVEALED BY CARMA, Niklaus M Dollhopf, Brett A. McGuire, Brandon Carroll, Anthony Remijan

WI13 **5:11 – 5:26**
THE CURIOUS CASE OF NH_2OH: HUNTING A DIRECT AMINO ACID PRECURSOR SPECIES IN THE INTERSTELLAR MEDIUM, Brett A. McGuire, Brandon Carroll, Niklaus M Dollhopf, Nathan Crockett, Geoffrey Blake, Anthony Remijan

WI14 **5:28 – 5:43**
NEW RESULTS FROM THE CARMA LARGE-AREA STAR FORMATION SURVEY (CLASSY) , Robert J Harris, Leslie Looney, Dominique M. Segura-Cox, Manuel Fernandez-Lopez, Lee Mundy, Shaye Storm, Maxime Rizzo, Katherine Lee, Héctor Arce

WI15 **5:45 – 6:00**
ADMIT: ALMA DATA MINING TOOLKIT, Douglas Friedel, Leslie Looney, Lisa Xu, Marc W. Pound, Peter J. Teuben, Kevin P. Rauch, Lee Mundy, Jeffrey S. Kern

WJ. Non-covalent interactions

Wednesday, June 24, 2015 – 1:30 PM

Room: 217 Noyes Laboratory

Chair: Wolfgang Jäger, University of Alberta, Edmonton, AB, Canada

WJ01 **1:30 – 1:45**
FORMATION OF COMPLEXES c-C_3H_6···MCl (M = Ag or Cu) AND THEIR CHARACTERIZATION BY BROADBAND ROTATIONAL SPECTROSCOPY, Daniel P. Zaleski, John Connor Mullaney, Nick Walker, Anthony Legon

WJ02 **1:47 – 2:02**
ROTATIONAL SPECTROSCOPY OF MONOFLUOROETHANOL AGGREGATES WITH ITSELF AND WITH WATER, Javix Thomas, Wenyuan Huang, Xunchen Liu, Wolfgang Jäger, Yunjie Xu

WJ03 **2:04 – 2:19**
O-TOLUIC ACID MONOMER AND MONOHYDRATE: ROTATIONAL SPECTRA, STRUCTURES, AND ATMOSPHERIC IMPLICATIONS, Elijah G Schnitzler, Brandi L M Zenchyzen, Wolfgang Jäger

WJ04 **2:21 – 2:36**
A ROVIBRATIONAL ANALYSIS OF THE WATER BENDING VIBRATION IN OC-H_2O AND A MORPHED POTENTIAL OF THE COMPLEX, Luis A. Rivera-Rivera, Sean D. Springer, Blake A. McElmurry, Igor I Leonov, Robert R. Lucchese, John W. Bevan, L. H. Coudert

WJ05 **2:38 – 2:53**
THE MICROWAVE SPECTRUM AND UNEXPECTED STRUCTURE OF THE BIMOLECULAR COMPLEX FORMED BETWEEN ACETYLENE AND (*Z*)-1-CHLORO-2-FLUOROETHYLENE, Nazir D. Khan, Helen O. Leung, Mark D. Marshall

WJ06 **2:55 – 3:10**
CHLORINE NUCLEAR QUADRUPOLE HYPERFINE STRUCTURE IN THE VINYL CHLORIDE–HYDROGEN CHLORIDE COMPLEX, Helen O. Leung, Mark D. Marshall, Joseph P. Messinger

WJ07 **3:12 – 3:27**
ELECTRONIC COMMUNICATION IN COVALENTLY *vs.* NON-COVALENTLY BONDED POLYFLUORENE SYSTEMS: THE ROLE OF THE COVALENT LINKER., Brandon Uhler, Neil J Reilly, Marat R Talipov, Maxim Ivanov, Qadir Timerghazin, Rajendra Rathore, Scott Reid

Intermission

WJ08 **3:46 – 4:01**
A GENERAL TRANSFORMATION TO CANONICAL FORM FOR POTENTIALS IN PAIRWISE INTERMOLECULAR INTERACTIONS, Jay R. Walton, Luis A. Rivera-Rivera, Robert R. Lucchese, John W. Bevan

WJ09 **4:03 – 4:18**
THREE-DIMENSIONAL WATER NETWORKS SOLVATING AN EXCESS POSITIVE CHARGE: NEW INSIGHTS INTO THE MOLECULAR PHYSICS OF ION HYDRATION, Conrad T. Wolke, Joseph Fournier, Gary H Weddle, Evangelos Miliordos, Sotiris Xantheas, Mark Johnson

WJ10 **4:20 – 4:35**
MATRIX ISOLATION INFRARED SPECTROSCOPY OF A SERIES OF 1:1 PHENOL-WATER COMPLEXES , Pujarini Banerjee, Tapas Chakraborty

WJ11 **4:37 – 4:52**
MATRIX ISOLATION IR SPECTROSCOPY AND QUANTUM CHEMISTRY STUDY OF 1:1 Π-HYDROGEN BONDED COMPLEXES OF BENZENE WITH A SERIES OF FLUOROPHENOLS, Pujarini Banerjee, Tapas Chakraborty

WJ12 **4:54 – 5:09**
MATRIX ISOLATION IR SPECTROSCOPY OF 1:1 COMPLEXES OF ACETIC ACID AND TRIHALOACETIC ACIDS WITH WATER AND BENZENE, Pujarini Banerjee, Tapas Chakraborty

WJ13 **5:11 – 5:21**
SPECTROSCOPIC INVESTIGATION OF THE EFFECTS OF ENVIRONMENT ON NEWLY-DEVELOPED NEAR INFRARED EMITTING DYES, Louis E. McNamara, Nalaka Liyanage, Jared Delcamp, Nathan I Hammer

WJ14 **5:23 – 5:38**
SPECTROSCOPIC SIGNATURES AND STRUCTURAL MOTIFS IN ISOLATED AND HYDRATED XANTHINE AND ITS METHYLATED DERIVATIVES, Vipin Bahadur Singh

RA. Metal containing

Thursday, June 25, 2015 – 8:30 AM
Room: 116 Roger Adams Lab

Chair: Damian L Kokkin, Arizona State University, Tempe, Arizona, USA

RA01 **8:30 – 8:45**
HYPERFINE RESOLVED PURE ROTATIONAL SPECTROSCOPY OF ScN, YN, AND BaNH ($X^1\Sigma^+$): INSIGHT INTO METAL-NITROGEN BONDING, Lindsay N. Zack, Matthew Bucchino, Justin Young, Marshall Binns, Phillip M. Sheridan, Lucy Ziurys

RA02 **8:47 – 9:02**
THE SUBMILLIMETER/THz SPECTRUM OF AlH ($X^1\Sigma^+$), CrH ($X^6\Sigma^+$), and SH^+ ($X^3\Sigma^-$), DeWayne T Halfen, Lucy Ziurys

RA03 **9:04 – 9:19**
FORMATION OF M-C≡C-Cl (M = Ag or Cu) AND CHARACTERIZATION BY ROTATIONAL SPECTROSCOPY, Daniel P. Zaleski, David Peter Tew, Nick Walker, Anthony Legon

RA04 **9:21 – 9:36**
$(CH_3)_3N\cdots AgI$ AND $H_3N\cdots AgI$ STUDIED BY BROADBAND ROTATIONAL SPECTROSCOPY AND *AB INITIO* CALCULATIONS, Dror M. Bittner, Daniel P. Zaleski, Susanna L. Stephens, Nick Walker, Anthony Legon

RA05 **9:38 – 9:53**
MICROWAVE SPECTRA AND GEOMETRIES OF $C_2H_2\cdots AgI$ and $C_2H_4\cdots AgI$, Susanna L. Stephens, David Peter Tew, Nick Walker, Anthony Legon

Intermission

RA06 **10:12 – 10:27**
EVALUATION OF THE EXOTHERMICITY OF THE CHEMI-IONIZATION REACTION $Sm + O \rightarrow SmO^+ + e^-$, Richard M Cox, JungSoo Kim, Peter Armentrout, Joshua Bartlett, Robert A. VanGundy, Michael Heaven, Joshua J. Melko, Shaun Ard, Nicholas S. Shuman, Albert Viggiano

RA07 **10:29 – 10:44**
The PERMANENT ELECTRIC DIPOLE MOMENT AND HYPERFINE INTERACTION IN GOLD SULFIDE, AuS , Ruohan Zhang, Damian L Kokkin, Thomas D. Varberg, Timothy Steimle

RA08 **10:46 – 11:01**
HIGH-ACCURACY CALCULATION OF Cu ELECTRIC-FIELD GRADIENTS: A REVISION OF THE Cu NUCLEAR QUADRUPOLE MOMENT VALUE , Lan Cheng, Devin A. Matthews, Jürgen Gauss, John F. Stanton

RA09 **11:03 – 11:18**
CATION-π AND CH-π INTERACTIONS IN THE COORDINATION AND SOLVATION OF Cu^+ $(ACETYLENE)_n$ (n=1-6) COMPLEXES INVESTIGATED VIA INFRARED PHOTODISSOCIATION SPECTROSCOPY , Antonio David Brathwaite, Richard S. Walters, TIMOTHY B WARD, Michael A Duncan

RA10 **11:20 – 11:35**
ANION PHOTOELECTRON SPECTROSCOPIC STUDIES OF $NbCr(CO)_n^-$ (n = 2,3) HETEROBIMETALLIC CARBONYL COMPLEXES, Melissa A. Baudhuin, Praveenkumar Boopalachandran, Doreen Leopold

RA11 **11:37 – 11:52**
MASS-ANALYZED THRESHOLD IONIZATION SPECTROSCOPY OF CYCLIC $La(C_5H_6)$ FORMED BY La ATOM ACTIVATION OF PENTYNE AND PENTADIENE, Wenjin Cao, Yuchen Zhang, Dong-Sheng Yang

RB. Mini-symposium: Accelerator-Based Spectroscopy

Thursday, June 25, 2015 – 8:30 AM

Room: 100 Noyes Laboratory

Chair: Gert von Helden, Fritz Haber Institute - MPG, Berlin, Germany

RB01 *INVITED TALK* **8:30 – 9:00**
PROBING INTRA- AND INTER- MOLECULAR INTERACTIONS VIA IRMPD EXPERIMENTS AND COMPUTATIONAL CHEMISTRY, Scott Hopkins, Terry McMahon

RB02 **9:05 – 9:20**
EXPLORING CONFORMATION SELECTIVE FAR INFRARED ACTION SPECTROSCOPY OF ISOLATED MOLECULES AND SOLVATED CLUSTERS, Daniël Bakker, Anouk Rijs, Jérôme Mahé, Marie-Pierre Gaigeot

RB03 **9:22 – 9:37**
FIRST INFRARED PREDISSOCIATION SPECTRA OF He-TAGGED PROTONATED PRIMARY ALCOHOLS AT 4 K, Alexander Stoffels, Britta Redlich, J. Oomens, Oskar Asvany, Sandra Brünken, Pavol Jusko, Sven Thorwirth, Stephan Schlemmer

RB04 **9:39 – 9:49**
METAL ION INDUCED PAIRING OF CYTOSINE BASES: FORMATION OF I-MOTIF STRUCTURES IDENTIFIED BY IR ION SPECTROSCOPY, Juehan Gao, Giel Berden, J. Oomens

RB05 **9:51 – 10:06**
MOLECULAR PROPERTIES OF THE "ANTI-AROMATIC" SPECIES CYCLOPENTADIENONE, C_5H_5=0, Thomas Ormond, Barney Ellison, John W Daily, John F. Stanton, Musahid Ahmed, Timothy S. Zwier, Patrick Hemberger

Intermission

RB06 **10:25 – 10:40**
HIGH-RESOLUTION SYNCHROTRON INFRARED SPECTROSCOPY OF THIOPHOSGENE: THE ν_1, ν_5, $2\nu_4$, and $\nu_2 + 2\nu_6$ bands , Bob McKellar, Brant E Billinghurst

RB07 **10:42 – 10:57**
THE SOLEIL VIEW ON SULFUR OXIDES: THE S_2O BENDING MODE ν_2 AT 380 cm^{-1} AND ITS ANALYSIS USING AN AUTOMATED SPECTRAL ASSIGNMENT PROCEDURE (ASAP), Marie-Aline Martin-Drumel, Christian Endres, Oliver Zingsheim, T. Salomon, Jennifer van Wijngaarden, Olivier Pirali, Sébastien Gruet, Frank Lewen, Stephan Schlemmer, Michael C McCarthy, Sven Thorwirth

RB08 **10:59 – 11:14**
THE SOLEIL VIEW ON SULFUR RICH OXIDES: THE ν_3 MODE OF S_2O REVISITED, Sven Thorwirth, Marie-Aline Martin-Drumel, Christian Endres, Oliver Zingsheim, T. Salomon, Jennifer van Wijngaarden, Olivier Pirali, Sébastien Gruet, Frank Lewen, Stephan Schlemmer, Michael C McCarthy

RB09 **11:16 – 11:31**
FT-IR MEASUREMENTS OF NH_3 LINE INTENSITIES IN THE 60 – 550 CM^{-1} USING SOLEIL/AILES BEAMLINE, Keeyoon Sung, Shanshan Yu, John Pearson, Laurent Manceron, F. Kwabia Tchana, Olivier Pirali

RB10 ***Post-Deadline Abstract*** **11:33 – 11:48**
THE H_2O-CH_3F COMPLEX: A COMBINED MICROWAVE AND INFRARED SPECTROSCOPIC STUDY SUPPORTED BY STRUCTURE CALCULATIONS, Sharon Priya Gnanasekar, Manuel Goubet, Elangannan Arunan, Robert Georges, Pascale Soulard, Pierre Asselin, T. R. Huet, Olivier Pirali

RC. Mini-symposium: Spectroscopy in the Classroom

Thursday, June 25, 2015 – 8:30 AM

Room: B102 Chemical and Life Sciences

Chair: Kristopher J Ooms, The King's University, Edmonton, Alberta, Canada

RC01 *INVITED TALK* **8:30 – 9:00**
DIRECT DIGITAL SYNTHESIS CHIRPED PULSE MICROWAVE SPECTROMETERS FOR THE CLASSROOM AND RESEARCH, Geoffrey Blake, Brandon Carroll, Ian A Finneran

RC02 **9:05 – 9:20**
A SIMPLE, COST EFFECTIVE RAMAN-FLUORESCENCE SPECTROMETER FOR USE IN LABORATORY AND FIELD EXPERIMENTS, Frank E Marshall, Michael A Pride, Michellle Rojo, Katelyn R. Brinker, Zachary Walker, Michael Storrie-Lombardi, Melanie R. Mormile, G. S. Grubbs II

RC03 **9:22 – 9:37**
LIF AND RAMAN SPECTROSCOPY IN UNDERGRADUATE LABS USING GREEN DIODE-PUMPED SOLID-STATE LASERS, Jeffrey A. Gray

RC04 **9:39 – 9:54**
SPECFITTER: A LEARNING ENVIRONMENT FOR THE ROTATIONAL SPECTROSCOPIST, Yoon Jeong Choi, Weixin Wu, A. J. Minei, S. A. Cooke

RC05 **9:56 – 10:11**
APPLICATIONS OF GROUP THEORY: INFRARED AND RAMAN SPECTRA OF THE ISOMERS OF *cis*- AND *trans*-1,2-DICHLOROETHYLENE, Norman C. Craig

Intermission

RC06 **10:30 – 10:40**
INFRARED ANALYSIS OF COMBUSTION PRODUCTS AND INTERMEDIATES OF HYDROCARBON COMBUSTION FOR SEVERAL SPECIES, Allen White, Rebecca Devasher

RC07 **10:42 – 10:52**
CHIRPED-PULSE MICROWAVE SPECTROSCOPY IN THE UNDERGRADUATE CHEMISTRY CURRICULUM, Sydney A Gaster, Taylor M Hall, Sean Arnold, Gordon G Brown

RC08 **10:54 – 11:04**
USB SPECTROMETERS AND THE TEMPERATURE OF THE SUN: MEASURING BLACK BODY RADIATION IN THE PALM OF YOUR HAND, Daniel P. Zaleski, Benjamin R Horrocks, Nick Walker

RC09 **11:06 – 11:16**
VIBRATION-ROTATION ANALYSIS OF THE $^{13}CO_2$ ASYMMETRIC STRETCH FUNDAMENTAL BAND IN AMBIENT AIR FOR THE PHYSICAL CHEMISTRY TEACHING LABORATORY, David A Dolson, CATHERINE B ANDERS

RC10 **11:18 – 11:33**
UTILIZING SPECTROSCOPIC RESEARCH TOOLS AND SOFTWARE IN THE CLASSROOM, G. S. Grubbs II

RD. Astronomy

Thursday, June 25, 2015 – 8:30 AM

Room: 274 Medical Sciences Building

Chair: Brett A. McGuire, California Institute of Technology, Pasadena, CA, USA

RD01 **8:30 – 8:40**
NEW INSTRUMENTAL TOOLS FOR ADVANCED ASTROCHEMICAL APPLICATIONS, Amanda Steber, Sabrina Zinn, Melanie Schnell, Anouk Rijs

RD02 **8:42 – 8:57**
DOPPLER AND SUB-DOPPLER MILLIMETER AND SUB-MILLIMETER WAVE SPECTROSCOPY OF KEY ASTRONOMICAL MOLECULES: HNC AND CS, Oliver Zingsheim, Thomas Schmitt, Frank Lewen, Stephan Schlemmer, Sven Thorwirth

RD03 **8:59 – 9:14**
MILLIMETRE-WAVE SPECTRUM OF ISOTOPOLOGUES OF ETHANOL FOR RADIO ASTRONOMY, Adam Walters, Mirko Schäfer, Matthias H. Ordu, Frank Lewen, Stephan Schlemmer, Holger S. P. Müller

RD04 **9:16 – 9:31**
TERAHERTZ SPECTROSCOPY OF DEUTERATED METHYLENE BI-RADICAL, CD_2, Hiroyuki Ozeki, Stephane Bailleux

RD05 **9:33 – 9:48**
THZ SPECTROSCOPY OF D_2H^+, Shanshan Yu, John Pearson, Takayoshi Amano

RD06 **9:50 – 10:05**
THZ SPECTROSCOPY OF $^{12}CH^+$, $^{13}CH^+$, AND $^{12}CD^+$, Shanshan Yu, Brian Drouin, John Pearson, Takayoshi Amano

RD07 **10:07 – 10:22**
ROTATIONAL SPECTROSCOPY OF VIBRATIONALLY EXCITED N_2H^+ and N_2D^+ UP TO 2 THZ, Shanshan Yu, John Pearson, Brian Drouin, Timothy J Crawford, Adam M Daly, Ben Elliott, Takayoshi Amano

Intermission

RD08 **10:41 – 10:56**
NEW ACCURATE WAVENUMBERS OF $H^{35}Cl^+$ AND $H^{37}Cl^+$ ROVIBRATIONAL TRANSITIONS IN THE $v = 0 - 1$ BAND OF THE $^2\Pi$ STATE., Jose Luis Domenech, Maite Cueto, Victor Jose Herrero, Isabel Tanarro, Jose Cernicharo

RD09 **10:58 – 11:13**
OSCILLATOR STRENGTHS AND PREDISSOCIATION RATES FOR $W - X$ BANDS AND THE 4P5P COMPLEX IN $^{13}C^{18}O$, Michele Eidelsberg, Jean Louis Lemaire, Steven Federman, Glenn Stark, Alan Heays, Lisseth Gavilan, James R Lyons, Peter L Smith, Nelson de Oliveira, Denis Joyeux

RD10 **11:15 – 11:30**
LINE STRENGTHS OF ROVIBRATIONAL AND ROTATIONAL TRANSITIONS IN THE $X^2\Pi$ GROUND STATE OF OH, James S.A. Brooke, Peter F. Bernath, Colin Western, Chris Sneden, Gang Li, Iouli E Gordon

RD11 ***Post-Deadline Abstract*** **11:32 – 11:42**
CLASS I METHANOL MASER CONDITIONS NEAR SNRS, Bridget C. McEwen, Ylva M. Pihlström, Loránt O. Sjouwerman

RD12 ***Post-Deadline Abstract*** **11:44 – 11:59**
THE MISSING LINK: ROTATIONAL SPECTRUM AND GEOMETRICAL STRUCTURE OF DISILICON CARBIDE, Si_2C, Michael C McCarthy, Joshua H Baraban, Bryan Changala, John F. Stanton, Marie-Aline Martin-Drumel, Sven Thorwirth, Neil J Reilly, Carl A Gottlieb

RE. Instrument/Technique Demonstration

Thursday, June 25, 2015 – 8:30 AM

Room: 217 Noyes Laboratory

Chair: Arthur Suits, Wayne State University, Detroit, MI, USA

RE01 **8:30 – 8:45**
OPTIMIZATION OF EXTREME ULTRAVIOLET LIGHT SOURCE FROM HIGH HARMONIC GENERATION FOR CONDENSED-PHASE CORE-LEVEL SPECTROSCOPY, Ming-Fu Lin, Max A Verkamp, Elizabeth S Ryland, Kristin Benke, Kaili Zhang, Michaela Carlson, Josh Vura-Weis

RE02 **8:47 – 9:02**
DEVELOPMENT OF TWO-PHOTON PUMP POLARIZATION SPECTROSCOPY PROBE TECHNIQUE (TPP-PSP) FOR MEASUREMENTS OF ATOMIC HYDROGEN . , Aman Satija, Robert P. Lucht

RE03 **9:04 – 9:19**
DEVELOPMENT OF COMBINED DUAL-PUMP VIBRATIONAL AND PURE-ROTATIONAL COHERENT ANTI-STOKES RAMAN SCATTERING TECHNIQUE., Aman Satija, Robert P. Lucht

RE04 **9:21 – 9:36**
VELOCITY MAP IMAGING STUDY OF THE PHOTOINITIATED CHARGE-TRANSFER DISSOCIATION OF $Cu^+(C_6H_6)$ AND $Ag^+(C_6H_6)$, Jon Maner, Daniel Mauney, Michael A Duncan

Intermission

RE05 **9:55 – 10:10**
MID-IR CAVITY RINGDOWN SPECTROSCOPY FOR ATMOSPHERIC ETHANE ABUNDANCE MEASUREMENTS, Linhan Shen, Thinh Quoc Bui, Lance Christensen, Mitchio Okumura

RE06 **10:12 – 10:27**
STRONG THERMAL NONEQUILIBRIUM IN HYPERSONIC CO AND CH_4 PROBED BY CRDS, Maud Louviot, Nicolas Suas-David, Vincent Boudon, Robert Georges, Michael Rey, Samir Kassi

RE07 **10:29 – 10:44**
ROTATIONALLY-RESOLVED INFRARED SPECTROSCOPY OF THE ν_{16} BAND OF 1,3,5-TRIOXANE, Bradley M. Gibson, Nicole Koeppen, Benjamin J. McCall

RE08 **10:46 – 11:01**
IMPROVING SNR IN TIME-RESOLVED SPECTROSCOPIES WITHOUT SACRIFICING TEMPORAL-RESOLUTION: APPLICATION TO THE UV PHOTOLYSIS OF METHYL CYANOFORMATE, Michael J. Wilhelm, Jonathan M. Smith, HAI-LUNG Dai

RE09 *Post-Deadline Abstract* **11:03 – 11:18**
LASER-INDUCED PLASMAS IN AMBIENT AIR FOR INCOHERENT BROADBAND CAVITY-ENHANCED ABSORPTION SPECTROSCOPY , Albert A Ruth, Sophie Dixneuf, Johannes Orphal

RF. Atmospheric science

Thursday, June 25, 2015 – 1:30 PM

Room: 116 Roger Adams Lab

Chair: Joseph Hodges, National Institute of Standards and Technology, Gaithersburg, MD, USA

RF01 **1:30 – 1:40**
PHOTOACOUSTIC SPECTROSCOPY OF THE OXYGEN A-BAND, Elizabeth M Lunny, Thinh Quoc Bui, Caitlin Bray, Priyanka Rupasinghe, Mitchio Okumura

RF02 **1:42 – 1:57**
HIGH PRESSURE OXYGEN A-BAND SPECTRA, Brian Drouin, Keeyoon Sung, Shanshan Yu, Elizabeth M Lunny, Thinh Quoc Bui, Mitchio Okumura, Priyanka Rupasinghe, Caitlin Bray, David A. Long, Joseph Hodges, David Robichaud, D. Chris Benner, V. Malathy Devi, Jiajun Hoo

RF03 **1:59 – 2:14**
COLLISION-DEPENDENT LINE AREAS IN THE $a^1\Delta_g \leftarrow X^3\Sigma_g^-$ BAND OF MOLECULAR OXYGEN, Vincent Sironneau, Adam J. Fleisher, Joseph Hodges

RF04 **2:16 – 2:31**
ANOMALOUS CENTRIFUGAL DISTORTION IN HDO AND SPECTROSCOPIC DATA BASES, L. H. Coudert

RF05 **2:33 – 2:48**
SPEED-DEPENDENT BROADENING AND LINE-MIXING IN CH_4 PERTURBED BY AIR NEAR 1.64 μm FOR THE FRENCH/GERMAN CLIMATE MISSION MERLIN, Thibault Delahaye, Thi Ngoc Ha Tran, Zachary Reed, Stephen E Maxwell, Joseph Hodges

RF06 **2:50 – 3:05**
MID INFRARED DUAL FREQUENCY COMB SPECTROMETER FOR THE DETECTION OF METHANE IN AMBIENT AIR , Hans Schuessler, Feng Zhu, Alexander Kolomenskii

RF07 **3:07 – 3:22**
IMPROVED OZONE AND CARBON MONOXIDE PROFILE RETRIEVALS USING MULTISPECTRAL MEASUREMENTS FROM NASA “A TRAIN”, NPP, AND TROPOMI SATELLITES, Dejian Fu

RF08 **3:24 – 3:39**
TEMPERATURE DEPENDENCES OF AIR-BROADENING AND SHIFT PARAMETERS IN THE ν_3 BAND OF OZONE, Mary Ann H. Smith, V. Malathy Devi, D. Chris Benner

Intermission

RF09 **3:58 – 4:13**
MICROWAVE OPTICAL DOUBLE RESONANCE STUDIES OF PERTURBATIONS IN THE SO $A^3\Pi$ STATE, Andrew Richard Whitehill, Alexander W. Hull, Trevor J. Erickson, Jun Jiang, Carrie Womack, Barratt Park, Shuhei Ono, Robert W Field

RF10 **4:15 – 4:30**
VALIDATION OF A NEW HNO_3 LINE PARAMETERS AT 7.6 μm USING LABORATORY INTENSITY MEASUREMENTS AND MIPAS SATELLITE SPECTRA, Marco Ridolfi, Agnes Perrin, Jean-Marie Flaud, Jean Vander Auwera, Massimo Carlotti

RF11 **4:32 – 4:47**
ROTATIONAL SPECTROSCOPY OF NEWLY DETECTED ATMOSPHERIC OZONE DEPLETERS: CF_3CH_2Cl, CF_3CCl_3, AND CF_2ClCCl_3, Zbigniew Kisiel, Ewa Białkowska-Jaworska, Lech Pszczółkowski, Iciar Uriarte, Patricia Ecija, Francisco J. Basterretxea, Emilio J. Cocinero

RF12 **4:49 – 5:04**
CHIRPED PULSE AND CAVITY FT MICROWAVE SPECTROSCOPY OF THE FORMIC ACID – TRIMETHYLAMINE WEAKLY BOUND COMPLEX, Becca Mackenzie, Chris Dewberry, Ken Leopold

RF13 **5:06 – 5:21**
FORMIC SULFURIC ANHYDRIDE: A NEW CHEMICAL SPECIES WITH POSSIBLE IMPLICATIONS FOR ATMOSPHERIC AEROSOL, Becca Mackenzie, Chris Dewberry, Ken Leopold

RF14 **5:23 – 5:38**
ROTATIONAL SPECTROSCOPY OF METHYL VINYL KETONE, Olena Zakharenko, R. A. Motiyenko, Juan-Ramon Aviles Moreno, T. R. Huet

RF15 **5:40 – 5:55**
THE MILLIMETER-WAVE SPECTRUM OF METHACROLEIN. TORSION-ROTATION-VIBRATION EFFECTS IN THE EXCITED STATES, Olena Zakharenko, R. A. Motiyenko, Juan-Ramon Aviles Moreno, T. R. Huet

RG. Vibrational structure/frequencies

Thursday, June 25, 2015 – 1:30 PM

Room: 100 Noyes Laboratory

Chair: John F. Stanton, The University of Texas, Austin, TX, USA

RG01 **1:30 – 1:45**
ALKYL CH STRETCH VIBRATIONS AS A PROBE OF LOCAL ENVIRONMENT AND STRUCTURE, Edwin Sibert, Daniel P. Tabor, Nathanael Kidwell, Jacob C. Dean, Timothy S. Zwier

RG02 **1:47 – 2:02**
COMPUTING THE VIBRATIONAL ENERGIES OF CH_2O AND CH_3CN WITH PHASE-SPACED LOCALIZED FUNCTIONS AND AN ITERATIVE EIGENSOLVER, James Brown, Tucker Carrington

RG03 **2:04 – 2:19**
A MULTILAYER SUM-OF-PRODUCTS METHOD FOR COMPUTING VIBRATIONAL SPECTRA WITHOUT STORING FULL-DIMENSIONAL VECTORS OR MATRCIES, Phillip Thomas, Tucker Carrington

RG04 **2:21 – 2:36**
QUANTUM MONTE CARLO ALGORITHMS FOR DIAGRAMMATIC VIBRATIONAL STRUCTURE CALCULATIONS, Matthew Hermes, So Hirata

RG05 **2:38 – 2:53**
DIAGRAMMATIC VIBRATIONAL COUPLED-CLUSTER, Jacob A Faucheaux, So Hirata

RG06 **2:55 – 3:10**
VIBRATIONAL JAHN-TELLER EFFECT IN NON-DEGENERATE ELECTRONIC STATES, Mahesh B. Dawadi, Bishnu P Thapaliya, Ram Bhatta, David S. Perry

RG07 **3:12 – 3:27**
SPECTRAL SIGNATURES AND STRUCTURAL MOTIFS IN NEUTRAL AND PROTONATED HISTAMINE: A COMPUTATIONAL STUDY , Santosh Kumar Srivastava, Vipin Bahadur Singh

RG08 **3:29 – 3:39**
ANALOG OF DUSCHINSKY MATRIX AND CO-ASSIGNMENT OF FREQUENCIES IN DIFFERENT ELECTRONIC STATES, Yurii Panchenko, Alexander Abramenkov

Intermission

RG09 **3:58 – 4:13**
HIGH RESOLUTION INFRARED SPECTRA OF TRIACETYLENE, Kirstin D Doney, Dongfeng Zhao, Harold Linnartz

RG10 **4:15 – 4:30**
INFRARED AND ULTRAVIOLET SPECTROSCOPY OF GAS-PHASE IMIDAZOLIUM AND PYRIDINIUM IONIC LIQUIDS., Justin W. Young, Ryan S Booth, Christopher Annesley, Jaime A. Stearns

RG11 **4:32 – 4:47**
GROUND AND EXCITED STATE ALKYL CH STRETCH IR SPECTRA OF STRAIGHT-CHAIN ALKYLBENZENES, Daniel M. Hewett, Joseph A. Korn, Timothy S. Zwier

RG12 **4:49 – 5:04**
ASYMMETRY OF $M^+(H_2O)RG$ COMPLEXES, (M=V, Nb) REVEALED WITH INFRARED SPECTROSCOPY, TIMOTHY B WARD, Evangelos Miliordos, Sotiris Xantheas, Michael A Duncan

RG13 **5:06 – 5:21**
INFRARED SPECTROSCOPY OF PROTONATED ACETYLACETONE AND MIXED ACETYLACETONE/WATER CLUSTERS, Daniel Mauney, Jon Maner, David C McDonald II, Michael A Duncan

RG14 **5:23 – 5:38**
HEAVY ATOM VIBRATIONAL MODES AND LOW-ENERGY VIBRATIONAL AUTODETACHMENT IN NITROMETHANE ANIONS, Michael C Thompson, Joshua H Baraban, John F. Stanton, J. Mathias Weber

RG15 ***Post-Deadline Abstract*** **5:40 – 5:55**
OBSERVATION OF DIPOLE-BOUND STATE AND HIGH-RESOLUTION PHOTOELECTRON IMAGING OF COLD ACETATE ANIONS, Guo-Zhu Zhu, Dao-Ling Huang, Lai-Sheng Wang

RH. Clusters/Complexes

Thursday, June 25, 2015 – 1:30 PM

Room: B102 Chemical and Life Sciences

Chair: Galen Sedo, University of Virginia's College at Wise, Wise, VA, USA

RH01 **1:30 – 1:45**
CHIRPED PULSE AND CAVITY FT MICROWAVE SPECTROSCOPY OF THE HCCH-2,6-DIFLUOROPYRIDINE WEAKLY BOUND COMPLEX, Chris Dewberry, Becca Mackenzie, Ken Leopold

RH02 **1:47 – 1:57**
MICROWAVE SPECTRUM, VAN DER WAALS BOND LENGTH, AND ^{131}Xe QUADRUPOLE COUPLING CONSTANT OF Xe-SO_3, Chris Dewberry, Anna Huff, Becca Mackenzie, Ken Leopold

RH03 **1:59 – 2:14**
DIMETHYL SULFIDE-DIMETHYL ETHER AND ETHYLENE OXIDE-ETHYLENE SULFIDE COMPLEXES INVESTIGATED BY FOURIER TRANSFORM MICROWAVE SPECTROSCOPY AND AB INITIO CALCULATION, Yoshiyuki Kawashima, Yoshio Tatamitani, Takayuki Mase, Eizi Hirota

RH04 **2:16 – 2:31**
INTERNAL DYNAMICS IN $SF_6 \cdots NH_3$ OBSERVED BY BROADBAND ROTATIONAL SPECTROSCOPY, Dror M. Bittner, Daniel P. Zaleski, Susanna L. Stephens, Nick Walker, Anthony Legon

RH05 **2:33 – 2:48**
EVIDENCE FOR A COMPLEX BETWEEN THF AND ACETIC ACID FROM BROADBAND ROTATIONAL SPECTROSCOPY, Daniel P. Zaleski, Dror M. Bittner, John Connor Mullaney, Susanna L. Stephens, Adrian King, Matthew Habgood, Nick Walker

RH06 **2:50 – 3:00**
THE ROTATIONAL SPECTRUM OF PYRIDINE-FORMIC ACID, Lorenzo Spada, Qian Gou, Barbara Michela Giuliano, Walther Caminati

Intermission

RH07 **3:19 – 3:34**
FOURIER-TRANSFORM MICROWAVE AND MILLIMETERWAVE SPECTROSCOPY OF THE H_2-HCN MOLECULAR COMPLEX, Keiichi Tanaka, Kensuke Harada, Yoshihiro Sumiyoshi, Masakazu Nakajima, Yasuki Endo

RH08 **3:36 – 3:51**
MICROWAVE SPECTROSCOPY OF THE CYCLOPENTANOL - WATER DIMER, Brandon Carroll, Ian A Finneran, Geoffrey Blake

RH09 **3:53 – 4:08**
HYDROGEN-BONDING AND HYDROPHOBIC INTERACTIONS IN THE ETHANOL-WATER DIMER, Ian A Finneran, Brandon Carroll, Marco A. Allodi, Geoffrey Blake

RH10 **4:10 – 4:25**
THE INFLUENCE OF FLUORINATION ON STRUCTURE OF THE TRIFLUOROACETONITRILE WATER COMPLEX, Wei Lin, Anan Wu, Xin Lu, Daniel A. Obenchain, Stewart E. Novick

RH11 **4:27 – 4:42**
THE POSITION OF DEUTERIUM IN THE HOD – N_2O AS DETERMINED BY STRUCTURAL AND NUCLEAR QUADRUPOLE COUPLING CONSTANTS, Daniel A. Obenchain, Derek S. Frank, Stewart E. Novick, William Klemperer

RH12 **4:44 – 4:59**
THE CP-FTMW SPECTROSCOPY AND ASSIGNMENT OF THE MONO- AND DIHYDRATE COMPLEXES OF PERFLUOROPROPIONIC ACID, G. S. Grubbs II, Daniel A. Obenchain, Derek S. Frank, Stewart E. Novick, S. A. Cooke, Agapito Serrato III, Wei Lin

RH13 **5:01 – 5:16**
HYDROGEN BONDING IN 4-AMINOPHENYL ETHANOL: A COMBINED IR-UV DOUBLE RESONANCE AND MICROWAVE STUDY, Caitlin Bray, Cara Rae Rivera, E. A. Arsenault, Daniel A. Obenchain, Stewart E. Novick, Joseph L. Knee

RH14 **5:18 – 5:28**
THEORETICAL STUDY OF THE EFFECT OF DOPING CLUSTERS (ZNO) 6 BY THE SELENIUM USING THE DFT, Nour el Houda Bensiradj, Ourida Ouamerali

RH15 ***Post-Deadline Abstract*** **5:30 – 5:45**
BORONYL MIMICS GOLD: A PHOTOELECTRON SPECTROSCOPY STUDY, Tian Jian, Gary Lopez, Lai-Sheng Wang

RI. Astronomy

Thursday, June 25, 2015 – 1:30 PM

Room: 274 Medical Sciences Building

Chair: Harshal Gupta, California Institute of Technology, Pasadena, CA, USA

RI01 **1:30 – 1:45**
THE COMPLETE, TEMPERATURE RESOLVED SPECTRUM OF METHYL FORMATE BETWEEN 214 AND 265 GHZ, James P. McMillan, Sarah Fortman, Christopher F. Neese, Frank C. De Lucia

RI02 **1:47 – 2:02**
ROTATIONAL SPECTROSCOPY OF 4-HYDROXY-2-BUTYNENITRILE, R. A. Motiyenko, L. Margulès, J.-C. Guillemin

RI03 **2:04 – 2:19**
TIME-DOMAIN TERAHERTZ SPECTROSCOPY OF ISOLATED PAHS , Brandon Carroll, Marco A. Allodi, Brett A. McGuire, Sergio Ioppolo, Geoffrey Blake

RI04 **2:21 – 2:36**
HIGH-RESOLUTION IR ABSORPTION SPECTROSCOPY OF POLYCYCLIC AROMATIC HYDROCARBONS: SHINING LIGHT ON THE INTERSTELLAR 3 MICRON EMISSION BANDS, Elena Maltseva, Alessandra Candian, Xander Tielens, Annemieke Petrignani, J. Oomens, Wybren Jan Buma

RI05 **2:38 – 2:53**
EXPLORING MOLECULAR COMPLEXITY WITH ALMA (EMoCA): HIGH-ANGULAR-RESOLUTION OBSERVATIONS OF SAGITTARIUS B2(N) AT 3 mm, Holger S. P. Müller, Arnaud Belloche, Karl M. Menten, Robin T. Garrod

RI06 **2:55 – 3:10**
FIRST SPECTROSCOPIC STUDIES AND DETECTION IN SgrB2 OF ^{13}C-DOUBLY SUBSTITUED ETHYL CYANIDE, L. Margulès, R. A. Motiyenko, J.-C. Guillemin, Holger S. P. Müller, Arnaud Belloche

RI07 **3:12 – 3:27**
MILLIMETERWAVE SPECTROSCOPY OF ETHANIMINE AND PROPANIMINE AND THEIR SEARCH IN ORION, L. Margulès, R. A. Motiyenko, J.-C. Guillemin, Jose Cernicharo

Intermission

RI08 **3:46 – 4:01**
FURTHER STUDIES OF λ 5797.1 DIFFUSE INTERSTELLAR BAND, Takeshi Oka, L. M. Hobbs, Daniel E. Welty, Donald G. York, Julie Dahlstrom, Adolf N. Witt

RI09 **4:03 – 4:18**
LABORATORY OPTICAL SPECTROSCOPY OF THE PHENOXY RADICAL AS A DIFFUSE INTERSTELLAR BANDS CANDIDATE, Mitsunori Araki, Yuki MATSUSHITA, Koichi Tsukiyama

RI10 **4:20 – 4:35**
INVESTIGATION OF CARBONACEOUS INTERSTELLAR DUST ANALOGUES BY INFRARED SPECTROSCOPY: EFFECTS OF ENERGETIC PROCESSING, Belén Maté, Miguel Jiménez-Redondo, Isabel Tanarro, Victor Jose Herrero

RI11 **4:37 – 4:52**
REACTIONS OF GROUND STATE NITROGEN ATOMS N(^{4}S) WITH ASTROCHEMICALLY-RELEVANT MOLECULES ON INTERSTELLAR DUSTS, Lahouari Krim, Sendres Nourry

RI12 **4:54 – 5:09**
STABILITY OF GLYCINE TO ENERGETIC PROCESSING UNDER ASTROPHYSICAL CONDITIONS INVESTIGATED VIA INFRARED SPECTROSCOPY, Belén Maté, Victor Jose Herrero, Isabel Tanarro, Rafael Escribano

RI13 **5:11 – 5:21**
MILLIMETER AND SUBMILLIMETER STUDIES OF INTERSTELLAR ICE ANALOGUES, AJ Mesko, Ian C Wagner, Houston Hartwell Smith, Stefanie N Milam, Susanna L. Widicus Weaver

RI14 **5:23 – 5:38**
UNTANGLING MOLECULAR SIGNALS OF ASTROCHEMICAL ICES IN THE THz: DISTINGUISHING AMORPHOUS, CRYSTALLINE, AND INTRAMOLECULAR MODES WITH BROADBAND THz SPECTROSCOPY, Brett A. McGuire, Sergio Ioppolo, Xander de Vries, Marco A. Allodi, Brandon Carroll, Geoffrey Blake

RI15 **5:40 – 5:55**
QUANTUM CHEMICAL STUDY OF THE REACTION OF C^+ WITH INTERSTELLAR ICE: PREDICTIONS OF VIBRATIONAL AND ELECTRONIC SPECTRA OF REACTION PRODUCTS, David E. Woon

RJ. Cold/Ultracold/Matrices/Droplets

Thursday, June 25, 2015 – 1:30 PM

Room: 217 Noyes Laboratory

Chair: Gary E. Douberly, The University of Georgia, Athens, GA, USA

RJ01 **1:30 – 1:45**
IR SPECTRA OF COLD PROTONATED METHANE, Oskar Asvany, Koichi MT Yamada, Sandra Brünken, Alexey Potapov, Stephan Schlemmer

RJ02 **1:47 – 2:02**
PROGRESS ON OPTICAL ROTATIONAL COOLING OF SiO+, Patrick R Stollenwerk, Yen-Wei Lin, Brian C. Odom

RJ03 **2:04 – 2:19**
THE OPTICAL BICHROMATIC FORCE IN MOLECULAR SYSTEMS, Leland M. Aldridge, Scott E. Galica, Edward E. Eyler

RJ04 **2:21 – 2:36**
A NEW EQUATION OF STATE FOR SOLID *para*-HYDROGEN, Lecheng Wang, Robert J. Le Roy, Pierre-Nicholas Roy

RJ05 **2:38 – 2:53**
INFRARED SPECTROSCOPY OF NOH SUSPENDED IN SOLID PARAHYDROGEN: PART TWO, Morgan E. Balabanoff, Fredrick M. Mutunga, David T. Anderson

RJ06 **2:55 – 3:10**
HIGH RESOLUTION INFRARED SPECTROSCOPY OF CH_3F-(*ortho*-$H_2)_n$ CLUSTER IN SOLID *para*-H_2, Hiroyuki Kawasaki, Asao Mizoguchi, Hideto Kanamori

Intermission

RJ07 **3:29 – 3:44**
REACTIVE INTERMEDIATES IN ^{4}He NANODROPLETS: INFRARED LASER STARK SPECTROSCOPY OF DIHYDROXYCARBENE, Bernadette M. Broderick, Christopher P. Moradi, Gary E. Douberly, Laura McCaslin, John F. Stanton

RJ08 **3:46 – 4:01**
INFRARED LASER STARK SPECTROSCOPY OF THE PRE-REACTIVE Cl···HCl COMPLEX FORMED IN SUPERFLUID ^{4}He DROPLETS, Christopher P. Moradi, Gary E. Douberly

RJ09 **4:03 – 4:13**
HELIUM NANODROPLET INFRARED SPECTROSCOPY OF THE TROPYL RADICAL, Matin Kaufmann, Bernadette M. Broderick, Gary E. Douberly

RJ10 **4:15 – 4:30**
MICROSOLVATION STUDIES IN HELIUM NANODROPLETS, Gerhard Schwaab, Matin Kaufmann, Daniel Leicht, Raffael Schwan, Theo Fischer, Devendra Mani, Martina Havenith

RJ11 **4:32 – 4:47**
INFRARED SPECTRA OF THE CO_2-H_2O, CO_2-$(H_2O)_2$, and $(CO_2)_2$-H_2O COMPLEXES ISOLATED IN SOLID NEON BETWEEN 90 AND 5300 cm^{-1}, Benoît Tremblay, Pascale Soulard

RJ12 **4:49 – 4:59**
MATRIX ISOLATION AND COMPUTATIONAL STUDY OF [2C, 2N, X] (X=S, SE) ISOMERS, Tamas Voros, Gyorgy Tarczay

RJ13 **5:01 – 5:16**
MATRIX ISOLATION SPECTROSCOPY AND PHOTOCHEMISTRY OF TRIPLET 1,3-DIMETHYLPROPYNYLIDENE (MeC_3Me), Stephanie N. Knezz, Terese A Waltz, Benjamin C. Haenni, Nicola J. Burrmann, Robert J. McMahon

RJ14 **5:18 – 5:33**
EVIDENCE OF INTERNAL ROTATION IN THE O-H STRETCHING REGION OF THE 1:1 METHANOL-BENZENE COMPLEX IN AN ARGON MATRIX, Jay Amicangelo, Ian Campbell, Joshua Wilkins

FA. Electronic structure, potential energy surfaces

Friday, June 26, 2015 – 8:30 AM

Room: 116 Roger Adams Lab

Chair: Timothy Steimle, Arizona State University, Tempe, AZ, USA

FA01 **8:30 – 8:45**
CHARACTERIZATION OF THE $1\,^5\Pi_u$ - $1\,^5\Pi_g$ BAND OF C_2 BY TWO-COLOR RESONANT FOUR-WAVE MIXING AND LIF, Peter Radi

FA02 **8:47 – 9:02**
SIGN CHANGES IN THE ELECTRIC DIPOLE MOMENT OF EXCITED STATES IN RUBIDIUM-ALKALINE EARTH DIATOMIC MOLECULES, Johann V. Pototschnig, Florian Lackner, Andreas W. Hauser, Wolfgang E. Ernst

FA03 **9:04 – 9:19**
HIGH RESOLUTION VELOCITY MAP IMAGING PHOTOELECTRON SPECTROSCOPY OF THE BERYLLIUM OXIDE ANION, BEO-, Amanda Reed, Kyle Mascaritolo, Michael Heaven

FA04 **9:21 – 9:36**
ELECTRONIC AUTODETACHMENT SPECTROSCOPY AND IMAGING OF THE ALUMINUM MONOXIDE ANION, ALO-, Amanda Reed, Kyle Mascaritolo, Adrian Gardner, Michael Heaven

FA05 **9:38 – 9:53**
SPECTROSCOPY OF THE LOW-ENERGY STATES OF BaO^+, Joshua Bartlett, Robert A. VanGundy, Michael Heaven

Intermission

FA06 **10:12 – 10:27**
THE OPTICAL SPECTRUM OF SrOH RE-VISITED: ZEEMAN EFFECT, HIGH-RESOLUTION SPECTROSCOPY AND FRANCK-CONDON FACTORS. , Trung Nguyen, Damian L Kokkin, Timothy Steimle, Ivan Kozyryev, John M. Doyle

FA07 **10:29 – 10:44**
SPECTROSCOPIC ACCURACY IN QUANTUM CHEMISTRY: A BENCHMARK STUDY ON Na_3, Andreas W. Hauser, Johann V. Pototschnig, Wolfgang E. Ernst

FA08 **10:46 – 11:01**
ACCURATE FIRST-PRINCIPLES SPECTRA PREDICTIONS FOR ETHYLENE AND ITS ISOTOPOLOGUES FROM FULL 12D AB INITIO SURFACES, Thibault Delahaye, Michael Rey, Vladimir Tyuterev, Andrei V. Nikitin, Peter Szalay

FA09 **11:03 – 11:18**
HIGH-RESOLUTION LASER SPECTROSCOPY OF S_1-S_0 TRANSITION OF NAPHTHALENE: MEASUREMENT OF VIBRATIONALLY EXCITED STATES, Takumi Nakano, Ryo Yamamoto, Shunji Kasahara

FA10 **11:20 – 11:35**
HIGH-RESOLUTION LASER SPECTROSCOPY OF THE $S_1 \leftarrow S_0$ TRANSITION OF Cl-NAPHTHALENES, Shunji Kasahara, Ryo Yamamoto

FB. Spectroscopy as an analytical tool

Friday, June 26, 2015 – 8:30 AM

Room: 100 Noyes Laboratory

Chair: Christopher F. Neese, The Ohio State University, Columbus, OH, USA

FB01 **8:30 – 8:45**
CONTINUOUS MONITORING OF PHOTOLYSIS PRODUCTS BY THZ SPECTROSCOPY, Abdelaziz Omar, Arnaud Cuisset, Gaël Mouret, Francis Hindle, Sophie Eliet, Robin Bocquet

FB02 **8:47 – 9:02**
MEDIUM RESOLUTION CAVITY SPECTROSCOPY FOR THE STUDY OF LARGE MOLECULES, Satyakumar Nagarajan, Christopher F. Neese, Frank C. De Lucia

FB03 **9:04 – 9:19**
SUBMILLIMETER/INFRARED DOUBLE RESONANCE: REGIMES FOR MOLECULAR SENSORS, Sree Srikantaiah, Ivan Medvedev, Christopher F. Neese, Dane Phillips, Henry O. Everitt, Frank C. De Lucia

FB04 **9:21 – 9:36**
ROTATIONAL SPECTROSCOPY AS A TOOL TO INVESTIGATE INTERACTIONS BETWEEN VIBRATIONAL POLYADS IN SYMMETRIC TOP MOLECULES: LOW-LYING STATES $v_8 \leq 2$ OF METHYL CYANIDE, Holger S. P. Müller, Matthias H. Ordu, Frank Lewen, Linda Brown, Brian Drouin, John Pearson, Keeyoon Sung, Isabelle Kleiner, Robert Sams

FB05 **9:38 – 9:53**
VIBRATIONAL SUM FREQUENCY STUDY OF THE INFLUENCE OF WATER-IONIC LIQUID MIXTURES IN THE CO_2 ELECTROREDUCTION ON SILVER ELECTRODES, Natalia Garcia Rey, Dana Dlott

Intermission

FB06 **10:12 – 10:27**
ELUCIDATING THE COMPLEX LINESHAPES RESULTING FROM THE HIGHLY SENSITIVE, ION SELECTIVE, TECHNIQUE NICE-OHVMS, James N. Hodges, Brian Siller, Benjamin J. McCall

FB07 **10:29 – 10:44**
CHARACTERIZATION AND INFRARED EMISSION SPECTROSCOPY OF BALL PLASMOID DISCHARGES, Scott E. Dubowsky, Benjamin J. McCall

FB08 **10:46 – 11:01**
VUV FLUORESCENCE OF WATER & AMMONIA FOR SATELLITE THRUSTER PLUME CHARACTERIZATION., Justin W. Young, Christopher Annesley, Ryan S Booth, Jaime A. Stearns

FB09 **11:03 – 11:18**
REACTIONS OF 3-OXETANONE AT HIGH TEMPERATURES, Emily Wright, Brian Warner, Hannah Foreman, Kimberly N. Urness, Laura R. McCunn

FC. Comparing theory and experiment

Friday, June 26, 2015 – 8:30 AM

Room: B102 Chemical and Life Sciences

Chair: Edwin Sibert, The Univeristy of Wisconsin, Madison, WI, USA

FC01 **8:30 – 8:45**
VIBRATIONAL COUPLING IN SOLVATED FORM OF EIGEN PROTON, Jheng-Wei Li, Kaito Takahashi, Jer-Lai Kuo

FC02 **8:47 – 9:02**
BINDING BETWEEN NOBEL GAS ATOMS AND PROTONATED WATER MONOMER AND DIMER, Ying-Cheng Li, Jer-Lai Kuo

FC03 **9:04 – 9:19**
ANALYSIS OF HYDROGEN BONDING IN THE OH STRETCH REGION OF PROTONATED WATER CLUSTERS, Laura C. Dzugan, Anne B McCoy

FC04 **9:21 – 9:36**
SEMIEXPERIMENTAL STRUCTURE OF THE NON-RIGID BF_2OH MOLECULE BY COMBINING HIGH RESOLUTION INFRARED SPECTROSCOPY AND AB INITIO CALCULATIONS. , Natalja Vogt, Jean Demaison, Agnes Perrin, Hans Bürger

FC05 **9:38 – 9:48**
CONFORMATIONAL, VIBRATIONAL AND ELECTRONIC PROPERTIES OF C5H3XOS (X = H, F, Cl OR Br): HALOGEN AND SOLVENT EFFECTS , Mustafa Senyel, Gunes Esma, Cemal Parlak

FC06 **9:50 – 10:05**
COMBINED EXPERIMENTAL AND THEORETICAL STUDIES ON THE VIBRATIONAL AND ELECTRONIC SPECTRA OF 5-QUINOLINECARBOXALDEHYDE, Mustafa Kumru, Mustafa Kocademir, Tayyibe Bardakci

Intermission

FC07 **10:24 – 10:39**
COMBINED COMPUTATIONAL AND EXPERIMENTAL STUDIES OF THE DUAL FLUORESCENCE IN DIMETHYLAMINOBENZONITRILE (DMABN), Anastasia Edsell, Steven Shipman

FC08 **10:41 – 10:56**
MODELING SPIN-ORBIT COUPLING IN THE HALOCARBENES , Phalgun Lolur, Richard Dawes, Scott Reid, Silver Nyambo

FC09 **10:58 – 11:13**
GAS-PHASE CONFORMATIONS AND ENERGETICS OF PROTONATED 2′-DEOXYADENOSINE-5′-MONOPHOSPHATE AND ADENOSINE-5′-MONOPHOSPHATE: IRMPD ACTION SPECTROSCOPY AND THEORETICAL STUDIES, Ranran Wu, Y-W Nei, Chenchen He, Lucas Hamlow, Giel Berden, J. Oomens, M T Rodgers

FD. Atmospheric science

Friday, June 26, 2015 – 8:30 AM

Room: 274 Medical Sciences Building

Chair: Kyle N Crabtree, University of California, Davis, Davis, CA, USA

FD01 **8:30 – 8:45**
OBSERVATION OF THE SIMPLEST CRIEGEE INTERMEDIATE CH_2OO IN THE GAS-PHASE OZONOLYSIS OF ETHYLENE, Carrie Womack, Marie-Aline Martin-Drumel, Gordon G Brown, Robert W Field, Michael C McCarthy

FD02 **8:47 – 9:02**
HIGH-RESOLUTION SPECTRA OF CH_2OO : ASSIGNMENTS OF ν_5 AND $2\nu_9$ BANDS AND OVERLAPPED BANDS OF ICH_2OO, Yu-Hsuan Huang, Li-Wei Chen, Yuan-Pern Lee

FD03 **9:04 – 9:19**
DIRECT INFRARED IDENTIFICATION OF THE CRIEGEE INTERMEDIATES syn- and $anti$-CH_3CHOO AND THEIR DISTINCT CONFORMATION-DEPENDENT REACTIVITY, Hui-Yu Lin, Yu-Hsuan Huang, Xiaohong Wang, Joel Bowman, Yoshifumi Nishimura, Henry A Witek, Yuan-Pern Lee

FD04 **9:21 – 9:36**
THE $\tilde{A}$-$\tilde{X}$ ELECTRONIC TRANSITIONS OF THE CH_2BrOO AND CH_2ClOO RADICALS IN THE NEAR INFRARED REGION, Neal Kline, Meng Huang, Terry A. Miller

FD05 **9:38 – 9:53**
THE $\tilde{A}$-$\tilde{X}$ ELECTRONIC TRANSITION OF CH_2IOO RADICAL IN THE NEAR INFRARED REGION, Neal Kline, Meng Huang, Terry A. Miller, Phalgun Lolur, Richard Dawes

FD06 **9:55 – 10:10**
A THEORETICAL CHARACTERIZATION OF ELECTRONIC STATES OF CH_2IOO AND CH_2OO RADICALS RELEVANT TO THE NEAR IR REGION , Richard Dawes, Phalgun Lolur, Meng Huang, Neal Kline, Terry A. Miller

Intermission

FD07 **10:29 – 10:44**
JET-COOLED LASER-INDUCED FLUORESCENCE SPECTROSCOPY OF T-BUTOXY, Neil J Reilly, Lan Cheng, John F. Stanton, Terry A. Miller, Jinjun Liu

FD08 **10:46 – 11:01**
NITROSYL IODIDE, INO: MILLIMETER-WAVE SPECTROSCOPY GUIDED BY $AB\ INITIO$ QUANTUM CHEMICAL COMPUTATION, Stephane Bailleux, Denis Duflot, Shohei Aiba, Hiroyuki Ozeki

FD09 **11:03 – 11:18**
DISPERSED FLUORESCENCE SPECTROSCOPY OF JET-COOLED ISOBUTOXY, 2-METHYL-1-BUTOXY, AND ISOPENTOXY RADICALS , Md Asmaul Reza, Neil J Reilly, Jahangir Alam, Amy Mason, Jinjun Liu

FD10 **11:20 – 11:35**
PHOTODISSOCIATION OF METHYL ISOTHIOCYANATE STUDIED USING CHIRPED PULSE UNIFORM FLOW SPECTROSCOPY, Nuwandi M Ariyasingha, Lindsay N. Zack, Chamara Abeysekera, Baptiste Joalland, Arthur Suits

FD11 **11:37 – 11:52**
DISPERSED FLUORESCENCE SPECTROSCOPY OF JET-COOLED METHYLCYCLOHEXOXY RADICALS , Jahangir Alam, Md Asmaul Reza, Amy Mason, Jinjun Liu

FE. Small molecules
Friday, June 26, 2015 – 8:30 AM
Room: 217 Noyes Laboratory

Chair: Robert W Field, MIT, Cambridge, MA, USA

FE01 **8:30 – 8:45**
TOWARDS A GLOBAL FIT OF THE COMBINED MILLIMETER-WAVE AND HIGH RESOLUTION FTIR DATA FOR THE LOWEST EIGHT VIBRATIONAL STATES OF HYDRAZOIC ACID (HN_3) , Brent K. Amberger, R. Claude Woods, Brian J. Esselman, Robert J. McMahon

FE02 **8:47 – 9:02**
MILLIMETER-WAVE SPECTROSCOPY AND GLOBAL ANALYSIS OF THE LOWEST EIGHT VIBRATIONAL STATES OF DEUTERATED HYDRAZOIC ACID (DN_3), Brent K. Amberger, R. Claude Woods, Brian J. Esselman, Robert J. McMahon

FE03 **9:04 – 9:19**
SIMPLIFIED CARTESIAN BASIS MODEL FOR INTRAPOLYAD EMISSION INTENSITIES IN THE $\tilde{A} \rightarrow \tilde{X}$ BENT-TO-LINEAR TRANSITION OF ACETYLENE, Barratt Park, Adam H. Steeves, Joshua H Baraban, Robert W Field

FE04 **9:21 – 9:36**
OBSERVATION OF LEVEL-SPECIFIC PREDISSOCIATION RATES IN S_1 ACETYLENE, Catherine A. Saladrigas, Jun Jiang, Robert W Field

FE05 **9:38 – 9:53**
FULL DIMENSIONAL ROVIBRATIONAL VARIATIONAL CALCULATIONS OF THE S_1 STATE OF C_2H_2, Bryan Changala, Joshua H Baraban, John F. Stanton

Intermission

FE06 **10:12 – 10:27**
MILLIMETER-WAVE SPECTROSCOPY OF FORMYL AZIDE ($HC(O)N_3$), Nicholas A. Walters, Brent K. Amberger, Brian J. Esselman, R. Claude Woods, Robert J. McMahon

FE07 **10:29 – 10:44**
MILLIMETER-WAVE ROTATIONAL SPECTRUM OF DEUTERATED NITRIC ACID, Rebecca A.H. Butler, Camren Coplan, Doug Petkie, Ivan Medvedev, Frank C. De Lucia

FE08 **10:46 – 11:01**
THEORETICAL ANALYSIS OF THE RESONANCE FOUR-WAVE MIXING AMPLITUDES: A FULLY NON-DEGENERATE CASE., Alexander Kouzov

FE09 **11:03 – 11:18**
SAND IN THE LABORATORY. PRODUCTION AND INTERROGATION OF GAS PHASE SILICATES., Damian L Kokkin, Timothy Steimle

FE10 ***Post-Deadline Abstract*** **11:20 – 11:35**
IMPACT OF COMPLEX-VALUED ENERGY FUNCTION SINGULARITIES ON THE BEHAVIOUR OF RAYLEIGH-SCHRÖDINGER PERTURBATION SERIES. H_2CO MOLECULE VIBRATIONAL ENERGY SPECTRUM., Andrey Duchko, Alexandr Bykov

MA. Plenary

Monday, June 22, 2015 – 8:30 AM

Room: Foellinger Auditorium

Chair: Gregory S. Girolami, University of Illinois at Urbana-Champaign, Urbana, IL, USA

Welcome **8:30**
Phyllis M. Wise, Chancellor
University of Illinois at Urbana-Champaign

MA01 **8:40 – 9:20**

BREATHING EASIER THROUGH SPECTROSCOPY: STUDYING FREE RADICAL REACTIONS IN AIR POLLUTION CHEMISTRY

MITCHIO OKUMURA[a], *Division of Chemistry and Chemical Engineering, California Institute of Technology, Pasadena, CA, USA.*

Air pollution arises from the oxidation of volatile organic compounds emitted into the atmosphere from both anthropogenic and biogenic sources. Free radicals dominate the gas phase chemistry leading to the formation of tropospheric ozone, oxygenated organic molecules and organic aerosols, but this chemistry is complex. In this presentation, advances in our understanding of the spectroscopy and chemistry of atmospheric free radicals will be described that have come from exploiting the sensitivity and specificity of methods such as Cavity Ringdown Spectroscopy, Multiplexed Photoionization Mass Spectrometry and Cavity-Enhanced Frequency Comb Spectroscopy.

[a]This work was supported by the National Science Foundation and the National Aeronautics and Space Administration.

MA02 **9:25 – 10:05**

MOLECULAR ROTATION SIGNALS: MOLECULE CHEMISTRY AND PARTICLE PHYSICS

JENS-UWE GRABOW, *Institut für Physikalische Chemie und Elektrochemie, Gottfried-Wilhelm-Leibniz-Universität, Hannover, Germany.*

Molecules - large or small - are attractive academic resources, with numerous questions on their chemical behaviour as well as problems in fundamental physics now (or still) waiting to be answered: Targeted by high-resolution spectroscopy, a rotating molecular top can turn into a laboratory for molecule chemistry or a laboratory for particle physics.

Once successfully entrained (many species - depending on size and chemical composition - have insufficient vapour pressures or are of transient nature, such that specifically designed pulsed-jet sources are required for their transfer into the gas phase or in-situ generation) into the collision-free environment of a supersonic-jet expansion, each molecular top comes with its own set of challenges, theoretically and experimentally: Multiple internal interactions are causing complicated energy level schemes and the resulting spectra will be rather difficult to predict theoretically. Experimentally, these spectra are difficult to assess and assign. With today's broad-banded chirp microwave techniques, finding and identifying such spectral features have lost their major drawback of being very time consuming for many molecules. For other molecules, the unrivalled resolution and sensitivity of the narrow-banded impulse microwave techniques provide a window to tackle - at the highest precision available to date – fundamental questions in physics, even particle physics – potentially beyond the standard model.

Molecular charge distribution, properties of the chemical bond, details on internal dynamics and intermolecular interaction, the (stereo-chemical) molecular structure (including the possibility of their spatial separation) as well as potential evidence for tiny yet significant interactions encode their signature in pure molecular rotation subjected to time-domain microwave spectroscopic techniques. Ongoing exciting technical developments promise rapid progress. We present recent examples from Hannover, new directions, and an outlook at the future of molecular rotation spectroscopy.

Intermission

MA03 **10:35 – 11:15**

IT IS WATER WHAT MATTERS: THz SPECTROSCOPY AS A TOOL TO STUDY HYDRATION DYNAMICS

MARTINA HAVENITH, *Physikalische Chemie II, Ruhr University Bochum, Bochum, Germany.*

Terahertz absorption spectroscopy has turned out to be a new powerful tool to study biomolecular hydration. The development of THz technology helped to fill the experimental gap in this frequency range. These experimental advances had to go hand in hand with the development of theoretical concepts that have been developed in the recent years to describe the underlying solute-induced sub-picosecond dynamics of the hydration shell. This frequency range covers the rattling modes of the ion with its hydration cage and allowed to derive major conclusions on the molecular picture of ion hydration, a key issue in chemistry. By a combination of experiment and theory it is now possible to rigorously dissect the THz spectrum of a solvated biomolecule into the distinct solute, solvent and solute-solvent coupled contributions Moreover, we highlight recent results that show the significance of hydrogen bond dynamics for molecular recognition. In all of these examples, a gradient of water motion toward functional sites of proteins is observed, the so-called hydration funnel. The efficiency of the coupling at THz frequencies is explained in terms of a two-tier (short- and long-range) solute-solvent interaction.

MA04 **11:20 – 12:00**

CPUF: CHIRPED-PULSE MICROWAVE SPECTROSCOPY IN PULSED UNIFORM SUPERSONIC FLOWS

ARTHUR SUITS, CHAMARA ABEYSEKERA, LINDSAY N. ZACK, BAPTISTE JOALLAND, NUWANDI M ARIYASINGHA, *Department of Chemistry, Wayne State University, Detroit, MI, USA*; BARRATT PARK, ROBERT W FIELD, *Department of Chemistry, MIT, Cambridge, MA, USA*; IAN SIMS, *Institut de Physique de Rennes, Université de Rennes 1, Rennes, France.*

Chirped-pulse Fourier-transform microwave spectroscopy has stimulated a resurgence of interest in rotational spectroscopy owing to the dramatic reduction in spectral acquisition time it enjoys when compared to cavity-based instruments. This suggests that it might be possible to adapt the method to study chemical reaction dynamics and even chemical kinetics using rotational spectroscopy. The great advantage of this would be clear, quantifiable spectroscopic signatures for polyatomic products as well as the possibility to identify and characterize new radical reaction products and transient intermediates. To achieve this, however, several conditions must be met: 1) products must be thermalized at low temperature to maximize the population difference needed to achieve adequate signal levels and to permit product quantification based on the rotational line strength; 2) a large density and volume of reaction products is also needed to achieve adequate signal levels; and 3) for kinetics studies, a uniform density and temperature is needed throughout the course of the reaction. These conditions are all happily met by the uniform supersonic flow produced from a Laval nozzle expansion. In collaboration with the Field group at MIT we have developed a new instrument we term a CPUF (Chirped-pulse/Uniform Flow) spectrometer in which we can study reaction dynamics, photochemistry and kinetics using broadband microwave and millimeter wave spectroscopy as a product probe. We will illustrate the performance of the system with a few examples of photodissociation and reaction dynamics, and also discuss a number of challenges unique to the application of chirped-pulse microwave spectroscopy in the collisional environment of the flow. Future directions and opportunities for application of CPUF will also be explored.

MF. Mini-symposium: High-Precision Spectroscopy

Monday, June 22, 2015 – 1:30 PM

Room: 116 Roger Adams Lab

Chair: Michael Heaven, Emory University, Atlanta, GA, USA

MF01 *Journal of Molecular Spectroscopy Review Lecture* **1:30 – 2:00**

PHYSICS BEYOND THE STANDARD MODEL FROM MOLECULAR HYDROGEN SPECTROSCOPY

WIM UBACHS, EDCEL JOHN SALUMBIDES, JULIJA BAGDONAITE, *Department of Physics and Astronomy, VU University , Amsterdam, Netherlands.*

The spectrum of molecular hydrogen can be measured in the laboratory to very high precision using advanced laser and molecular beam techniques, as well as frequency-comb based calibration [1,2]. The quantum level structure of this smallest neutral molecule can now be calculated to very high precision, based on a very accurate (10^{-15} precision) Born-Oppenheimer potential [3] and including subtle non-adiabatic, relativistic and quantum electrodynamic effects [4]. Comparison between theory and experiment yields a test of QED, and in fact of the Standard Model of Physics, since the weak, strong and gravitational forces have a negligible effect. Even fifth forces beyond the Standard Model can be searched for [5]. Astronomical observation of molecular hydrogen spectra, using the largest telescopes on Earth and in space, may reveal possible variations of fundamental constants on a cosmological time scale [6]. A study has been performed at a 'look-back' time of 12.5 billion years [7]. In addition the possible dependence of a fundamental constant on a gravitational field has been investigated from observation of molecular hydrogen in the photospheres of white dwarfs [8]. The latter involves a test of the Einsteins equivalence principle.

[1] E.J. Salumbides et al., Phys. Rev. Lett. 107, 143005 (2011).
[2] G. Dickenson et al., Phys. Rev. Lett. 110, 193601 (2013).
[3] K. Pachucki, Phys. Rev. A82, 032509 (2010).
[4] J. Komasa et al., J. Chem. Theory Comp. 7, 3105 (2011).
[5] E.J. Salumbides et al., Phys. Rev. D87, 112008 (2013).
[6] F. van Weerdenburg et al., Phys. Rev. Lett. 106, 180802 (2011).
[7] J. Badonaite et al., Phys. Rev. Lett. 114, 071301 (2015).
[8] J. Bagdonaite et al., Phys. Rev. Lett. 113, 123002 (2014).

MF02 **2:05 – 2:20**

PRECISION SPECTROSCOPY ON HIGHLY-EXCITED VIBRATIONAL LEVELS OF H_2

MING LI NIU, EDCEL JOHN SALUMBIDES, WIM UBACHS, *Department of Physics and Astronomy, VU University , Amsterdam, Netherlands.*

The ground electronic energy levels of H_2 have been used as a benchmark system for the most precise comparisons between *ab initio* calculations and experimental investigations. Recent examples include the determinations of the ionization energy [1], fundamental vibrational energy splitting [2], and rotational energy progression extending to $J = 16$ [3]. In general, the experimental and theoretical values are in excellent agreement with each other. The energy calculations, however, reduce in accuracy with the increase in rotational and vibrational excitation, limited by the accuracy of non-Born Oppenheimer corrections, as well as the higher-order QED effects. While on the experimental side, it remains difficult to sufficiently populate these excited levels in the ground electronic state.

We present here our high-resolution spectroscopic study on the X $^1\Sigma_g^+$ electronic ground state levels with very high vibrational quanta ($\nu = 10, 11, 12$). Vibrationally-excited H_2 are produced from the photodissociation of H_2S [4], and subsequently probed by a narrowband pulsed dye laser system. The experimental results are consistent with and more accurate than the best theoretical values [5]. These vibrationally-excited level energies are also of interest to studies that extract constraints on the possible new interactions that extend beyond the Standard Model [6].

[1] J. Liu et al., J. Chem. Phys. 130, 174306 (2009).
[2] G. Dickenson et al., Phys. Rev. Lett. 110, 193601 (2013).
[3] E.J. Salumbides et al., Phys. Rev. Lett. 107, 143005 (2011).
[4] J. Steadman and T. Baer, J. Chem. Phys. 91, 6113 (1989).
[5] J. Komasa et al., J. Chem. Theory Comp. 7, 3105 (2011).
[6] E.J. Salumbides et al., Phys. Rev. D 87, 112008 (2013).

MF03 **2:22 – 2:37**

BOUNDS ON THE NUMBER AND SIZE OF EXTRA DIMENSIONS FROM MOLECULAR SPECTROSCOPY

EDCEL JOHN SALUMBIDES, *Department of Physics and Astronomy, VU University , Amsterdam, Netherlands*; BERT SCHELLEKENS, *Theoretical Physics, Nikhef, Amsterdam, Netherlands*; BEATRIZ GATO-RIVERA, *Instituto de Fisica Fundamental, CSIC, Madrid, Spain*; WIM UBACHS, *Department of Physics and Astronomy, VU University , Amsterdam, Netherlands.*

Modern string theories, which seek to produce a consistent description of physics beyond the Standard Model that also includes the gravitational interaction, appear to be most consistent if a large number of dimensions are postulated. For example the mysterious M-theory, which generalizes all consistent versions of superstring theories, require 11 dimensions. We demonstrate that investigations of quantum level energies in simple molecular systems provide a testing ground to constrain the size of compactified extra dimensions, for example those proposed in the ADD [1] and RS scenarios [2]. This is made possible by the recent progress in precision metrology with ultrastable lasers on energy levels in neutral molecular hydrogen (H_2, HD and D_2) [3] and the molecular hydrogen ions (H_2^+, HD^+ and D_2^+) [4]. Comparisons between experiment and quantum electrodynamics calculations for these molecular systems can be interpreted in terms of probing large extra dimensions, under which conditions gravity will become much stronger. Molecules are a probe of space-time geometry at typical distances where chemical bonds are effective, i.e. at length scales of an Å.

[1] N. Arkani-Hamed, S. Dimopoulos and G. Dvali, Phys. Lett. B 429, 263 (1998)
[2] L. Randall and R. Sundrum, Phys. Rev. Lett. 83, 3370 (1999).
[3] G. Dickenson et al., Phys. Rev. Lett. 110, 193601 (2013).
[4] J. C. J. Koelemeij et al., Phys. Rev. Lett. 98, 173002 (2007).

MF04 **2:39 – 2:54**

CONTINUOUS SUPERSONIC EXPANSION DISCHARGE SOURCE FOR HIGH-PRECISION MID-INFRARED SPECTROSCOPY OF COLD MOLECULAR IONS

COURTNEY TALICSKA, MICHAEL PORAMBO, *Department of Chemistry, University of Illinois at Urbana-Champaign, Urbana, IL, USA*; BENJAMIN J. McCALL, *Departments of Chemistry and Astronomy, University of Illinois at Urbana-Champaign, Urbana, IL, USA.*

The low temperatures and pressures of the interstellar medium provide an ideal environment for gas phase ion-neutral reactions that play an essential role in the chemistry of the universe. High-precision laboratory spectra of molecular ions are necessary to facilitate new astronomical discoveries and provide a deeper understanding of interstellar chemistry, but forming ions in measurable quantities in the laboratory has proved challenging. Even when cryogenically cooled, the high temperatures and pressures of typical discharge cells lead to diluted and congested spectra from which extracting chemical information is difficult. Here we overcome this challenge by coupling an electric discharge to a continuous supersonic expansion source to form ions cooled to low temperatures. The ion production abilities of the source have been demonstrated previously as ion densities on the order of 10^{10}-10^{12} cm^{-3} have been observed for $H_3{}^+$.[a] With a smaller rotational constant and the expectation that it will be formed with comparable densities, $HN_2{}^+$ is used as a reliable measure of the cooling abilities of the source. Ions are probed through the use of a widely tunable mid-infrared (3-5 μm) spectrometer based on light formed by difference frequency generation and noise-immune cavity-enhanced optical heterodyne molecular spectroscopy (NICE-OHMS).[b] To improve the sensitivity of the instrument the discharge is electrically modulated and the signal is fed into a lock-in amplifier before being recorded by a custom data acquisition program. Rovibrational transitions of $H_3{}^+$ and $HN_2{}^+$ have been recorded, giving rotational temperatures of 80-120 K and 35-40 K, respectively. With verification that the source is producing rotationally cold ions, we move toward the study of primary ions of more astronomical significance, including H_2CO^+.

[a] K. N. Crabtree, C. A. Kaufman, and B. J. McCall, *Rev. Sci. Instrum.* **81**, 086103 (2010).
[b] M. W. Porambo, B. M. Siller, J. M. Pearson, and B. J. McCall, *Opt. Lett.* **37**, 4422 (2012)

MF05 **2:56 – 3:11**

PROGRESS TOWARDS A HIGH-PRECISION INFRARED SPECTROSCOPIC SURVEY OF THE H_3^+ ION

ADAM J. PERRY, JAMES N. HODGES, CHARLES R. MARKUS, G. STEPHEN KOCHERIL, PAUL A. JENKINS II, *Department of Chemistry, University of Illinois at Urbana-Champaign, Urbana, IL, USA*; BENJAMIN J. McCALL, *Departments of Chemistry and Astronomy, University of Illinois at Urbana-Champaign, Urbana, IL, USA.*

The trihydrogen cation, H_3^+, represents one of the most important and fundamental molecular systems. Having only two electrons and three nuclei, H_3^+ is the simplest polyatomic system and is a key testing ground for the development of new techniques for calculating potential energy surfaces and predicting molecular spectra. Corrections that go beyond the Born-Oppenheimer approximation, including adiabatic, non-adiabatic, relativistic, and quantum electrodynamic corrections are becoming more feasible to calculate[abcd]. As a result, experimental measurements performed on the H_3^+ ion serve as important benchmarks which are used to test the predictive power of new computational methods.

By measuring many infrared transitions with precision at the sub-MHz level it is possible to construct a list of the most highly precise experimental rovibrational energy levels for this molecule. Until recently, only a select handful of infrared transitions of this molecule have been measured with high precision ($\sim$ 1 MHz)[e]. Using the technique of Noise Immune Cavity Enhanced Optical Heterodyne Velocity Modulation Spectroscopy, we are aiming to produce the largest high-precision spectroscopic dataset for this molecule to date. Presented here are the current results from our survey along with a discussion of the combination differences analysis used to extract the experimentally determined rovibrational energy levels.

[a] O. Polyansky, *et al., Phil. Trans. R. Soc. A* (2012), **370**, 5014.
[b] M. Pavanello, *et al., J. Chem. Phys.* (2012), **136**, 184303.
[c] L. Diniz, *et al., Phys. Rev. A* (2013), **88**, 032506.
[d] L. Lodi, *et al., Phys. Rev. A* (2014), **89**, 032505.
[e] J. Hodges, *et al., J. Chem. Phys* (2013), **139**, 164201.

Intermission

MF06 **3:30 – 3:45**

HIGH PRECISION INFRARED SPECTROSCOPY OF OH^+

CHARLES R. MARKUS, ADAM J. PERRY, JAMES N. HODGES, G. STEPHEN KOCHERIL, PAUL A. JENKINS II, *Department of Chemistry, University of Illinois at Urbana-Champaign, Urbana, IL, USA*; BENJAMIN J. McCALL, *Departments of Chemistry and Astronomy, University of Illinois at Urbana-Champaign, Urbana, IL, USA.*

The molecular ion OH^+ is of significant importance to interstellar chemistry. OH^+ is a key intermediate in the formation of water, and the ratios of OH^+ to H_2O^+ and H_3O^+ have been used to calculate the cosmic ray ionization rates in diffuse molecular clouds.[abc] To improve on previous spectroscopic work, the sensitive technique Noise Immune Cavity Enhanced Optical Heterodyne Velocity Modulation Spectroscopy (NICE-OHVMS) has been used to record rovibrational transitions of OH^+. Previously this approach has been used to investigate HCO^+, H_3^+, CH_5^+, and HeH^+.[de] Using an optical frequency comb for precise frequency calibration, the OH^+ line centers have been determined with $\sim$MHz uncertainties. Here the most precise and accurate list of rovibrational transitions of OH^+ is presented. These values can then be used to empirically determine rotational transitions through combination difference analysis.

[a] F. Wyrowski *et al., Astron. Astrophys.*, (2010),**26**,5.
[b] E. González-Alfonso *et al., Astron. Astrophys.*, (2013), **550**, A25.
[c] N. Indriolo *et al., Astrophys. J.*, (2015), **800**, 40.
[d] J.N. Hodges *et al., Chem. Phys.*, (2013), **139**, 164201.
[e] A.J. Perry *et al., J. Chem. Phys.*, (2014), **141**, 101101.

MF07 **3:47 – 4:02**

TOWARD TWO-COLOR SUB-DOPPLER SATURATION RECOVERY KINETICS IN CN (X, v = 0, J)

HONG XU, DAMIEN FORTHOMME, *Department of Chemistry, Brookhaven National Laboratory, Upton, NY, USA*; TREVOR SEARS[a], GREGORY HALL, *Chemistry Department, Brookhaven National Laboratory, Upton, NY, USA*; PAUL DAGDIGIAN, *Department of Chemistry, Johns Hopkins University, Baltimore, MD, USA*.

Collision-induced rotational energy transfer among rotational levels of ground state CN (X $^2\Sigma^+$, v = 0) radicals has been probed by saturation recovery experiments, using high-resolution, polarized transient FM spectroscopy to probe the recovery of population and the decay of alignment following ns pulsed laser depletion of selected CN rotational levels. Despite the lack of Doppler selection in the pulsed depletion and the thermal distribution of collision velocities, the recovery kinetics are found to depend on the probed Doppler shift of the depleted signal. The observed Doppler-shift-dependent recovery rates are a measure of the velocity dependence of the inelastic cross sections, combined with the moderating effects of velocity-changing elastic collisions. New experiments are underway, in which the pulsed saturation is performed with sub-Doppler velocity selection. The time evolution of the spectral hole bleached in the initially thermal CN absorption spectrum can characterize speed-dependent inelastic collisions along with competing elastic velocity-changing collisions, all as a function of the initially bleached velocity group and rotational state. The initial time evolution of the depletion recovery spectrum can be compared to a stochastic model, using differential cross sections for elastic scattering as well as speed-dependent total inelastic cross sections, derived from ab initio scattering calculations. Progress to date will be reported.

Acknowledgments: Work at Brookhaven National Laboratory was carried out under Contract No. DE-AC02-98CH10886 and DE-SC0012704 with the U.S. Department of Energy and supported by its Office of Basic Energy Sciences, Division of Chemical Sciences, Geosciences and Biosciences.

---

[a]also: Chemistry Department, Stony Brook University, Stony Brook, NY 11794-3400

MF08 **4:04 – 4:19**

AN EMPIRICAL DIPOLE POLARIZABILITY FOR He FROM A FIT TO SPECTROSCOPIC DATA YIELDING ANALYTIC EMPIRICAL POTENTIALS FOR ALL ISOTOPOLOGUES OF HeH^+

YOUNG-SANG CHO, ROBERT J. LE ROY, *Department of Chemistry, University of Waterloo, Waterloo, ON, Canada*; NIKESH S. DATTANI[a], *Graduate School of Science, Department of Chemistry, Kyoto University, Kyoto, Japan.*

All available spectroscopic data for all stable isotopologues of HeH^+ are analyzed with a direct-potential-fit (DPF) procedure that uses least-squares fits to experimental data in order to optimize the parameters defining an analytic potential. Since the coefficient of the leading $(1/r^4)$ inverse-power term is $C_4 = \alpha^{\rm He}/2$, when treated as a free parameter in the fit, it provides an independent empirical estimate of the polarizability of the He atom. The fact that the present model for the long-range behaviour includes accurate theoretical C_6, C_7 and C_8 coefficients (which are held fixed in the fits) should make it possible to obtain a good estimate of this quantity.

The Boltzmann constant k_B, a fundamental constant that can define temperature, is directly related to the dipole polarizability α of a gas by the expression $k_B = \frac{\alpha}{3\epsilon_0}\left(\frac{\epsilon_r+2}{\epsilon_r-1}\right)\frac{p}{T}$, in which ϵ_0 is the permitivity of free space, and ϵ_r is the relative dielectric permitivity at pressure p and temperature T. If k_B can be determined with greater precision, it can be used to define temperature based on a fundamental constant, rather than based on the rather arbitrary triple point of water, which is only known to 5 digits of precision. α for He is known theoretically to 8 digits of precision, but an empirical value lags behind. This work, examines the question of how precisely $\alpha^{\rm He}$ can be determined from a DPF to spectroscopic HeH^+ data, where the limiting long-range tail of the analytic potential has the correct form implied by Rydberg theory: $\alpha^{\rm He}/2r^4$. Although the highest observed vibrational level is bound by over 1000 cm^{-1}, our current fits determine an empirical $C_4 = \alpha^{\rm He}/2$ with an uncertainty of only 0.6%. It has been shown that with more precise spectroscopic data near the dissociation, $\alpha^{\rm He}$ can be determined with high enough precision to determine a more precise k_B and hence redefine temperature more accurately[b].

---

[a]dattani.nike@gmail.com

[b]Dattani N S. & Puchalski M. (2015) *Physical Review Letters* (in press)

MF09 **4:21 – 4:36**

ANALYTIC EMPIRICAL POTENTIALS FOR BeH^+, BeD^+, AND BeT^+ INCLUDING UP TO 4TH ORDER QED IN THE LONG-RANGE, AND PREDICTIONS FOR THE HALO NUCLEONIC MOLECULES $^{11}BeH^+$ and $^{14}BeH^+$.

LENA C. M. LI CHUN FONG, *Department of Chemistry, University of Waterloo, Waterloo, ON, Canada*; GRZEGORZ LACH, *International Institute of Molecular and Cell Biology, Warsaw, Poland*; ROBERT J. LE ROY, *Department of Chemistry, University of Waterloo, Waterloo, ON, Canada*; NIKESH S. DATTANI[a], *Graduate School of Science, Department of Chemistry, Kyoto University, Kyoto, Japan.*

The 13.81(8)s half-life of the halo nucleonic atom ^{11}Be is orders of magnitude longer than those for any other halo nucleonic atom known, and makes Be-based diatomics the most promising candidates for the formation of the first halo nucleonic molecules. However, the $4e^-$ species LiH and BeH^+ are some of the first molecules for which the highest accuracy *ab initio* methods *are not* accessible, so empirical potential energy functions will be important for making predictions and for benchmarking how *ab initio* calculations break down at this transition from $3e^-$ to $4e^-$. BeH^+ is also very light, and has one of the most extensive data sets involving a tritium isotopologue, making it a very useful benchmark for studying Born-Oppenheimer breakdown. We therefore seek to determine an empirical analytic potential energy function for BeH^+ that has as much precision as possible. To this end, all available spectroscopic data for all stable isotopologues of BeH^+ are analyzed in a standard direct-potential-fit procedure that uses least-squares fits to optimize the parameters defining an analytic potential. The "Morse/Long-range" (MLR) model used for the potential energy function incorporates the inverse-power long-range tail required by theory, and the calculation of the leading long-range coefficients C_4, C_6, C_7, and C_8 include non-adiabatic terms, and up to 4th order QED corrections. As a by-product, we have calculated some fundamental properties of $1e^-$ systems with unprecedented precision, such as the dipole, quadrupole, octupole, non-adiabatic, and mixed higher order polarizabilities of hydrogen, deuterium, and tritium. We provide good first estimates for the transition energies for the halo nucleonic species $^{11}BeH^+$ and $^{14}BeH^+$.

[a]dattani.nike@gmail.com

MF10 **4:38 – 4:53**

ANALYTIC EMPIRICAL POTENTIAL AND ITS COMPARISON TO STATE OF THE ART *ab initio* CALCULATIONS FOR THE $6e^-$ EXCITED $b(1^3\Pi_u)$-STATE OF Li_2.

NIKESH S. DATTANI, *Graduate School of Science, Kyoto University, Kyoto, Japan*; ROBERT J. LE ROY, *Department of Chemistry, University of Waterloo, Waterloo, ON, Canada.*

Despite only having $6e^-$, the most sophisticated $Li_2(b, 1^3\Pi_u)$ calculation[a] has an r_e that disagrees with the empirical value by over 1500% of the latter's uncertainty, and energy spacings that disagree with those of the empirical potential by up to over 1.5cm^{-1}. The discrepancy here is far more than for the ground state of the $5e^-$ system BeH, for which the best *ab initio* calculation gives an r_e which disagrees with the empirical value by less than 200% of the latter's uncertainty[b]. In addition to this discrepancy, other reasons motivating the construction of an analytic empirical potential for $Li_2(b, 1^3\Pi_u)$ include (1) the fact that it is the most deeply bound Li_2 state, (2) it is the only Li_2 state out of the lowest five, for which no analytic empirical potential has yet been built, (3) the state it mixes with, the $A(1^1\Sigma_u)$-state, is one of the most thoroughly characterized molecular states, but has a small gap of missing data in part of the region where it mixes with the b-state, and (4) it is one of the states accessible by new ultra-high precision techniques based on photoassociation[c,d]. Finally (5) there is currently a discrepancy between the most sophisticated $3e^-$ *ab initio* calculation[e], and the most current empirical value[f], for the first $Li(^2S)$–$Li(^2P)$ interaction term (C_3), despite the latter being the most precise experimentally determined oscillator strength for any system, by an order of magnitude[e]. The b-state is one of the states that has this exact C_3 interaction term.

[a]Musial & Kucharski (2014) *J. Chem. Theor. Comp.* **10**, 1200.
[b]Dattani N. S. (2015) *J. Mol. Spec.* http://dx.doi.org/10.1016/j.jms.2014.09.005
[c]Semczuk M., Li X., Gunton W., Haw M., Dattani N. S., Witz J., Mills A., Jones D. J., Madison K. W. (2013) *Phys. Rev. A* **87**, 052505
[d]Gunton W., Semczuk M., Dattani N. S., Madison K. W. (2013) *Phys. Rev. A* **88**, 062510
[e]Tang L.-Y., Yan Z.-C., Shi T.-Y., Mitroy J (2011) *Phys. Rev. A* **84**, 052502.
[f]Le Roy R. J., Dattani N. S., Coxon J. A., Ross A. J., Crozet P., Linton C. (2009) *J. Chem Phys.* **131**, 204309

MF11 **4:55 – 5:10**

PRECISION SPECTROSCOPY OF TRAPPED HfF^+ WITH A COHERENCE TIME OF 1 SECOND

KEVIN COSSEL, WILLIAM CAIRNCROSS, MATT GRAU, DAN GRESH, YAN ZHOU, JUN YE, ERIC CORNELL, *JILA, National Institute of Standards and Technology and Univ. of Colorado Department of Physics, University of Colorado, Boulder, Boulder, CO, USA.*

Trapped molecular ions provide new systems for precision spectroscopy and tests of fundamental physics. For example, measurements of the permanent electric dipole moment of the electron (eEDM) test time-reversal symmetry[a]. Currently, we are using Ramsey spectroscopy between spin states of the metastable $^3\Delta_1$ state in trapped HfF^+ for a measurement of the eEDM[b,c]. We are regularly performing spectroscopy with a Ramsey time of 500 ms yielding what, to our knowledge, is the narrowest spectral line observed in a molecular system. Here, we will provide an overview of the experiment and the current eEDM results.

[a]The ACME Collaboration, *et al.*, *Science* **343**, 269 (2014)

[b]H. Loh, K. C. Cossel, M. C. Grau, K.-K. Ni, E. R. Meyer, J. L. Bohn, J. Ye, E. A. Cornell, *Science* **342**, 1220 (2013).

[c]A. E. Leanhardt, J. L. Bohn, H. Loh, M. C. Grau, P. Maletinski, E. R. Meyer, L. C. Sinclair, R. P. Stutz, E. A. Cornell, *J. Mol. Spec.* **270**, 1 (2011).

MF12 **5:12 – 5:27**

BROADBAND FREQUENCY COMB AND CW-LASER VELOCITY MODULATION SPECTROSCOPY OF ThF^+

DAN GRESH, KEVIN COSSEL, JUN YE, ERIC CORNELL, *JILA, National Institute of Standards and Technology and Univ. of Colorado Department of Physics, University of Colorado, Boulder, CO, USA.*

An experimental search for the permanent electric dipole moment of the electron (eEDM) is currently being performed using the metastable $^3\Delta_1$ state in trapped HfF^+ [(a)]. The use of ThF^+ could significantly increase the sensitivity due to the larger effective electric field and longer $^3\Delta_1$ state lifetime. Previous work by the Heaven group has identified several low-lying ThF^+ electronic states[b]; however, the ground state could not be conclusively assigned. In addition, transitions to intermediate electronic states have not been identified, but they are necessary for state detection, manipulation, and readout in an eEDM experiment. To date we have acquired 3700 cm^{-1} of densely-sampled ThF^+ spectra in the 695 – 1020 nm region with frequency comb[c] and cw-laser velocity modulation spectroscopy[d]. With high resolution, we have accurately fit more than 20 ThF^+ vibronic transitions, including electronic states spaced by the known X-a energy separation[b]. We will report on the ThF^+ ground state assignment and its implications for an eEDM experiment.

[a]H. Loh, K. C. Cossel, M. C. Grau, K.-K. Ni, E. R. Meyer, J. L. Bohn, J. Ye, E. A. Cornell, Science **342**, 1220 (2013).

[b]B. J. Barker, I. O. Antonov, M. C. Heaven, K. A. Peterson, J. Chem. Phys. **136**, 104305 (2012).

[c]L. C. Sinclair, K. C. Cossel, T. Coffey, J. Ye, E. A. Cornell, PRL **107**, 093002 (2011).

[d]K.C. Cossel *et. al.*, Chem. Phys. Lett. **546**, 1 (2012).

MF13 **5:29 – 5:44**

PURE MW DATA FOR $v = 0 - 6$ OF PbI GIVE VIBRATIONAL SPACINGS AND A FULL ANALYTIC POTENTIAL ENERGY FUNCTION

JI HO (CHRIS) YOO, *Department of Chemistry, University of Waterloo, Waterloo, ON, Canada*; COREY EVANS, *Department of Chemistry, University of Leicester, Leicester, United Kingdom*; NICK WALKER, *School of Chemistry, Newcastle University, Newcastle-upon-Tyne, United Kingdom*; ROBERT J. LE ROY, *Department of Chemistry, University of Waterloo, Waterloo, ON, Canada.*

At last year's ISMS meeting, Zaleski *et al.* reported new broadband MW spectroscopy measurements of pure rotational transitions in the $v = 0 - 6$ levels of the $^2\Pi_{1/2}$ ground electronic state of PbI.[a] The analysis presented at that time was a conventional v-level by v-level 'band-constant' analysis performed using the PGopher program.[b] That level-by-level PGopher analysis yielded values of B_v, D_v and five spin-splitting parameters for each vibrational level of each isotopologue. Ignoring the spin-splitting information, the B_v and D_v values were used to generate a set of synthetic pure $R(0)$ transitions for each level that were taken to represent the "mechanical" information about the molecule contained in these spectra. A standard direct-potential-fit (DPF) analysis[c] was then used to fit these data to an "Expanded Morse Oscillator" (EMO) potential function form. The well-depth parameter $\mathcal{D}_e$ was fixed at the literature value, while values of the equilibrium distance r_e and three EMO exponent-coefficient expansion 'potential shape' parameters are determined from the fits. The best fits to the data yield potentials whose fundamental vibrational spacings are in excellent agreement with experiment[d] together with reliable predictions for the first five overtone energies.

---

[a] D.P. Zaleski, H. Köckert, S.L. Stephens, N. Walker, L.-M. Dickens, and C. Evans, paper RE08 at the 69$^{\rm th}$ International Symposium on Molecular Spectroscopy, University of Illinois (2014).

[b] **PGopher** - *a Program for Simulating Rotational Structure*, C. M. Western, University of Bristol, http://pgopher.chm.bris.ac.uk

[c] **DPotFit 2.0**: *A Computer Program for fitting Diatomic Molecule Spectra to Potential Energy Functions*, R.J. Le Roy, J. Seto and Y. Huang, University of Waterloo Chemical Physics Research Report CP-667 (2013); see http://leroy.uwaterloo.ca/programs/.

[d] K. Ziebarth, R. Breidohr, O. Shestakov and E.H. Fink, *Chem. Phys. Lett.* **190**, 271 (1992).

MG. Structure determination

Monday, June 22, 2015 – 1:30 PM

Room: 100 Noyes Laboratory

Chair: Ha Vinh Lam Nguyen, Université Paris-Est Créteil, Créteil, France

MG01 **1:30 – 1:45**

DETECTION OF HSNO, A CRUCIAL INTERMEDIATE LINKING NO AND H_2S CHEMISTRIES

MARIE-ALINE MARTIN-DRUMEL, *Spectroscopy Lab, Harvard-Smithsonian Center for Astrophysics, Cambridge, MA, USA*; CARRIE WOMACK, *Department of Chemistry, MIT, Cambridge, MA, USA*; KYLE N CRABTREE, *Department of Chemistry, The University of California, Davis, CA, USA*; SVEN THORWIRTH, *I. Physikalisches Institut, Universität zu Köln, Köln, Germany*; MICHAEL C McCARTHY, *Atomic and Molecular Physics, Harvard-Smithsonian Center for Astrophysics, Cambridge, MA, USA.*

The simplest S-nitrosothiol, thionitrous acid (HSNO), is a reactive molecule of both biological and astronomical interest. Here we report the first detection of both *cis*- and *trans*-HSNO by means of Fourier-transform microwave spectroscopy and double resonance experiments. Surprisingly, HSNO is readily produced in a gas expansion of H_2S and NO, i.e. without applying any discharge. Once formed, HSNO appears quite stable, as evidenced by its high steady-state concentration. A precise empirical molecular equilibrium structure was derived from a combination of theory and experiment.

MG02 **1:47 – 2:02**

DETECTION AND STRUCTURAL CHARACTERIZATION OF NITROSAMIDE H_2NNO: A CENTRAL INTERMEDIATE IN deNO$_x$ PROCESSES

MICHAEL C McCARTHY, *Atomic and Molecular Physics, Harvard-Smithsonian Center for Astrophysics, Cambridge, MA, USA*; KELVIN LEE, *School of Chemistry, UNSW, Sydney, NSW, Australia*; JOHN F. STANTON, *Department of Chemistry, The University of Texas, Austin, TX, USA.*

H_2NNO plays a central role as the initial intermediate in the NH_2 + NO reaction. As the simplest N-nitrosamine, it is also the basis for understanding how specific substituents subtly change the structure of the NNO unit in larger nitrosamines, an important set of compounds that can be produced in foods by nitrites, but which are commonly carcinogenic to humans. Due to its perceived instability, H_2NNO has never been isolated and spectroscopically characterized in the gas phase, but, by means of Fourier transform microwave spectroscopy in combination with millimeter-wave double resonance techniques, the rotational spectrum of the normal and six of its rare isotopic species have been measured to high accuracy between 15 and 90 GHz. For each isotopic species, all three rotational constants have been determined to a fractional accuracy of 10 ppm or better; nitrogen quadrupole coupling constants have also been derived to very high accuracy. By correcting the experimental rotational constants for vibrational corrections calculated theoretically, a precise semi-experimental structure has been derived. These findings are consistent with new CCSD(T) calculations which predict that the equilibrium geometry of H_2NNO is planar, but that it possesses an extremely flat H_2N-X potential, like NH_3. Other aspects of this joint work, including the bond order inferred from the eQq(N) coupling constants, and the issue of planarity in substituted derivatives of the form R_1R_2NNO, will be discussed.

MG03 **2:04 – 2:19**

MICROWAVE SPECTRA OF 1- AND 2-BROMOBUTANE

SOOHYUN KA, JIHYUN KIM[a], HEESU JANG, JUNG JIN OH, *Research Institute of Global Environment, Sookmyung Women's University, Seoul, Korea.*

The rotational spectrum of 1-bromobutane measured by the 480 MHz bandwidth chirped-pulse Fourier transform microwave (CP-FTMW) spectroscopy. In this paper, the *ab initio* calculation and the analysis of rotational spectrum were performed, and the properties of gas molecule are reported.

1-bromobutane have five confomers; aa, ag, ga, gg, gg'. The transitions were assigned to three different conformers which are most stable forms; aa, ag, ga. The spectra for the normal isotopic species and ^{81}Br substitution were observed and assigned.

The rotational spectrum of 2-bromobutane has been observed in the frequency region 7-18 GHz. 2-bromobutane has the three possible conformers; G+, A, G-. The difference of their energy is very small, so the spectra of all conformers were found in the full range of our spectrum.

Consequentially, the rotational constants, nuclear quadrupole constants, and centrifugal distortion constants were determined and the dipole moment of the aa conformer with ^{79}Br were measured. All the experimental data is in good agreement with the calculated data.

[a]Current affiliation: Department of Chemistry, Texas A&M University, College Station, TX 77842-3012, USA

MG04 **2:21 – 2:36**

ACCURATE EQUILIBRIUM STRUCTURES FOR *trans*-HEXATRIENE BY THE MIXED ESTIMATION METHOD AND FOR THE THREE ISOMERS OF OCTATETRAENE FROM THEORY; STRUCTURAL CONSEQUENCES OF ELECTRON DELOCALIZATION

NORMAN C. CRAIG, *Department of Chemistry and Biochemistry, Oberlin College, Oberlin, OH, USA*; JEAN DEMAISON, *Université Lille 1, Laboratoire PhLAM, Villeneuve d'Ascq, France*; PETER GRONER, *Department of Chemistry, University of Missouri - Kansas City, Kansas City, MO, USA*; HEINZ DIETER RUDOLPH, *Department of Chemistry, Universität Ulm, Ulm, Germany*; NATALJA VOGT, *Section of Chemical Information Systems, Universität Ulm, Ulm, Germany.*

An accurate equilibrium structure of *trans*-hexatriene has been determined by the *mixed estimation* method with rotational constants from 8 deuterium and carbon isotopologues and high-level quantum chemical calculations. In the mixed estimation method bond parameters are fit concurrently to moments of inertia of various isotopologues and to theoretical bond parameters, each data set carrying appropriate uncertainties. The accuracy of this structure is 0.001 Å and 0.1°. Structures of similar accuracy have been computed for the cis,cis, trans,trans, and cis,trans isomers of octatetraene at the CCSD(T) level with a basis set of wCVQZ(ae) quality adjusted in accord with the experience gained with *trans*-hexatriene. The structures are compared with butadiene and with *cis*-hexatriene to show how increasing the length of the chain in polyenes leads to increased blurring of the difference between single and double bonds in the carbon chain. In *trans*-hexatriene r("C_1=C_2") = 1.339 Å and r("C_3=C_4") = 1.346 Å compared to 1.338 Å for the "double" bond in butadiene; r("C_2–C_3") = 1.449 Å compared to 1.454 Å for the "single" bond in butadiene. "Double" bonds increase in length; "single" bonds decrease in length.

MG05 **2:38 – 2:53**

RING PUCKERING POTENTIALS OF THREE FLUORINATED CYCLOPENTENES: C_5F_8, C_5HF_7, and $C_5H_2F_6$

E. A. ARSENAULT, B. E. LONG, WALLACE C. PRINGLE, *Department of Chemistry, Wesleyan University, Middletown, CT, USA*; YOON JEONG CHOI, S. A. COOKE, *Natural and Social Science, Purchase College SUNY, Purchase, NY, USA*; ESTHER J OCOLA, JAAN LAANE, *Department of Chemistry, Texas A & M University, College Station, TX, USA.*

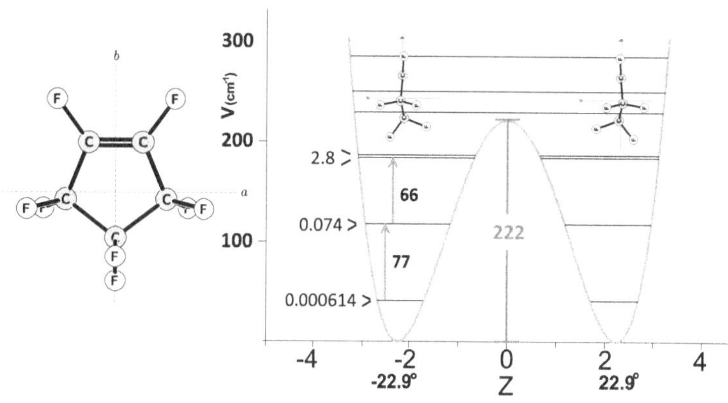

A systematic study on the ring puckering potentials of three fluorinated cyclopentenes has been performed using Fourier transform microwave spectroscopy in tandem with quantum chemical calculations. Spectra between 8 GHz and 16 GHz have been measured for octafluorocyclopentene, 1H-heptafluorocyclopentene, and 1H,2H-hexafluorocyclopentene, where the hydrogens sequentially replace the fluorines on the sp^2 hybridized carbons. Rotational constants and centrifugal distortion constants have been determined for the parent species and all ^{13}C isotopologues. In regards to the ring puckering, double minimum potential, both cross state and intra-state transitions were observed for all molecules except the 1H,2H-hexafluorocyclopentene. Experimental Coriolis coupling constants and ΔE_{01} values will be presented and discussed. The ring puckering barrier heights for C_5F_8, C_5HF_7, and $C_5H_2F_6$, have been calculated to be 222 cm^{-1}, 302 cm^{-1}, and 367 cm^{-1}, respectively.

MG06 **2:55 – 3:10**

CONFORMATIONAL TRANSFORMATION OF FIVE-MEMBERED RINGS: THE GAS PHASE STRUCTURE OF 2-METHYLTETRAHYDROFURAN

VINH VAN, *Institute for Physical Chemistry, RWTH Aachen University, Aachen, Germany*; HA VINH LAM NGUYEN, *CNRS et Universités Paris Est et Paris Diderot, Laboratoire Interuniversitaire des Systèmes Atmosphériques (LISA), Créteil, France*; WOLFGANG STAHL, *Institute for Physical Chemistry, RWTH Aachen University, Aachen, Germany.*

2-Methyltetrahydrofuran (2-MeTHF) is a promising environmentally friendly solvent and biofuel component which is derived from renewable resources[a]. Following the principles of Green Chemistry, 2-MeTHF has been evaluated in various fields like organometallics, metathesis, and biosynthesis on the way to more eco-friendly syntheses[b].

Cyclopentane as the prototype of five-membered rings is well-known to exist as twist or envelope structures. However, the conformational analysis of its heterocyclic derivative 2-methyl-tetrahydrothiophene (MTTP) yielded two stable twist conformers and two envelope transition states[c]. Here, we report on the heavy atom r_s structure of the oxygen-analog of MTTP, 2-MeTHF, studied by a combination of molecular beam Fourier transform microwave spectroscopy and quantum chemistry. One conformer of 2-MeTHF was observed and highly accurate molecular parameters were determined using the XIAM program[d]. In addition, all ^{13}C-isotopologues were assigned in natural abundance of 1%. A structural determination based on the r_s positions of all carbon atoms was achieved via Kraitchman's equations[e]. The methyl group in 2-MeTHF undergoes internal rotation and causes A–E splittings of the rotational lines. The barrier was calculated to be 1142 cm^{-1}at the MP2/6-311++G(d,p) level of theory, which is rather high. Accordingly, narrow A–E splittings could be observed for only a few transitions. However, the barrier height could be fitted while the angles between the internal rotor axis and the principal axes of inertia were taken from the experimental geometry.

[a] V. Pace, P. Hoyos, L. Castoldi, P. Domínguez de María, A. R. Alcántara, *ChemSusChem* **5** (2012), 1369 – 1379.
[b] a) D. F. Aycock, *Org. Process Res. Dev.* **11** (2007),156 – 159. b) M. Smoleń, M. Kędziorek, K. Grela, *Catal. Commun.* **44** (2014), 80 – 84.
[c] V. Van, C. Dindic, H.V.L. Nguyen, W. Stahl, *ChemPhysChem* **16** (2015), 291 – 294.
[d] H. Hartwig, H. Dreizler, *Z. Naturforsch. A* **51** (1996), 923 – 932.
[e] J. Kraitchman, *Am. J. Phys.* **21** (1953), 17 – 24.

MG07 **3:12 – 3:27**

ASSIGNMENT OF THE MICROWAVE SPECTRUM OF 1,2-DIFLUOROBENZENE$\cdots$HCCH: LESSONS LEARNED FROM ANALYSIS OF A DENSE BROADBAND SPECTRUM

ANURADHA AKMEEMANA, REBECCA D. NELSON, MIKAYLA L. GRANT, REBECCA A. PEEBLES, SEAN A. PEEBLES, *Department of Chemistry, Eastern Illinois University, Charleston, IL, USA*; JUSTIN M. KANG, *Department of Chemistry and Biochemistry, Oberlin College, Oberlin, OH, USA*; NATHAN A SEIFERT, BROOKS PATE, *Department of Chemistry, The University of Virginia, Charlottesville, VA, USA.*

Dimers of aromatic molecules with weak proton donors such as acetylene are prototypical systems for investigating weak CH$\cdots\pi$ interactions. A logical progression from our recent rotational spectroscopic studies of benzene...HCCH and fluorobenzene$\cdots$HCCH was to study 1,2-difluorobenzene(1,2-dfbz)$\cdots$HCCH, so the effect of increasing the number of electronegative substituents could be investigated. In this talk, structures of benzene, fluorobenzene, and 1,2-difluorobenzene complexed with HCCH will be compared, and the challenges and pitfalls encountered during assignment of the very rich chirped-pulse Fourier-transform microwave (CP-FTMW) spectrum will be discussed.

The spectrum of a mixture of 1,2-dfbz and HCCH in a neon carrier was initially recorded using the CP-FTMW spectrometer at the University of Virginia. Transitions matching the patterns and approximate rotational constants predicted for 1,2-dfbz$\cdots$HCCH were readily identified; however, efforts to fit the observed frequencies to an asymmetric top Hamiltonian were unsuccessful. A second CP-FTMW scan of only 1,2-dfbz monomer revealed that the transitions initially believed to be 1,2-dfbz$\cdots$HCCH were actually present in both scans. Subtraction of lines common to both data sets revealed a previously unidentified pattern of transitions that have now been confirmed by isotopic substitution to belong to 1,2-dfbz$\cdots$HCCH. The originally identified transitions are likely 1,2-dfbz$\cdots{}^{20}$Ne, which has a similar mass to the HCCH complex. Ab initio calculations for 1,2-dfbz$\cdots$HCCH and 1,2-dfbz$\cdots$Ne lead to several possible orientations for each dimer with similar energies and rotational constants, and efforts to improve the computational methods and to reliably identify stationary points on the dimer potential energy surfaces are ongoing.

MG08 **3:29 – 3:39**

STRUCTURE DETERMINATION AND CH$\cdots$F INTERACTIONS IN H_2C=CHF$\cdots H_2C=CF_2$ BY FOURIER-TRANSFORM MICROWAVE SPECTROSCOPY

RACHEL E. DORRIS, REBECCA A. PEEBLES, SEAN A. PEEBLES, *Department of Chemistry, Eastern Illinois University, Charleston, IL, USA.*

The structure of the weakly bound dimer between fluoroethylene (FE) and 1,1-difluoroethylene (DFE) has been determined using a combination of chirped-pulse and resonant-cavity Fourier-transform microwave spectroscopy over a 7.5 to 19 GHz range. The rotational constants of the most abundant isotopomer were determined to be A = 6601.14(35) MHz, B = 833.3336(5) MHz and C = 744.0217(5) MHz, and are in excellent agreement with *ab initio* predictions at the MP2/6-311++G(2d,2p) level. Observation of all four unique ^{13}C isotopologues in natural abundance allowed for a full structure determination, showing that the dimer takes on a planar configuration with the H–C–F end of FE aligned with one of the F–C=C–H sides of DFE, forming two inequivalent CH$\cdots$F contacts. The dipole moment components (μ_a = 0.9002(18) D, μ_b = 0.0304(80) D) were determined using Stark effect measurements and confirm the observed structure.

Intermission

MG09 **3:58 – 4:13**

MILLIMETER WAVE SPECTROSCOPY AND EQUILIBRIUM STRUCTURE DETERMINATION OF PYRIMIDINE (*m*-$C_4H_4N_2$)

ZACHARY N. HEIM, BRENT K. AMBERGER, BRIAN J. ESSELMAN, R. CLAUDE WOODS, ROBERT J. McMAHON, *Department of Chemistry, The Univeristy of Wisconsin, Madison, WI, USA.*

Pyrimidine, the *meta* substituted dinitrogen analog of benzene, has been studied in the mm-wave region from 260 – 360 GHz, expanding on previous studies up to 337 GHz.[abc] The spectra of all four of the singly-substituted ^{13}C and ^{15}N isotopologues were observed in natural abundance. Samples of deuterium enriched pyrimidine were synthesized, giving access to several deuterium-substituted isotopologues. The experimental rotational constants have been corrected for vibration-rotation coupling and electron mass. The vibration-rotation corrections were calculated with an anharmonic frequency calculation at the CCSD[T]/ANO1 level using CFOUR. An equilibrium structure determination has been performed using the corrected rotational constants with the xrefit module of CFOUR. Several vibrational satellites of pyrimidine have also been studied. Their rotational constants have been compared to those obtained computationally.

[a]Z. Kisiel, L. Pszczolkowski, I. R. Medvedev, M. Winnewisser, F. C. De Lucia, E. Herbst, J. Mol. Spectrosc. **233**, 231-243 (2005).
[b]G. L. Blackman, R. D. Brown, F. R. Burden, J. Mol. Spectrosc. **35**, 444-454 (1970).
[c]W. Caminati, D. Damiani, Chem. Phys. Lett. **179**, 460-462 (1991).

MG10 **4:15 – 4:30**

MILLIMETER-WAVE SPECTROSCOPY OF PHENYL ISOCYANATE

CARA E. SCHWARZ, BRENT K. AMBERGER, BENJAMIN C. HAENNI, BRIAN J. ESSELMAN, R. CLAUDE WOODS, ROBERT J. McMAHON, *Department of Chemistry, University of Wisconsin, Madison, WI, USA.*

Phenyl isocyanate (PhNCO) has been studied in the frequency range of 250-360 GHz, improving on rotational and centrifugal distortion constants based on previous spectroscopic studies between 4.7 and 40 GHz.[ab] Using the rigid rotor/centrifugal distortion model, many transitions have been assigned for the ground state (approximately 2200 transitions) and the fundamental of the -NCO torsional vibration (approximately 1500 transitions) for J values ranging between 140 and 210 and $K_{prolate}$ values from 0 to 42. Beyond these K values, these two spectra show effects of perturbations with other vibrational states. Vibrational energy levels and vibration-rotation interaction constants were predicted using CFOUR at the CCSD(T)/ANO0 level. The two lowest energy excited vibrational modes are predicted to have energies of 47 cm^{-1} (-NCO torsion) and 95 cm^{-1} (in-plane -NCO wag). Fermi resonance between the first overtone of the -NCO torsional vibration (94 cm^{-1}) and the fundamental of the in-plane -NCO wag has been observed in the spectra of these two states. Analysis for vibrationally excited states up to 190 cm^{-1} is in progress.

[a]A. Bouchy and G. Roussy, *Journal of Molecular Spectroscopy*. **65** (1977), 395-404.
[b]W. Kasten and H. Dreizler, *Z. Naturforsch.* **42a** (1987), 79-82.

MG11 **4:32 – 4:47**

BROADBAND MICROWAVE SPECTROSCOPY AS A TOOL TO STUDY THE STRUCTURES OF ODORANT MOLECULES AND WEAKLY BOUND COMPLEXES IN THE GAS PHASE

SABRINA ZINN[a], THOMAS BETZ, CHRIS MEDCRAFT, MELANIE SCHNELL, *MPSD, Max Planck Institute for the Structure and Dynamics of Matter, Hamburg, Germany.*

The rotational spectrum of *trans*-cinnamaldehyde ((2E)-3-phenylprop-2-enal) has been obtained with chirped-pulse microwave spectroscopy in the frequency range of 2 - 8.5 GHz. The odorant molecule is the essential component in cinnamon oil and causes the characteristic smell. In the measured high-resolution spectrum, we were able to assign the rotational spectra of two conformers of *trans*-cinnamaldehyde as well as all singly ^{13}C-substituted species of the lowest-energy conformer in natural abundance. Two different methods were used to determine the structure from the rotational constants, which will be compared within this contribution.

In addition, the current progress of studying ether-alcohol complexes, aiming at an improved understanding of the interplay between hydrogen bonding and dispersion interaction, will be reported. Here, a special focus is placed on the complexes of diphenylether with small aliphatic alcohols.

[a]The author thanks "The Hamburg Centre for Ultrafast Imaging" for financial support.

MG12 **4:49 – 5:04**

MICROWAVE SPECTRA OF 9-FLUORENONE AND BENZOPHENONE

CHANNING WEST, GALEN SEDO, *Department of Natural Sciences, University of Virginia's College at Wise, Wise, VA, USA*; JENNIFER VAN WIJNGAARDEN, *Department of Chemistry, University of Manitoba, Winnipeg, MB, Canada.*

The pure rotational spectra of 9-fluorenone ($C_{13}H_8O$) and benzophenone ($C_{13}H_{10}O$) were observed using chirped-pulse Fourier transform microwave spectroscopy (cp-FTMW). The 9-fluorenone spectrum was collected between 8 and 13 GHz, which allowed for the assignment of 124 rotational transitions. A separate spectrum spanning from 8 to 14 GHz was collected for benzophenone, allowing for the assignment of 133 rotational transitions. Both aromatic ketones exhibited strong b-type spectra with little to no centrifugal distortion, indicating highly rigid molecular structures. A comparison of the experimentally determined spectral constants of 9-fluorenone to those calculated using both ab initio and density functional theory strongly suggest the molecule conforms to a planar C_{2v} symmetric geometry as expected for its polycyclic structure. Whereas, a comparison of the experimental benzophenone constants to those predicted by theory suggests a molecule with non-planar C_2 symmetry, where the two phenyl groups are rotated approximately 32° out-of-plane to form a paddlewheel like geometry.

MG13 **5:06 – 5:21**

ASSESSING THE IMPACT OF BACKBONE LENGTH AND CAPPING AGENT ON THE CONFORMATIONAL PREFERENCES OF A MODEL PEPTIDE: CONFORMATION SPECIFIC IR AND UV SPECTROSCOPY OF 2-AMINOISOBUTYRIC ACID

JOSEPH R. GORD, DANIEL M. HEWETT, *Department of Chemistry, Purdue University, West Lafayette, IN, USA*; MATTHEW A. KUBASIK, *Department of Chemistry and Biochemistry, Fairfield University, Fairfield, CT, USA*; TIMOTHY S. ZWIER, *Department of Chemistry, Purdue University, West Lafayette, IN, USA.*

2-Aminoisobutyric acid (Aib) is an achiral, α-amino acid having two equivalent methyl groups attached to C_α. Extended Aib oligomers are known to have a strong preference for the adoption of a 3_{10}-helical structure in the condensed phase.[a] Here, we have taken a simplifying step and focused on the intrinsic folding propensities of Aib by looking at a series of capped Aib oligomers in the gas phase, free from the influence of solvent molecules and cooled in a supersonic expansion. Resonant two-photon ionization and IR-UV holeburning have been used to record single-conformation UV spectra using the Z-cap as the UV chromophore. Resonant ion-dip infrared (RIDIR) spectroscopy provides single-conformation IR spectra in the OH stretch and NH stretch regions. Data have been collected on a set of Z-$(\text{Aib})_n$-X oligomers with n = 1, 2, 4, 6 and X = -OH and -OMethyl. The impacts of these capping groups and differences in backbone length have been found to dramatically influence the conformational space accessed by the molecules studied here. Oligomers of n=4 have sufficient backbone length for a full turn of the 3_{10}-helix to be formed. Early interpretation of the data collected shows clear spectroscopic markers signaling the onset of 3_{10}-helix formation as well as evidence of structures incorporating C7 and C14 hydrogen bonded rings.

[a]Toniolo, C.; Bonora, G. M.; Barone, V.; Bavoso, A.; Benedetti, E.; Di Blasio, B.; Grimaldi, P.; Lelj, F.; Pavone, V.; Pedone, C., Conformation of Pleionomers of α-Aminoisobutyric Acid. *Macromolecules* **1985**, *18*, 895-902.

MG14 **5:23 – 5:38**

COMPARISON OF INTRAMOLECULAR FORCES IN DIPEPTIDES WITH TWO AROMATIC RINGS: DOES DISPERSION DOMINATE?

JESSICA A. THOMAS, *Department of Biology and Chemistry, Purdue University North Central, Westville, IN, USA.*

IR/UV double resonance spectroscopy has shown that the structure of the capped dipeptide Ac-Trp-Tyr-NH_2 is dominated by a hydrophobic interaction between the aromatic rings on the side chains. Using the same method, a similar molecule, Ac-Phe-Phe-NH_2, had three experimentally observed conformers including one similar to that of Ac-Trp-Tyr-NH_2. In this work, calculations were performed on additional dipeptides containing two aromatic rings to determine if the dispersion-dominated structure was among the lowest energy structures in all such cases.

The B3LYP-DCP method developed by DiLabio and Torres was used to calculate the conformations of each dipeptide. Appending dispersion correction potentials (DCP) to B3LYP input files improves results for systems containing dispersion interactions without significantly increasing the calculation time. This method was used first on Ac-Trp-Tyr-NH_2 and Ac-Phe-Phe-NH_2 to confirm that it successfully identified the experimentally observed structures among the lowest energy results and was then applied to other capped dipeptides containing two aromatic rings including Ac-Phe-Tyr-NH_2 and Ac-Tyr-Phe-NH_2.

MH. Linelists

Monday, June 22, 2015 – 1:30 PM

Room: B102 Chemical and Life Sciences

Chair: Shanshan Yu, California Institute of Technology, Pasadena, CA, USA

MH01 **1:30 – 1:45**

HITRAN IN THE XXIst CENTURY: BEYOND VOIGT AND BEYOND EARTH

LAURENCE S. ROTHMAN, IOULI E GORDON, CHRISTIAN HILL, ROMAN V KOCHANOV, PIOTR WCISLO, *Atomic and Molecular Physics, Harvard-Smithsonian Center for Astrophysics, Cambridge, MA, USA*; JONAS WILZEWSKI, *Department of Astronomy, Harvard University, Cambridge, MA, USA.*

The line-by-line portion of the most recent HITRAN2012 edition[a] contains spectroscopic parameters for 47 gases and associated isotopologues. Continuing the effort of the last five decades, our task has been to improve the accuracy of the existing parameters as well as to add new bands, molecules, and their isotopologues. In this talk we will briefly summarize some of the most important efforts of the past year.

Particular attention will be given to explaining the new development in providing line-shape information in HITRAN. There are two important directions in which the database is evolving with respect to line shapes. The first direction is that, apart from the Voigt profile parameters that were traditionally provided in HITRAN, we are able to add parameters associated with many "mainstream" line shapes, including Galatry, speed-dependent Voigt, and the HT profile[b] recently recommended by IUPAC[c]. As a test case, we created a first complete dataset of the HT parameters for every line of molecular hydrogen in the HITRAN database. Another important development is that in order to increase the potential of the HITRAN database in planetary sciences, experimental and theoretical line-broadening coefficients, line shifts and temperature-dependence exponents of molecules of planetary interest broadened by H_2, He, and CO_2 have been assembled from available peer-reviewed sources. The collected data were used to create semi-empirical models for calculating relevant parameters for every line of the studied molecules in HITRAN.

This work has been supported by NASA Aura Science Team Grant NNX14AI55G and NASA Planetary Atmospheres Grant NNX13AI59G.

---

[a]L.S. Rothman, et al. "The HITRAN 2012 molecular spectroscopic database," JQSRT 130, 4–50 (2013).

[b]N.H. Ngo, et al. "An isolated line-shape model to go beyond the Voigt profile in spectroscopic databases and radiative transfer codes," JQSRT 129, 89–100 (2013).

[c]J. Tennyson, et al. "Recommended isolated-line profile for representing high-resolution spectroscopic transitions," Pure Appl.Chem. 86, 1931–1943 (2014).

MH02 **1:47 – 2:02**

HITRANonline: A NEW STRUCTURE AND INTERFACE FOR HITRAN LINE LISTS AND CROSS SECTIONS

CHRISTIAN HILL, LAURENCE S. ROTHMAN, IOULI E GORDON, ROMAN V KOCHANOV, PIOTR WCISLO, *Atomic and Molecular Physics, Harvard-Smithsonian Center for Astrophysics, Cambridge, MA, USA*; JONAS WILZEWSKI, *Department of Astronomy, Harvard University, Cambridge, MA, USA.*

We present **HITRAN***online*, an online interface to the internationally-recognised HITRAN molecular spectroscopic database[1], and describe the structure of its relational database backend[2].

As the amount and complexity of spectroscopic data on molecules used in atmospheric modelling has increased, the existing 160-character, text-based format has become inadequate for its description. For example, line shapes such as the Hartmann-Tran profile[3] require up to six parameters for their full description (each with uncertainties and references), data is available on line-broadening by species other than "air" and "self" and more than the current maximum of 10 isotopologues of some molecules (for example, CO_2) can be important for accurate radiative-transfer modelling. The new relational database structure overcomes all of these limitations as well as allowing for better data provenance through "timestamping" of transitions and a direct link between items of data and their literature sources.

To take full advantage of this new database structure, the online interface **HITRAN***online*, available at www.hitran.org, provides a user-friendly way to make queries of HITRAN data with the option of returning it in a customizable format with user-defined fields and precisions. Binary formats such as HDF-5 are also supported. In addition to the data, each query also produces its own bibliography (in HTML and BibTeX formats), "README" documentation and interactive graph for easy visualization.

1. L. S. Rothman *et al.*, *JSQRT* **130**, 4-50 (2013).
2. C. Hill, I. E. Gordon, L. S. Rothman, J. Tennyson, *JQSRT***130**, 51-61 (2013).
3. N. H. Ngo, D. Lisak, H. Tran, J.-M. Hartmann, *JQSRT* **129**, 89–100, (2013); erratum: *JQSRT* **134**, 105 (2014).

This work has been supported by NASA Aura Science Team Grant NNX14AI55G and NASA Planetary Atmospheres Grant NNX13AI59G.

MH03 **2:04 – 2:19**

WORKING WITH HITRAN DATABASE USING HAPI: HITRAN APPLICATION PROGRAMMING INTERFACE

ROMAN V KOCHANOV[a], CHRISTIAN HILL, PIOTR WCISLO, IOULI E GORDON, LAURENCE S. ROTHMAN, *Atomic and Molecular Physics, Harvard-Smithsonian Center for Astrophysics, Cambridge, MA, USA*; JONAS WILZEWSKI, *Department of Astronomy, Harvard University, Cambridge, MA, USA.*

A HITRAN Application Programing Interface (HAPI) has been developed to allow users on their local machines much more flexibility and power. HAPI is a programming interface for the main data-searching capabilities of the new "HITRANonline" web service (http://www.hitran.org). It provides the possibility to query spectroscopic data from the HITRAN[b] database in a flexible manner using either functions or query language. Some of the prominent current features of HAPI are: a) Downloading line-by-line data from the HITRANonline site to a local machine b) Filtering and processing the data in SQL-like fashion c) Conventional Python structures (lists, tuples, and dictionaries) for representing spectroscopic data d) Possibility to use a large set of third-party Python libraries to work with the data e) Python implementation of the HT lineshape[c] which can be reduced to a number of conventional line profiles f) Python implementation of total internal partition sums (TIPS-2011[d]) for spectra simulations g) High-resolution spectra calculation accounting for pressure, temperature and optical path length h) Providing instrumental functions to simulate experimental spectra i) Possibility to extend HAPI's functionality by custom line profiles, partitions sums and instrumental functions

Currently the API is a module written in Python and uses Numpy library providing fast array operations. The API is designed to deal with data in multiple formats such as ASCII, CSV, HDF5 and XSAMS.

This work has been supported by NASA Aura Science Team Grant NNX14AI55G and NASA Planetary Atmospheres Grant NNX13AI59G.

[a] QUAMER, Tomsk State University, Tomsk, Russia
[b] L.S. Rothman et al. JQSRT, Volume 130, 2013, Pages 4-50
[c] N.H. Ngo et al. JQSRT, Volume 129, November 2013, Pages 89–100
[d] A. L. Laraia at al. Icarus, Volume 215, Issue 1, September 2011, Pages 391–400

MH04 **2:21 – 2:31**

GPU ACCELERATED INTENSITIES: A NEW METHOD OF COMPUTING EINSTEIN-A COEFFICIENTS

AHMED FARIS AL-REFAIE, *Department of Physics and Astronomy, University College London, Gower Street, London WC1E 6BT, United Kingdom*; SERGEI N. YURCHENKO, *Department of Physics and Astronomy, University College London, Gower Street, London WC1E 6BT, United Kingdom*; JONATHAN TENNYSON, *Department of Physics and Astronomy, University College London, London, IX, United Kingdom.*

Abstract

The use of variational nuclear motion calculations to produce comprehensive molecular line lists is now becoming common. In order to produce high quality and complete line-lists in particular applicable to high temperatures requires large amounts of computational resources. The more accuracy required, the larger the problem and the more computational resources needed. The two main bottlenecks in the production of these line-lists are solving the eigenvalue problem and the computation of the Einstein-A coefficients. From the project's recently released line-lists, the number of transitions can reach up to 10 billion evaluated by the combination of millions of eigenvalues and eigenvectors corresponding to individual energy states. For line-lists of this size, the evaluation of Einstein-A coefficients take up the vast majority of computational time compared to solving the eigenvalue problem. Recently, as part of the ExoMol [1] project, we have developed a new program called **G**PU **A**ccelerated **IN**tensities (GAIN) that utilises the highly parallel Graphics Processing Units (GPU) in order to accelerate the evaluation of the Einstein-A coefficients. Speed-ups of up to 70x can be achieved on a single GPU and can be further improved by utilising multiple GPUs. The GPU hardware, its limitations and how the problem was implemented to exploit parallelism will be discussed.

References

[1] J. Tennyson and S. N. Yurchenko. ExoMol: molecular line lists for exoplanet and other atmospheres. *MNRAS*, 425:21–33, 2012.

MH05 **2:33 – 2:48**

LINE SHAPE PARAMETERS FOR NEAR INFRARED CO_2 BANDS IN THE 1.61 AND 2.06 MICRON SPECTRAL REGIONS

V. MALATHY DEVI, D. CHRIS BENNER, *Department of Physics, College of William and Mary, Williamsburg, VA, USA*; KEEYOON SUNG, LINDA BROWN, TIMOTHY J CRAWFORD, *Jet Propulsion Laboratory, California Institute of Technology, Pasadena, CA, USA*; MARY ANN H. SMITH, *Science Directorate, NASA Langley Research Center, Hampton, VA, USA*; ARLAN MANTZ, *Department of Physics, Astronomy and Geophysics, Connecticut College, New London, CT, USA.*

Accurate spectroscopic measurements of self- and air-broadened Lorentz half-width and pressure-shift coefficients and their temperature dependence exponents are crucial for the Orbiting Carbon Observatory (OCO-2) mission.[a] We therefore analyzed 73 high-resolution high signal-to-noise spectra of CO_2 and CO_2+air for OCO-2 channels at 1.61 and 2.06 μm. These spectra were recorded at various spectral resolutions (0.004-0.013 cm^{-1}) using two spectrometers (the Kitt Peak FTS in Arizona and the Bruker 125HR FTS at the Jet Propulsion Laboratory in Pasadena, California). Six different absorption cells with path lengths between 0.2 and 121 m were used with gas samples at a range of temperatures (170-297 K). The gas pressures ranged from (0.3-898 Torr for pure sample and 26-924 Torr for mixtures of CO_2 and air with CO_2 volume mixing ratios between 0.01 and 0.4. The cold sample spectra were acquired using a short 0.2038 m straight pass celland a multipass Herriott cell having a 20.941 m total path A multispectrum fitting technique was employed to fit all the spectra simultaneously with a non-Voigt line shape profile including speed dependence and full line mixing. Examples of fitted spectra and retrieved parameters in both CO_2 band regions will be shown. Comparisons of some of the results with other published values will be provided.[b]

[a] D. Crisp, B.M. Fisher, C. O'Dell, et.al., Atmos. Meas. Tech. Discuss 4 (2011) 1-59.

[b] Research described in this paper are performed at the College of William and Mary, Jet Propulsion Laboratory, California Institute of Technology, NASA Langley Research Center and Connecticut College under contracts and cooperative agreements with the National Aeronautics and Space Administration.

MH06 **2:50 – 3:05**

RELIABLE IR LINE LISTS FOR SO_2 AND CO_2 ISOTOPOLOGUES COMPUTED FOR ATMOSPHERIC MODELING ON VENUS AND EXOPLANETS

XINCHUAN HUANG, *Carl Sagan Center, SETI Institute, Moutain View, CA, USA*; DAVID SCHWENKE, *MS 258-2, NAS Facility, NASA Ames Research Center, Moffett Field, CA, USA*; TIMOTHY LEE, *Space Science and Astrobiology Division, NASA Ames Research Center, Moffett Field, CA, USA*; ROBERT R. GAMACHE, *Department of Environmental, Earth, and Atmospheric Sciences, University of Massachusetts, Lowell, MA, USA.*

For SO_2 atmospheric characterization in Venus and other Exoplanetary environments, recently we presented Ames-296K line lists for 626 (upgraded) and other 4 symmetric isotopologues: 636, 646, 666 and 828. For CO_2, we reported Ames-296K (1E-42 cm/molecule) and Ames-1000K (1E-36 cm/molecule) IR line lists up to E'=18000 cm^{-1} for 13 CO_2 isotopologues, including symmetric species 626, 636, 646, 727, 737, 828, 838, and asymmetric species 627, 628, 637, 638, 728, 738. CO_2 line shape parameters were also determined for four different temperature ranges: Mars, Earth, Venus, and higher temperatures. General line position prediction accuracy up to 5000 cm^{-1} (SO_2) or 13000 cm^{-1} (CO_2) is 0.01 – 0.02 cm^{-1}. Most transition intensity deviations are less than 5-10%, when compare to experimentally measured quantities. With such prediction accuracy, these SO_2 and CO_2 isotopologue lists are the best available alternative for those wide spectra region missing from spectroscopic databases such as HITRAN and CDMS. For example, only very limited data exist for SO_2 646/636 and no data at all for other minor isotopologues. They should greatly facilitate spectroscopic analyses in future laboratory or astronomical observations. Our line list work are based on "Best Theory + Reliable High-Resolution Experiment" strategy, i.e. using an ab initio potential energy surface refined with selected reliable high resolution experimental data, and high quality CCSD(T)/aug-cc-pVQ(or Q+d)Z dipole moment surfaces. Note that we have solved a convergence defect on SO_2 Ames-1 PES and further improved the quality and completeness of the Ames-296K SO_2 list by including most recent experimental data into the refinement. We will compare the Ames-296K SO_2 and CO_2 lists to latest experiments and HITRAN/CDMS models. We expect more interactions between experimental and theoretical efforts. Currently the Ames-296K lists are available at http://huang.seti.org/.

MH07 **3:07 – 3:22**

LASER SPECTROSCOPIC STUDY OF CaH IN THE $B^2\Sigma^+$ AND $D^2\Sigma^+$ STATES

KYOHEI WATANABE, KANAKO UCHIDA, KAORI KOBAYASHI, FUSAKAZU MATSUSHIMA, YOSHIKI MORIWAKI, *Department of Physics, University of Toyama, Toyama, Japan.*

Calcium hydride is one of the abundant molecules in the stellar environment, and is considered as a probe of stellar analysis[a]. Ab initio calculations have shown that the electronic excited states of CaH have complex potential curves. It is suggested that the $B^2\Sigma^+$ state has an interesting double minimum potential due to the avoided crossing[b]. Such a potential leads to drastic change of the rotational constants when the vibrational energy level goes across the potential barrier. Spectroscopic studies on CaH began in the 1920's[c], and many studies have been carried out since then. Bell et al. extensively assigned the $D^2\Sigma^+ - X^2\Sigma^+$ bands in the UV region[d]. Bernath's group has observed transitions in the IR and visible regions and identified their upper states as the $A^2\Sigma^+$, $B^2\Sigma^+$ and $E^2\Sigma^+$ states[efgh]. We have carried out a laser induced fluorescence (LIF) study in the UV region between 360 and 430 nm. We have produced CaH by using laser ablation of a calcium target in a hydrogen gas environment, then molecules have been excited by a second harmonic pulse of dye laser and the fluorescence from molecules have been detected through a monochromator. Detection of the $D^2\Sigma^+$-$X^2\Sigma^+$ bands already identified by Bell et al. indicates the production of CaH. In addition, many other bands have been also found and a few bands have been assigned by using the combination differences, the lower state of these bands have been confirmed to the vibrational ground state of $X^2\Sigma^+$ state. We have tentatively assigned these bands as the $B^2\Sigma^+ - X^2\Sigma^+$ transition. We will discuss the assignment of these bands, together with the rotational constants comparing with those calculated from the ab initio potential.

[a]B. Barbuy, R. P. Schiavon, J. Gregorio-Hetem, P. D. Singh C. Batalha , *Astron. Astrophys. Sippl. Ser.* **101**, 409 (1993).
[b]P. F. Weck and P. C .Stabcil, *J. Chem. Phys.* **118**, 9997 (2003).
[c]R. S. Mulliken, *Phys. Rev.* **25**, 509 (1925).
[d]G. D. Bell, M, Herman, J. W. C. Johns, and E. R. Peck, *Physica Scripta* **20**, 609 (1979).
[e]A. Shayesteh, K. A. Walker, I. Gordon, D. R. T. Appadoo, and P. F. Bernath, *J. Mol. Struct.* **695-696**, 23 (2004).
[f]R. S. Ram, K. Tereszchuk, I. E. Gordon, K. A. Walker, and P. F. Bernath, *J. Mol. Spec.* **266**, 86 (2011).
[g]G. Li, J. J. Harrison, R. S. Ram, C. M. Western, and P. F. Bernath *Quant. Spectrosc. Rad. Transfer.* **113**, 67 (2012).
[h]A. Shayesteh, R. S. Ram, and P. F. Bernath, *J. Mol. Spec.* **288**, 46 (2013).

Intermission

MH08 **3:41 – 3:56**

ADDITIONAL MEASUREMENTS AND ANALYSES OF $H_2^{17}O$ AND $H_2^{18}O$

JOHN PEARSON, SHANSHAN YU, *Jet Propulsion Laboratory, California Institute of Technology, Pasadena, CA, USA*; ADAM WALTERS, *IRAP, Université de Toulouse 3 - CNRS - OMP, Toulouse, France.*

Historically the analysis of the spectrum of water has been a balance between the quality of the data set and the applicability of the Hamiltonian to a highly non-rigid molecule. Recently, a number of different non-rigid analysis approaches have successfully been applied to ^{16}O water resulting in a self-consistent set of transitions and energy levels to high J which allowed the spectrum to be modeled to experimental precision[ab]. The data set for ^{17}O and ^{18}O water was previously reviewed and many of the problematic measurements identified[c], but Hamiltonian modeling of the remaining data resulted in significantly poorer quality fits than that for the ^{16}O parent. As a result, we have made additional microwave measurements and modeled the existing ^{17}O and ^{18}O data sets with an Euler series model[d]. This effort has illuminated a number of additional problematic measurements in the previous data sets and has resulted in analyses of ^{17}O and ^{18}O water that are of similar quality to the ^{16}O analysis. We report the new lines, the analyses and make recommendations on the quality of the experimental data sets.

[a]SS. Yu, J.C. Pearson, B.J. Drouin *et al. J. Mol. Spectrosc.* **279**, 16-25 (2012)
[b]J. Tennyson, P.F. Bernath, L.R. Brown *et al. J. Quant. Spectrosc. Rad. Trans.* **117**, 29-58 (2013)
[c]J. Tennyson, P.F. Bernath, L.R. Brown *et al. J. Quant. Spectrosc. Rad. Trans.* **110**, 573-596 (2009)
[d]H.M. Pickett, J.C. Pearson, C.E. Miller *J. Mol. Spectrosc.* **233**, 174-179 (2005)

MH09 **3:58 – 4:13**

EXPERIMENTAL LINE LISTS OF HOT METHANE

ROBERT J. HARGREAVES, PETER F. BERNATH, *Department of Chemistry and Biochemistry, Old Dominion University, Norfolk, VA, USA*; JEREMY BAILEY, *School of Physics, University of New South Wales, New South Wales, Australia*; MICHAEL DULICK, *Department of Chemistry and Biochemistry, Old Dominion University, Norfolk, VA, USA.*

Line lists of CH_4 at high temperatures (up to 900°C) have been produced between 2500 and 5000 cm^{-1}. This spectral range contains the pentad and octad regions, and includes numerous fundamental, overtone and hot bands. Our method makes use of a quartz sample cell that is heated by a tube furnace. Four spectra are then recorded at each temperature using a Fourier transform infrared spectrometer at high resolution (0.02 cm^{-1}). By combining these four spectra at each temperature, the emission and absorption from the cell and molecules are accounted for, and we obtain the true transmission spectrum of hot CH_4. Analysis of this series of spectra enables the production of line lists that include positions, intensities and empirical lower state energies.

We also compare our line lists to the best available theoretical line lists at high temperatures. Whilst our experimental line lists contain fewer lines than theoretical line lists, we are able to demonstrate the quality of our observed spectra by considering our observations as absorption cross sections. This is important at elevated temperatures, when numerous blended lines appear as a continuum.

MH10 **4:15 – 4:30**

EXPERIMENTAL TRANSMISSION SPECTRA OF HOT AMMONIA IN THE INFRARED

CHRISTOPHER A. BEALE, *Department of Ocean, Earth and Atmospheric Sciences, Old Dominion University, Norfolk, VA, USA*; ROBERT J. HARGREAVES, MICHAEL DULICK, PETER F. BERNATH, *Department of Chemistry and Biochemistry, Old Dominion University, Norfolk, VA, USA.*

High resolution absorption spectra of hot ammonia have been recorded in the 2400–5500 cm^{-1} region and the line lists are presented. This extends our previous work on ammonia in the 740–4000 cm^{-1} region[a,b] and utilizes our improved cell design that has been successfully applied to methane in a similar spectral region. Transmission spectra were acquired for seven temperatures up to 700°C using a Bruker IFS 125HR Fourier transform spectrometer and empirical lower state energies are obtained from the temperature dependence of intensities. Applications of our spectra and line lists include modeling of brown dwarfs and (exo)planetary atmospheres.

[a] R.J. Hargreaves, G. Li and P.F. Bernath. 2011, ApJ, 735, 111
[b] R.J. Hargreaves, G. Li and P.F. Bernath. 2011, JQSRT, 113, 670

MH11 **4:32 – 4:47**

HYPERSONIC POST-SHOCK CAVITY RING-DOWN SPECTROSCOPY

NICOLAS SUAS-DAVID, *IPR UMR6251, CNRS - Université Rennes 1, Rennes, France*; SAMIR KASSI, *UMR5588 LIPhy, Université Grenoble 1/CNRS, Saint Martin D'heres, France*; ABDESSAMAD BENIDAR, ROBERT GEORGES, *IPR UMR6251, CNRS - Université Rennes 1, Rennes, France.*

A highly sensitive experimental set-up (α_{min} = 10^{-10} cm^{-1}) has been developed to produce high-temperature infrared spectra of methane in the Tetradecad polyad region (1.67 μm) using cw-CRDS. A continuous flow of methane admixed to argon is initially heated at 1000 – 1500 K and then accelerated to hypersonic speeds in a vacuum chamber before being abruptly stopped by the impact on a planar screen set perpendicular to the flow axis, forming a stationary shock wave detached from the screen (bow shock). The CRD optical beam probes the very hot subsonic zone behind the shock where the gas temperature is close to the stagnation one. Computational Fluid Dynamics calculations have been performed to characterize the post-shock structure of the flow. Spectra reveal a series of new hot bands of fundamental interest for the modeling of highly excited levels of methane.

MH12 **4:49 – 5:04**

CH_3D NEAR INFRARED CAVITY RING-DOWN SPECTRUM REANALYSIS AND IR-IR DOUBLE RESONANCE

SHAOYUE YANG, GEORGE SCHWARTZ, *Department of Physics, The University of Virginia, Charlottesville, VA, USA*; KEVIN LEHMANN, *Departments of Chemistry and Physics, University of Virginia, Charlottesville, VA, USA.*

As one of the most important hydrocarbon prototype molecules, CH_3D's overtone band in near infrared region has not been well studied. Various methods were used to help identifying transitions from previous cavity ring down spectrum of CH_3D in the near infrared region. Symmetric top molecules' Hamiltonian diagonal terms for the ground state, perpendicular state and parallel state were simulated by software PGopher. Combination differences were used to find possible pairs of transitions starting from adjacent ground state and ending in same excited states. Also we introduced our temperature controlled spectrum setup for ground state energy and rotational quanta prediction from temperature dependence, and proven to be working well for lower J levels for CH_4. At last, we set up a double resonance system, using two lasers (3.3 and 1.65 μm, respectively) to excite transitions from the same ground state, to provide strong proof for the lower state quanta.

MH13 **5:06 – 5:21**

AYTY: A NEW LINE-LIST FOR HOT FORMALDEHYDE

AHMED FARIS AL-REFAIE, *Department of Physics and Astronomy, University College London, Gower Street, London WC1E 6BT, United Kingdom*; SERGEI N. YURCHENKO, *Department of Physics and Astronomy, University College London, Gower Street, London WC1E 6BT, United Kingdom*; JONATHAN TENNYSON, *Department of Physics and Astronomy, University College London, London, IX, United Kingdom*; ANDREY YACHMENEV, *Department of Physics and Astronomy, University College London, Gower Street, London WC1E 6BT, United Kingdom.*

Abstract

The ExoMol [1] project aims at providing spectroscopic data for key molecules that can be used to characterize the atmospheres of exoplanets and cool stars. Formaldehyde (H_2CO) is of growing importance in studying and modelling terrestrial atmospheric chemistry and dynamics. It also has relevance in astrophysical phenomena that include interstellar medium abundance, proto-planetary and cometary ice chemistry and masers from extra-galactic sources. However there gaps in currently available absolute intensities and a lack of higher rotational excitations that makes it unfeasible to accurately model high temperature systems such as hot Jupiters. Here we present **AYTY** [2], a new line list for formaldehyde applicable to temperatures up to 1500 K. AYTY contains almost 10 million states reaching rotational excitations up to $J = 70$ and over 10 billion transitions at up to 10 000 cm^{-1}. The line list was computed using the variational ro-vibrational solver TROVE with a refined *ab-initio* potential energy surface and dipole moment surface.

References

[1] J. Tennyson and S. N. Yurchenko. *MNRAS*, 425:21–33, 2012.

[2] A. F. Al-Refaie, S. N. Yurchenko, A. Yachmenev, and J. Tennyson. *MNRAS*, 2015.

MH14 **5:23 – 5:38**

THE MICROWAVE SPECTROSCOPY OF AMINOACETONITRILE IN THE VIBRATIONAL EXCITED STATE

CHIHO FUJITA, HIROYUKI OZEKI, *Department of Environmental Science, Toho University, Funabashi, Japan*; KAORI KOBAYASHI, *Department of Physics, University of Toyama, Toyama, Japan.*

Aminoacetonitrile (NH_2CH_2CN) is a potential precursor of the simplest amino acid, glycine and was detected toward SgrB2(N). [a] It is expected that the strongest transitions will be found in the terahertz region so that we have extended measurements up to 1.3 THz. [b] This study gave an accurate prediction of aminoacetonitrile up to 2 THz which is useful for astronomically search. This molecule has a few low-lying vibrational excited states and the pure rotational transitions in these vibrational excited states are expected to found. [c] We found a series of transitions with intensity of about 30%. Eighty-eight spectral lines including both *a*-type and *b*-type transitions were recorded in the frequency region of 400 - 450 GHz, and centrifugal distortion constants up to the sextic term were determined. Perturbation was recognized. We will report the current status of the analysis.

[a] A. Belloche, K. M. Menten, C. Comito, H. S. P. Müller, P. Schilke, J. Ott, S. Thorwirth, and C. Hieret, 2008, *Astronom. & Astrophys.* **482**, 179 (2008).

[b] Y. Motoki, Y. Tsunoda, H. Ozeki, and K. Kobayashi, *Astrophys. J. Suppl. Ser.* **209**, 23 (2013).

[c] B. Bak, E. L. Hansen, F. M. Nicolaisen, and O. F. Nielsen, *Can. J. Phys.* **53**, 2183 (1975).

MI. Ions

Monday, June 22, 2015 – 1:30 PM

Room: 274 Medical Sciences Building

Chair: Mark Johnson, Yale University, New Haven, CT, USA

MI01 **1:30 – 1:45**

ROTATIONAL ACTION SPECTROSCOPY VIA STATE-SELECTIVE HELIUM ATTACHMENT

LARS KLUGE, ALEXANDER STOFFELS[a], SANDRA BRÜNKEN, OSKAR ASVANY, STEPHAN SCHLEMMER, *I. Physikalisches Institut, Universität zu Köln, Köln, Germany.*

Helium atoms can attach to molecular cations via ternary collision processes forming weakly bound ($\approx$ 1 kcal/mol) He-M^+ complexes. We developed a novel sensitive action spectroscopic scheme for molecular ions based on an observed rotational state dependency of the He attachment process [1]. A detailed account of the underlying kinetics will be presented on the example of the CD^+ ion, where our studies indicate a decrease of around 50% for the rotational state dependent ternary He attachment rate coefficient of the $J = 1$ level with respect to the $J = 0$ level. Experiments are performed on mass-selected ions stored in a temperature-variable ($T \geq 3.9$ K) cryogenic rf 22-pole ion trap in the presence of a high number density of He ($\approx 10^{15}$ cm^{-3}) [2]. Rotational spectra of the bare ions are recorded by measuring the change in the number of formed He-M^+ complexes after a certain storage time as a function of excitation wavelength. Here we will also present the first measurements of the rotational ground state transitions of CF^+ ($J = 1 - 0$, hfs resolved) and NH_3D^+ ($J_K = 1_0 - 0_0$), recorded in this way.

[1] Brünken et al., ApJL **783**, L4 (2014)
[2] Asvany et al., Applied Physics B **114**, 203 (2014)

[a] also at: Institute for Molecules and Materials (IMM), Radboud University Nijmegen, Nijmegen, Netherlands

MI02 **1:47 – 2:02**

SYMMETRY BEYOND PERTURBATION THEORY: FLOPPY MOLECULES AND ROTATION-VIBRATION STATES

HANNO SCHMIEDT, STEPHAN SCHLEMMER, *I. Physikalisches Institut, University of Cologne, Cologne, Germany*; PER JENSEN, *Fachbereich C-Physikalische und Theoretische Chemie, Bergische Universität Wuppertal, D-42097 Wuppertal, Germany.*

In the customary approach to the theoretical description of the nuclear motion in molecules, the molecule is seen as a near-static structure rotating in space. Vibrational motion causing small structural deformations induces a perturbative treatment of the rotation-vibration interaction, which fails in fluxional molecules, where *all* vibrational motions are large compared to the linear extension of the molecule. An example is protonated methane (CH_5^+) [a]. For this molecule, customary theory fails to simulate reliably even the low-energy spectrum. Within the traditional view of rotation and vibration being near-separable, rotational and vibrational wavefunctions can be symmetry classified separately in the molecular symmetry (MS) group [b]. In the present contribution we discuss a fundamental group theoretical approach to the problem of determining the symmetries of molecular rotation-vibration states. We will show that all MS groups discussed so far are subgroups of the special orthogonal group in three dimensions SO(3)[c]. This leads to a group theoretical foundation of the technique of equivalent rotations [d]. The MS group of protonated methane (G_{240}) represents, to the best of our knowledge, the first example of an MS group which is not a subgroup of SO(3) (nor of O(3) nor of SU(2)). Because of this, a separate symmetry classification of vibrational and rotational wavefunctions becomes impossible in this MS group, consistent with the fact that a decoupling of vibrational and rotational motion is impossible. We want to discuss the consequences of this. In conclusion, we show that the prototypical floppy molecule CH_5^+ represents a new class of molecules, where usual group theoretical methods for determining selection rules and spectral assignments fail so that new methods have to be developed.

[a] P. Kumar and D. Marx, Physical Chemistry Chemical Physics **8**, 573 (2006); Z. Jin, B. J. Braams, and J. M. Bowman, The Journal of Physical Chemistry A **110**, 1569 (2006); A. S. Petit, J. E. Ford, and A. B. McCoy, The Journal of Physical Chemistry A **118**, 7206 (2014).
[b] P.R. Bunker and P. Jensen, *Molecular Symmetry and Spectroscopy* (NRC Research Press, Ottawa, Canada, 1998).
[c] Being precise, we must include O(3) and SU(2), but our theory can be easily extended to these two groups.
[d] H. Longuet-Higgins, Molecular Physics **6**, 445 (1963).

MI03 **2:04 – 2:19**

STUDYING ROTATION/TORSION COUPLING IN H_5^+ USING DIFFUSION MONTE CARLO

MELANIE L. MARLETT, ZHOU LIN, ANNE B McCOY, *Department of Chemistry and Biochemistry, The Ohio State University, Columbus, OH, USA.*

H_5^+ is a highly fluxional intermediate found in interstellar clouds. The rotational/torsional couplings in this molecule are of great interest due to the unusually large coupling between these modes. However, theoretical studies of highly fluxional molecules like H_5^+ are challenging due to the lack of a good zero-order model. In order to better understand the rotation/vibration interaction, a method has been developed to model the rotational/torsional motions. This method is based upon diffusion Monte Carlo (DMC). In this approach, the vibrational contribution to the wavefunction is modeled using standard DMC approaches, while the rotational/torsional contribution is treated as a set of coefficients that are assigned to the various rotational/torsional state vectors. The potential portion of the Hamiltonian is expressed as a low-order expansion in terms of the torsion angle between the two outer H_2 units. The expansion coefficients are evaluated at each time step for each walker and depend on the $3N - 7$ other internal coordinates. The transition frequencies obtained from this method for $J \leq 1$ agree well with results obtained using other methods such as fixed-node diffusion Monte Carlo.[a] This new method is advantageous over the fixed-node approach because it allows for multiple state calculations at once which saves on computation time.

[a] Sarka, J.; Fábri, C.; Szidarovszky, T.; Császár, A.G.; Lin Z.; McCoy, A.B., "Modeling Rotations, Vibrations, and Rovibrational Couplings in Astructural Molecules - A Case Study Based on the H_5^+ Molecular Ion.", accepted by Mol. Phys.

MI04 **2:21 – 2:36**

HIGH-J ROTATIONAL LINES OF ^{13}C ISOTOPOLOGUES OF HCO^+ MEASURED BY USING EVENSON-TYPE TUNABLE FIR SPECTROMETER

MARI SUZUKI, RYO OISHI, YOSHIKI MORIWAKI, FUSAKAZU MATSUSHIMA, *Department of Physics, University of Toyama, Toyama, Japan*; TAKAYOSHI AMANO, *Jet Propulsion Laboratory, California Institute of Technology, Pasadena, CA, USA.*

Frequencies of high-J rotational lines of HCO^+ and its isotopologues have been measured precisely by using an Evenson-type spectrometer in Toyama. The tunable far-infrared spectrometer (TuFIR in short) is based on synthesizing terahertz radiation from two mid-infrared CO_2 laser lines and one microwave source. Study of the isotopologues containing H or D, ^{12}C, and ^{16}O were reported last year. In the present work, isotopologues of H or D, ^{13}C, and ^{16}O have been studied. The HCO^+ ions are produced by discharging a ^{13}CO, H_2 (or D_2), and Ar mixture in an extended negative glow discharge cell cooled with liquid nitrogen. Because the low-J rotational lines have been investigated by other groups, our present study was focussed mainly to the measurements of higher-J rotational lines. Currently we have observed the lines J + 1 ← J(J=11, 13-21) for $H^{13}CO^+$, and J + 1 ← J (J=13-18, 20-22, 24-25) for $D^{13}CO^+$. Molecular contstants for these isotopologues (B, D, H, L) have been modified. From the analysis of the intensity of each rotational line, we estimate the rotational temperature to be as low as 140K. This low temperature makes it difficult to measure yet higher-J lines. Measurement of other isotopogues such as those containing oxygen isotopes is now in preparation.

MI05 **2:38 – 2:48**

UV-UV HOLE-BURNING SPECTROSCOPY OF A PROTONATED ADENINE DIMER IN A COLD QUADRUPOLE ION TRAP

HYUK KANG, *Department of Chemistry, Ajou University, Suwon, Korea.*

A novel method for double-resonance photofragmentation spectroscopy in a cold quadrupole ion trap has been developed and utilized to differentiate the structures of a cold protonated adenine dimer. A burn laser generates a population hole of a certain conformer of the dimer stored in a cold quadrupole ion trap, and an auxiliary dipolar RF ejects the photofragments by the burn laser from the trap. A probe laser detects depletion of a certain conformer by the burn laser, and a conformer-specific UV or IR spectrum of a cold ion is obtained by scanning the wavelength of the burn or the probe laser. This simple and versatile method is applicable to any type of double-resonance photofragmentation spectroscopy in a cold quadrupole ion trap. To demonstrate its capability, it was applied to UV-UV hole-burning spectroscopy of a protonated adenine dimer. It is proved that a cold protonated adenine dimer has at least two hydrogen-bonding geometries and each has multiple electronically excited states with significantly different spectral bandwidths, possibly due to different excited state dynamics.

MI06 **2:50 – 3:05**

SPECTROSCOPIC INVESTIGATION OF PROTON-COUPLED ELECTRON TRANSFER IN WATER OXIDATION CATALYZED BY A RUTHENIUM COMPLEX, $[Ru(tpy)(bpy)(H_2O)]^{2+}$

ERIN M. DUFFY, BRETT MARSH, JONATHAN VOSS, ETIENNE GARAND, *Department of Chemistry, University of Wisconsin, Madison, WI, USA.*

The splitting of H_2O into H_2 and O_2 is an attractive option for alternative energy, but the oxygen evolution step poses a significant challenge. A decades-long effort to produce a suitable water oxidation catalyst (WOC) has made progress on this front, but the precise reaction mechanism of these catalysts is still not well understood. One of the most extensively studied WOCs is $[Ru(tpy)(bpy)(H_2O)]^{2+}$ (tpy = 2,2':6,2"-terpyridine, bpy = 2,2'-bipyridine). Presented here are gas-phase infrared spectra of water clusters of $[Ru(tpy)(bpy)(OH_2)]^{2+}$ and the first intermediate of the catalytic cycle, $[Ru(tpy)(bpy)(OH)]^{2+}$. In particular, the O-H stretches are used as a probe of solvation strength, and trends in their spectral shifts are examined as a function of cluster size. With the aid of density functional theory (DFT) calculations, these spectra reveal structural changes induced by solvation that provide clear evidence for proton-coupled electron transfer (PCET), in support of proposed mechanisms.

MI07 **3:07 – 3:22**

PROBING SOLVATAION SHELLS OF $Ni(H_2O)_m{}^{2+}$ (m=4-10) AND $NiOH(H_2O)_n{}^{+}$ (n=2-5) WITH CRYOGENIC ION VIBRATIONAL SPECTROSCOPY.

JONATHAN VOSS, BRETT MARSH, JIA ZHOU, ETIENNE GARAND, *Department of Chemistry, University of Wisconsin, Madison, WI, USA.*

The solvation of metal cations, a process that dictates chemistry in both catalytic and biological systems, has been well studied using gas-phase spectroscopy. However, until recently the solvation of cation-anion pairs has been poorly explored. Here we present gas-phase spectra of $Ni(H_2O)_m{}^{2+}$ (m=4-10) and $NiOH(H_2O)_n{}^{+}$ (n=2-5) obtained via cryogenic ion vibrational spectroscopy (CIVS). Our results indicate that as cluster size decreases, the $NiOH(H_2O)_n{}^{+}$ moiety becomes more favorable over the $Ni(H_2O)_m{}^{2+}$ moiety. Analysis of the spectral data in conjunction with density functional theory calculations shows that both species have a 1st solvation shell consisting of six lingands. However, the $NiOH(H_2O)_n{}^{+}$ clusters show evidence of strong interactions between a first solvation shell water ligand and the OH^- group of the metal, similar to the interactions previously observed in $CaOH(H_2O)_n{}^{+}$ and $MgOH(H_2O)_n{}^{+}$.

MI08 **3:24 – 3:39**

MICROSOLVATION OF THE $Mg_2SO_4{}^{2+}$ CATION: CRYOGENIC VIBRATIONAL SPECTROSCOPY OF $(Mg^{2+})_2SO_4{}^{2-}(H_2O)_{n=4-11}$

PATRICK J KELLEHER, JOSEPH W DePALMA, *Department of Chemistry, Yale University, New Haven, CT, USA*; CHRISTOPHER J JOHNSON, *Department of Chemistry, Stony Brook University, Stony Brook, NY, USA*; JOSEPH FOURNIER, MARK JOHNSON, *Department of Chemistry, Yale University, New Haven, CT, USA.*

Cryogenic ion vibrational predissociation (CIVP) spectroscopy was used to examine the onset of solvation upon the incremental addition of water molecules to the $Mg_2SO_4{}^{2+}(H_2O)_n$ cation (n = 4 – 11). D_2 predissociation spectra are reported for each cluster over the range 1000-3800 cm^{-1}. Initially, the Mg^{2+} atoms each interact with two oxygen atoms on the sulfate anion in a bifurcated arrangement. The breaking of this motif occurs upon addition of the eighth water molecule as evidenced by splitting of the water bend, and broad absorption in the 3000-3400 cm^{-1} range indicative of hydrogen bonding between the water molecules and sulfate ion.

Intermission

MI09 **3:58 – 4:13**

CAPTURE AND STRUCTURAL DETERMINATION OF ACTIVATED INTERMEDIATES IN NICKEL CATALYZED CO_2 REDUCTION

STEPHANIE CRAIG, FABIAN MENGES, ARRON WOLK, JOSEPH FOURNIER, *Department of Chemistry, Yale University, New Haven, CT, USA*; NIKLAS TÖTSCH, *Physikalische Chemie II, Ruhr University Bochum, Bochum, Germany*; MARK JOHNSON, *Department of Chemistry, Yale University, New Haven, CT, USA.*

The catalyzed reduction of CO_2 is an important step in the conversion of this small molecule into liquid fuels. Nickel 1,4,8,11-tetraazacyclotetradecane, Ni(cyclam), is a well-known catalyst for the reduction of CO_2 in solution. Cryogenic ion vibrational predissociation (CIVP) spectroscopy of CO_2-messenger tagged ions cooled in a temperature controlled ion trap was used to study the starting Ni^{2+}(cyclam) reactant, and possible reaction intermediates and products in the gas phase. Additionally, parental CO_2 reduction was observed on the Ni(I) species $[Ni(bipyridine-(NMe)_2)]_2$(diphenyldiacetylene).

MI10 **4:15 – 4:30**

THRESHOLD IONIZATION SPECTROSCOPIC CHARACTERIZATION OF La ATOM REACTION WITH ISOPRENE

WENJIN CAO, DONG-SHENG YANG, *Department of Chemistry, University of Kentucky, Lexington, KY, USA.*

The reaction between La atom and isoprene ($CH_2 = CHC(CH_3) = CH_2$) was investigated in a supersonic molecular source. $La(C_2H_2)$, $La(C_3H_4)$, and $La(C_5H_8)$ were observed by time-of-flight mass spectrometry, and their structures and electronic states were characterized by mass-analyzed threshold ionization spectroscopy. Both $La(C_2H_2)$ and $La(C_3H_4)$ are three-membered metallacycles formed by the C-C bond cleavage and hydrogen migration. $La(C_2H_2)$ has a C_{2v} structure, whereas $La(C_3H_4)$ has a C_s structure. $La(C_5H_8)$ was identified as lanthano-methylcyclobutene ($La(CH_2C(CH_3) = CHCH_2)$) (C_1) formed by association and double-bond migration. All three complexes have a doublet ground state with the highest occupied molecular orbital being largely a La 6s character. Ionization removes the metal based electron, and the resultant ion has a similar structure to the neutral complex.

MI11 **4:32 – 4:47**

Ce-PROMOTED BOND ACTIVATION OF ETHYLENE PROBED BY MASS-ANALYZED THRESHOLD IONIZATION SPECTROSCOPY

YUCHEN ZHANG, SUDESH KUMARI, WENJIN CAO, DONG-SHENG YANG, *Department of Chemistry, University of Kentucky, Lexington, KY, USA.*

$Ce(C_2H_2)$ and $Ce(C_4H_6)$ complexes were observed in the reaction of Ce atom with ethylene in a supersonic molecular beam source and investigated by mass-analyzed threshold ionization spectroscopy (MATI) and theoretical calculations. Preliminary data analysis shows that $Ce(C_2H_2)$ has a triangle structure (C_{2v}) with Ce binding to C_2H_2 in a two-fold mode and $Ce(C_4H_6)$ has a five-membered metallacyclic structure (C_s) with Ce binding to the two terminal carbon atoms of butadiene. The ground states of both species are triplets with a $4f^16s^1$ Ce-based electron configuration and those of the corresponding ions are doublets from the removal of the $6s^1$ electron. The $Ce(C_2H_2)$ complex is formed by ethylene dehydrogenation, whereas $Ce(C_4H_6)$ by ethylene dehydrogenation and carbon-carbon bond coupling. The MATI spectra of $Ce(C_2H_2)$ and $Ce(C_4H_6)$ are rather similar to those of the corresponding La complexes previously observed by our group, except that the spectra of the Ce complexes exhibit two electronic transitions with almost identical vibrational intervals. This observation suggests that the existence of a 4f electron results in an increased complexity of the electronic spectra and states of the lanthanide hydrocarbons.

MI12 **4:49 – 5:04**

STRUCTURE DETERMINATION OF CISPLATIN-AMINO ACID ANALOGUES BY INFRARED MULTIPLE PHOTON DISSOCIATION ACTION SPECTROSCOPY

CHENCHEN HE, XUN BAO, YANLONG ZHU, STEPHEN STROBEHN, BETT KIMUTAI, Y-W NEI, C S CHOW, M T RODGERS, *Department of Chemistry, Wayne State University, Detroit, MI, USA*; JUEHAN GAO, J. OOMENS, *Institute for Molecules and Materials (IMM), Radboud University Nijmegen, Nijmegen, Netherlands.*

To gain a better understanding of the binding mechanism and assist in the optimization of relevant drug and chemical probe design, both experimental and theoretical studies were performed on a series of amino acid-linked cisplatin derivatives, including glycine-, lysine-, and ornithine-linked cisplatin, Gplatin, Kplatin, and Oplatin, respectively. Cisplatin, the first FDA-approved platinum-based anticancer drug, has been widely used in cancer chemotherapy. Its pharmacological mechanism has been identified as its ability to coordinate to genomic DNA, and guanine is its major target. In previous reports, cisplatin was successfully utilized as a chemical probe to detect solvent accessible sites in ribosomal RNA (rRNA). Among the amino-acid-linked cisplatin derivatives, Oplatin exhibits preference for adenine over guanine. The mechanism behind its different selectivity compared to cisplatin may relate to its potential of forming a hydrogen bond between the carboxylate group in Pt (II) complex and the 6-amino moiety of adenosine stabilizes A-Oplatin products. Tandem mass spectrometry analysis also indicates that different coordination sites of Oplatin on adenosine affect glycosidic bond stability.

Infrared multiple photon dissociation (IRMPD) action spectroscopy experiments were performed on all three amino acid-linked cisplatin to characterize their structures. An extensive theoretical study has been performed on Gplatin to guide the selection of the most effective theory and basis set based on its geometric information. The results for Gplatin provide the foundation for characterization of the more complex amino acid-linked cisplatin derivatives, Oplatin and Kplatin. Structural and energetic information elucidated for these compounds, particularly Oplatin reveal the reason for its alternative selectivity compared to cisplatin.

MI13 **5:06 – 5:21**

STRUCTUAL EFFECTS OF CYTIDINE 2′ RIBOSE MODIFICATIONS AS DETERMINED BY IRMPD ACTION SPECTROSCOPY

LUCAS HAMLOW, CHENCHEN HE, LIN FAN, RANRAN WU, BO YANG, M T RODGERS, *Department of Chemistry, Wayne State University, Detroit, MI, USA*; GIEL BERDEN, J. OOMENS, *Institute for Molecules and Materials (IMM), Radboud University Nijmegen, Nijmegen, Netherlands.*

Modified nucleosides, both naturally occurring and synthetic play an important role in understanding and manipulating RNA and DNA. Naturally occurring modified nucleosides are commonly found in functionally important regions of RNA and also affect antibiotic resistance or sensitivity. Synthetic modifications of nucleosides such as fluorinated and arabinosyl nucleosides have found uses as anti-virals and chemotherapy agents. Understanding the effect that modifications have on structure and glycosidic bond stability may lend insight into the functions of these modified nucleosides.

Modifications such as the naturally occurring 2′-O-methylation and the synthetic 2′-fluorination are believed to help stabilize the nucleoside through the glycosidic bond stability and intramolecular hydrogen bonding. Changing the sugar from ribose to arabinose alters the stereochemistry at the 2′ position and thus shifts the 3D orientation of the 2′-hydroxyl group, which also affects intramolecular hydrogen bonding and glycosidic bond stability. The structures of 2′-deoxy-2′-fluorocytidine, 2′-O-methylcytidine and cytosine arabinoside are examined in the current work by measuring the infrared spectra in the IR fingerprint region using infrared multiple photon dissociation (IRMPD) action spectroscopy. The structures accessed in the experiments were determined via comparison of the measured IRMPD action spectra to the theoretical linear IR spectra determined by density functional theory and molecular modeling for the stable low-energy structures. Although glycosidic bond stability cannot be quantitatively determined from this data, complementary TCID studies will establish the effect of these modifications. Comparison of these modified nucleosides with their RNA and DNA analogues will help elucidate differences in their intrinsic chemistry.

MI14 **5:23 – 5:38**

GAS-PHASE CONFORMATIONS AND ENERGETICS OF SODIUM CATIONIZED 2′-DEOXYGUANOSINE AND GUANOSINE: IRMPD ACTION SPECTROSCOPY AND THEORETICAL STUDIES

YANLONG ZHU, LUCAS HAMLOW, CHENCHEN HE, XUN BAO, M T RODGERS, *Department of Chemistry, Wayne State University, Detroit, MI, USA*; JUEHAN GAO, J. OOMENS, *Institute for Molecules and Materials (IMM), Radboud University Nijmegen, Nijmegen, Netherlands.*

In living systems, the local structures of DNA and RNA are influenced by protonation, deprotonation and noncovalent binding interactions with cations. In order to determine the effects of Na^+ cationization on the gas-phase structures of 2′-deoxyguanosine, $[dGuo+Na]^+$, and guanosine, $[Guo+Na]^+$, infrared multiple photon dissociation (IRMPD) action spectra of these two sodium cationized DNA and RNA mononucleosides are measured over the range extending from 500 to 1850 cm^{-1} using the FELIX free electron laser. Complementary electronic structure calculations are performed to determine the stable low-energy conformations of these complexes. Geometry optimizations and frequency analyses of these species are performed at the B3LYP/6-31G* level of theory, whereas single-point energies are calculated at the B3LYP/6-311+G(2d,2p) level of theory to determine the relative stabilities of these conformations. Comparison of the measure IRMPD action spectra and computed linear IR spectra enable the conformations accessed in the experiments to be elucidated. In both cases, preferential binding of the Na^+ cation to O6 and N7 positions of the nucleobase is observed. Present results for the sodium cationized nucleosides are compared to results for the analogous protonated forms of these nucleosides to elucidate the effects of multiple chelating interactions with the sodium cation to hydrogen bonding interactions in the protonated systems on the structures and stabilities of these nucleosides.

MI15 **5:40 – 5:55**

UNRAVELING PROTON TRANSFER IN STEPWISE HYDRATED N-HETEROCYCLIC ANIONS

JOHN T. KELLY, NATHAN I HAMMER, *Chemistry and Biochemistry, University of Mississippi, Oxford, MS, USA*; KIT BOWEN, *Department of Chemistry, Johns Hopkins University, Baltimore, MD, USA*; GREGORY S. TSCHUMPER, *Chemistry and Biochemistry, University of Mississippi, Oxford, MS, USA.*

Depending upon the number and location of nitrogen atoms in a N-heterocyclic azabenzene, the addition of a single water molecule can result in a positive electron affinity. The transfer of a proton from a solvating water azine base can be induced by excess electron attachment. Here we explore this phenomenon through the use of photoelectron spectroscopy and electronic structure theory. Carefully calibrated density functional theory (DFT) computations indicate that the excess electron predominantly resides in a π^* orbital of the heterocycle.

MJ. Small molecules

Monday, June 22, 2015 – 1:30 PM

Room: 217 Noyes Laboratory

Chair: Leah C O'Brien, Southern Illinois University, Edwardsville, IL, USA

MJ01 1:30 – 1:45

DEPERTURBATION ANALYSIS FOR THE $a^3\Pi$ AND $c^3\Sigma^-$ STATES OF C_2

JIAN TANG, WANG CHEN, KENTAROU KAWAGUCHI, *Graduate School of Natural Science and Technology , Okayama University, Okayama, Japan.*

In the last symposium and a recent paper[a], we reported a simultaneous analysis for the Phillips and Ballik-Ramsay band systems with a deperturbation treatment for the $X^1\Sigma^+$ and $b^3\Sigma^-$ states of C_2 and also, for the first time, the observation of the forbidden transitions between the singlet and triplet states of C_2. In the present study, we consider the interaction between the $a^3\Pi$ and $c^3\Sigma^-$ states to remove some anomalies in the higher order constants of the $a^3\Pi$ state presented in the previous work. The local interaction between the $a^3\Pi$ v=7 and $c^3\Sigma^-$ v=1 states was considered in a recent analysis[b] for the perturbation of the spectrum. We consider the interaction between all the vibrational levels of the two electronic states with a set of Dunham-like constants. The progress and results will be presented.

[a] W. Chen, K. Kawaguchi, P. F. Bernath, and J. Tang, J. Chem. Phys. 142, 064317 (2015).
[b] M. Nakajima and Y. Endo, J. Mol. Spectrosc. 302, 9 (2014).

MJ02 1:47 – 2:02

HIGH – RESOLUTION LASER SPECTROSCOPY OF THE $A\,^3\Pi_1 \leftarrow X\,^1\Sigma^+$ SYSTEM OF ICl IN 0.7 μm REGION.

NOBUO NISHIMIYA, TOKIO YUKIYA, MASAO SUZUKI, *Faculty of Engineering, Tokyo Polytechnic University, Atsugi, Japan*; ROBERT J. LE ROY, *Department of Chemistry, University of Waterloo, Waterloo, ON, Canada.*

Spectroscopic data for the $A\,^3\Pi_1$ and $X\,^1\Sigma^+$ states of $I^{35/37}Cl$ have been obtained by many researchers using grating spectrometers and Fourier-transform infrared spectrometers.[a,b] In a previous paper[c] we reported the measurement of doppler limited electronic vib-rotational absorption lines of the $A\,^3\Pi_1 \leftarrow X\,^1\Sigma^+$ system of $I^{35/37}Cl$ using a source modulation method, and new Mass-reduced Dunham coefficients were reported for the X-state. However, it is becoming increasingly common to analyse diatomic molecule spectroscopic data using the "direct-potential-fit" (DPF) method in which observed transition energies are fitted to simulated spectra generated from analytic models for the potential energy function(s). This method tends to require fewer fitting parameters than traditional Dunham analyses, as well as having more robust extrapolation properties in both the v and J domains. The present work combines all available previously reported data for the $A\,^3\Pi_1$ and $X\,^1\Sigma^+$ states with new measurements up to $v' = 10$ in the 0.7μm region obtained with a tone burst method using a Ti:Sapphire Ring Laser (M Squared Ltd SolsTis CW with Tera scan) in the the first DPF analysis reported for this system. The results of this study and our new fully analytic potential energy functions for the $A\,^3\Pi_1$ and $X\,^1\Sigma^+$ states of ICl will be presented.

[a] J.A. Coxon, R.M. Gordon and M.A. Wickramaaratchi, J. Mol. Spectr. **79** (1983) 363, 380.
[b] H. Hedderich P.F. Bernath and G.A. McRae J. Mol. Spectr. **155** (1992) 384.
[c] T.Yukiya, N. Nishimiya and M. Suzuki, J. Mol. Spectr. **269** (2011) 193.

MJ03 **2:04 – 2:19**

HIGH RESOLUTION LASER SPECTROSCOPY FOR ABSORPTION TO LEVELS LYING NEAR THE DISSOCIATION LIMIT OF THE $A\,^3\Pi_1$ STATE OF IBr

TOKIO YUKIYA, NOBUO NISHIMIYA, MASAO SUZUKI, *Faculty of Engineering, Tokyo Polytechnic University, Atsugi, Japan*; ROBERT J. LE ROY, *Department of Chemistry, University of Waterloo, Waterloo, ON, Canada.*

Spectroscopic data involving levels lying near the dissociation limit are very important for determining accurate molecular well depths and full potential energy curves. In previous work, we have reported the potential functions and values of parameters $\mathcal{D}_e$ and r_e for the $A\,^3\Pi_1$ and $X\,^1\Sigma^+$ states of IBr.[a] That study used data extending to $v'(A) = 29$ and determined anomalous fluctuations in the v–dependence of the first differences of $\Delta B_v = B_{v+1} - B_v$ for levels $v' = 27 - 29$ of the $A\,^3\Pi_1$ state which, surprisingly, seems to have been smoothly accounted by a fitted potential energy function that shows no visually perceptible irregularities. In the present work, a Ti:Sapphire ring laser(M SQUARED LASERS Ltd. SolsTiS CW with Tera–scan) has been introduced to probe the 0.7μm region closer to the dissociation limit and examine whether the anomalous ΔB_v behaviour expends further up the well. The results of this study will be presented.

[a]T.Yukiya, N. Nishimiya, M. Suzuki and R.J. Le Roy, paper MG03 at the 69th International Symposium on Molecular Spectroscopy, University of Illinois (2014)

MJ04 **2:21 – 2:36**

THE NEAR-INFRARED SPECTRUM OF NiCl: ANALYSES OF THE (0,1), (1,0), & (2,1) BANDS OF SYSTEM G AND THE (1,0) BAND OF SYSTEM H

JACK C HARMS, COURTNEY N GIPSON, ETHAN M GRAMES, JAMES J O'BRIEN, *Chemistry and Biochemistry, University of Missouri, St. Louis, MO, USA*; LEAH C O'BRIEN, *Department of Chemistry, Southern Illinois University, Edwardsville, IL, USA.*

The near-infrared spectrum of nickel chloride, NiCl, has been recorded at high resolution using intracavity laser absorption spectroscopy. The NiCl molecules were produced in a plasma discharge of a nickel hollow cathode from a trace amount of CCl_4 using Ar as the sputter gas. Spectra were collected from 12,490-12,660 cm^{-1}and 13,200-13,350 cm^{-1}as a series of overlapping 5 cm^{-1}scans. The (0,1), (1,0), and (2,1) bands of the [13.0] $^2\Pi_{3/2}$-X $^2\Pi_{3/2}$ transition, System G, were observed at 12,537 cm^{-1}, 13,352 cm^{-1}, and 13,318 cm^{-1}, respectively. The (1,0) band of the [12.3] $^2\Sigma^-$-X $^2\Pi_{3/2}$ transition, System H, was observed at 12,645 cm^{-1}. Analyses of these bands will be presented.

MJ05 **2:38 – 2:53**

ANALYSIS OF EMISSION SPECTRA OF YTTRIUM MONOIODIDE PRODUCED BY THE PHOTODISSOCIATION OF YI_3

WENTING WENDY CHEN, THOMAS C. GALVIN, THOMAS J. HOULAHAN, JR., J. GARY EDEN, *Department of Electrical and Computer Engineering, University of Illinois at Urbana-Champaign, Urbana, IL, USA.*

Emission spectra of yttrium monoiodide (YI) spanning the 250 - 940 nm spectral region were generated by the photodissociation of yttrium tri-iodide under photoexcitation at 248 nm (KrF laser). Fluorescent spectra in the13,000 - 19,000 cm^{-1} and 24,000 - 40,000 cm^{-1} regions will be first reported. New vibrational transitions of YI in the 20,000 - 25,000 cm^{-1} interval will be presented as well.

MJ06 **2:55 – 3:05**

GENERATION OF VIBRATIONALLY EXCITED HCP FROM A STABLE SYNTHETIC PRECURSOR

ALEXANDER W. HULL, JUN JIANG, TREVOR J. ERICKSON, CARRIE WOMACK, MATTHEW NAVA, CHRISTOPHER CUMMINS, ROBERT W FIELD, *Department of Chemistry, MIT, Cambridge, MA, USA.*

HCP belongs to a class of reactive small molecules with much interest to spectroscopists. It bears certain similarities to HCN, including a strong Ã(bent) - X̃(linear) ultraviolet transition, associated with the HCP-HPC isomerization pathway. HCP has traditionally been generated by the *in situ* reaction of PH_3 and acetylene. In this talk, we will discuss a recently developed synthetic precursor molecule, 1,1-((triphenylphosphoranylidene)methyl)-9,10-phosphanoanthracene. At temperatures above 200 degrees Celsius, this precursor is thought to release HCP in a vibrationally excited state. We will present preliminary spectra on this system obtained by LIF and chirped pulse millimeter wave spectroscopy.

Intermission

MJ07 **3:24 – 3:39**

DPF ANALYSES YIELD FULLY ANALYTIC POTENTIALS FOR THE $B\,^1\Pi_u$ "BARRIER" STATES OF Rb_2 and Li_2 AND AN IMPROVED GROUND-STATE WELL DEPTH FOR Rb_2

KAI SLAUGHTER, *Department of Chemistry, University of Waterloo, Waterloo, ON, Canada*; NIKESH S. DATTANI, *Graduate School of Science, Department of Chemistry, Kyoto University, Kyoto, Japan*; CLAUDE S. AMIOT, *Laboratoire Aimé Cotton, CNRS, Orsay, France*; AMANDA J. ROSS, *UMR 5306, ILM University Lyon 1 and CNRS, Villeurbanne, France*; ROBERT J. LE ROY[a], *Department of Chemistry, University of Waterloo, Waterloo, ON, Canada.*

Determining full model potential energy functions for molecular states that have a 'natural' rotationless barrier which protrudes above the potential asymptote, such as the $B\,^1\Pi_u$ states of alkali dimers, is a challenging problem. The present work extends our previous Direct-Potential-Fit (DPF) analysis of data for the $B\,^1\Pi_u$ state of Li_2[b] by introducing a more sophisticated model for the long-range tail of the fully analytic 'Double Exponential Long-Range' (DELR) potential function form[a] that takes account of the interstate coupling that occurs near the asymptotes of $nS + nP$ alkali dimers.[c] This type of analysis is then applied to data for the $B\,^1\Pi_u$ state of Rb_2, and a concurrent extension of the DPF analysis of Seto and Le Roy[d] yields an improved fully analytic potential energy function for its ground $X\,^1\Sigma_g^+$ state. The effect of taking account of the long-range inter-state coupling on the shapes of the outer walls of the $B\,^1\Pi_u$ state potential functions for these two species will also be examined.

[a]leroy@uwaterloo.ca
[b] Y. Huang and R.J. Le Roy, *J. Chem. Phys.*, **119**, 7398 (2003)
[c] M. Aubert-Frécon and G. Hadinger and S. Magnier and S. Rousseau, *J. Mol. Spectosc.*, **288**, 182 (1998).
[d]J.Y. Seto and R.J. Le Roy, *J. Chem. Phys.*, **113**, 3067 (2000).

MJ08 **3:41 – 3:56**

LASER SPECTROSCOPY OF THE PHOTOASSOCIATION OF Rb–Ar AND Rb–Kr THERMAL PAIRS: STRUCTURE OF THE Rb–RARE GAS $A^2\Pi_{1/2}$ STATE NEAR THE CLASSICAL LIMIT

ANDREY E. MIRONOV, WILLIAM GOLDSHLAG, KYLE T RAYMOND, J. GARY EDEN, *Department of Electrical and Computer Engineering, University of Illinois at Urbana-Champaign, Urbana, IL, USA.*

A new laser spectroscopic technique has been demonstrated for examining the structure of alkali–rare gas diatomic electronic states near the classical limit. In two-color experiments, Rb–Ar or Rb–Kr thermal pairs are excited by free←free or bound←free transitions while monitoring the amplified spontaneous emission produced on the Rb D_1 or D_2 lines. Spectra observed lying within 10 cm^{-1} of the separated atom limit for the $A^2\Pi_{1/2}$ states of Rb–Ar and Rb–Kr will be presented and discussed.

MJ09 **3:58 – 4:13**

COLLISION-INDUCED ABSORPTION WITH EXCHANGE EFFECTS AND ANISOTROPIC INTERACTIONS: THEORY AND APPLICATION TO $H_2 - H_2$ and $N_2 - N_2$.

TIJS KARMAN, *Institute for Molecules and Materials (IMM), Radboud University Nijmegen, Nijmegen, Netherlands*; EVANGELOS MILIORDOS, KATHARINE HUNT, *Department of Chemistry, Michigan State University, East Lansing, MI, USA*; AD VAN DER AVOIRD, GERRIT GROENENBOOM, *Institute for Molecules and Materials (IMM), Radboud University Nijmegen, Nijmegen, Netherlands.*

Collision-induced absorption spectra can be calculated quantum mechanically and from first principles. However, such calculations are usually performed in the approximation of an isotropic interaction potential and neglecting exchange effects. We present theory for including exchange and anisotropic interactions in the calculation of collision-induced absorption spectra, and apply this method to the $H_2 - H_2$ and $N_2 - N_2$ systems. For $H_2 - H_2$, the isotropic interaction approximation is generally accurate, although significant effects of anisotropic interactions are observed in the far wing of the spectrum. For $N_2 - N_2$, anisotropic interactions increase the line strength at low energy by two orders of magnitude. The agreement with experimental data is reasonable in the isotropic interaction approximation, and improves when the full anisotropic potential is considered. The effect of the interaction anisotropy decreases at higher energy, which validates the usual isotropic interaction approximation as a high-temperature approximation for the calculation of collision-induced absorption spectra.

MJ10 ***Post-Deadline Abstract*** **4:15 – 4:30**

PHOTO-DISSOCIATION RESONANCES OF JET-COOLED NO_2 AT THE DISSOCIATION THRESHOLD BY CW-CRDS, CHALLENGING RRKM THEORIES

PATRICK DUPRÉ, *Laboratoire de Physico-Chimie de l'Atmosphère, Université du Littoral Côte d'Opale, Dunkerque, France.*

Around 398 nm, the jet-cooled NO_2 spectrum exhibits a well identified dissociation threshold (D_0). Combining LIF detection and continuous-wave absorption-based CRDS technique a frequency range of $\sim 25\,\mathrm{cm}^{-1}$ is analyzed at high resolution around D_0. In addition to the usual rovibronic transitions towards long-lived energy levels, ~ 115 wider resonances are observed. Over this energy range, the resonance widths spread from $\sim 0.006\,\mathrm{cm}^{-1}$ (~ 450 ps) to $\sim 0.7\,\mathrm{cm}^{-1}$ (~ 4 ps) with large fluctuations. At least two ranges of resonance width can be identified when increasing the excess energy. They are associated with the opening of the dissociation channels $NO_2 \rightarrow NO\left(X\,^2\Pi_{1/2},\, v=0,\, J=1/2\right) + O\left(^3P_2\right)$ and $NO_2 \rightarrow NO\left(X\,^2\Pi_{1/2},\, v=0,\, J=3/2\right) + O\left(^3P_2\right)$. Weighted mean unimolecular dissociation rate coefficients k_{uni} are calculated. The density of reactants (following the RRKM predictions) is deduced, and it will be discussed versus the density of transitions, the density of resonances and the density of vibronic levels. The data are analyzed in the light of time-resolved data previously reported. This analysis corroborates the existence of loose transition states along the reaction path close to the dissociation energy in agreement with the phase space theory predictions[a].

[a][Accpeted in J. Chem. Phys.]

MJ11 **4:32 – 4:47**

SELF- AND CO_2-BROADENED LINE SHAPE PARAMETERS FOR THE ν_2 AND ν_3 BANDS OF HDO

V. MALATHY DEVI, D. CHRIS BENNER, *Department of Physics, College of William and Mary, Williamsburg, VA, USA*; KEEYOON SUNG, LINDA BROWN, *Jet Propulsion Laboratory, California Institute of Technology, Pasadena, CA, USA*; ARLAN MANTZ, *Department of Physics, Astronomy and Geophysics, Connecticut College, New London, CT, USA*; MARY ANN H. SMITH, *Science Directorate, NASA Langley Research Center, Hampton, VA, USA*; ROBERT R. GAMACHE, *Department of Environmental, Earth, and Atmospheric Sciences, University of Massachusetts, Lowell, MA, USA*; GERONIMO L. VILLANUEVA, *Astrochemistry, NASA Goddard Space Flight Center, Greenbelt, MD, USA.*

Knowledge of CO_2-broadened HDO widths and their temperature dependence exponents are required to interpret atmospheric spectra of Mars and Venus. We therefore used nine high-resolution, high signal-to-noise spectra of HDO and HDO+CO_2 mixtures to obtain broadening coefficients for selected transitions of the ν_2 and ν_3 vibrational bands located at 7.13 and 2.70 μm, respectively. The gas samples were prepared by mixing equal amounts of high-purity distilled H_2O and a 99% enriched D_2O sample. Spectra at different temperatures (255-296 K) were obtained using a 20.38 cm long coolable cell[a] installed in the sample compartment of the Bruker 125HR Fourier transform spectrometer at the Jet Propulsion Laboratory, in Pasadena, CA. The retrieved parameters included accurate line positions, intensities, self- and CO_2-broadened half-width and pressure-shift coefficients and the temperature dependences of CO_2 broadened HDO. The spectroscopic parameters for many transitions were obtained simultaneously by multispectrum fitting[b] of all nine spectra in each band. A non-Voigt line shape with speed dependence was applied. Line mixing was also observed for several transition pairs. Preliminary results will be compared to other recent measurements reported in the literature.[c]

[a] K. Sung, A.W. Mantz, M.A.H. Smith, L.R. Brown, T.J. Crawford, V.M. Devi, D.C. Benner. J. Mol. Spectrosc. 162 (2010) 124-134.

[b] D.C. Benner, C.P. Rinsland, V. Malathy Devi, M.A.H. Smith, and D. Atkins. JQSRT 53 (1995) 705-721.

[c] Research described in this paper are performed at the College of William and Mary, Jet Propulsion Laboratory, California Institute of Technology, Connecticut College and NASA Langley Research Center under contracts and cooperative agreements with the National Aeronautics and Space Administration.

MJ12 **4:49 – 5:04**

DISPERSED FLUORESCENCE SPECTRA OF JET COOLED SiCN

MASARU FUKUSHIMA, TAKASHI ISHIWATA, *Information Sciences, Hiroshima City University, Hiroshima, Japan.*

The laser induced fluorescence (LIF) spectrum of $\tilde{A}\ ^2\Delta - \tilde{X}\ ^2\Pi$ transition was obtained for SiCN generated by laser ablation under supersonic free jet expansion. The vibrational structure of the dispersed fluorescence (DF) spectra from single vibronic levels (SVL's) was analyzed with consideration of Renner-Teller (RT) interaction. The usual analysis based on the perturbation approach[a], indicated considerably different spin splitting for the μ and κ levels of the $\tilde{X}\ ^2\Pi$ state of SiCN, in contrast to identical spin splitting for general species based on the usual RT analysis. Further analysis of the vibrational structure is being carried out via direct RT diagonalization.

[a] J. M. Brown and F. Jørgensen, Advances in Chemical Physics 52, 117 (1983).

MJ13 **5:06 – 5:21**

INTERNAL FORCE FIELD DETERMINATION OF $\tilde{C}^1$ B_2 STATE of SO_2

JUN JIANG, BARRATT PARK, CARRIE WOMACK, ROBERT W FIELD, *Department of Chemistry, MIT, Cambridge, MA, USA.*

The internal force field of $\tilde{C}^1B_2$ state of SO_2 is determined up to quartic terms. The fit incorporates observed vibrational energy levels of both $S^{16}O_2$ and $S^{18}O_2$ below 3000 cm^{-1}, as well as rotational information of both isotopologues. With inclusion of nine recently observed B_2 symmetry levels of $S^{16}O_2$ in the fit, the double-well potential in asymmetric stretching coordinate can be better characterized. By inspecting the wavefunctions, as well as the basis state distribution of the eigenvectors, we are able to give vibrational assignments to majority of states in this energy region, based on Kellman's semiclassical study on fermi-resonant systems. Our analysis calls into question the validity of previous assignments of several vibrational levels. In addition, the force field allows us to calculate Coriolis matrix elements between vibrational bands, and the calculated values agree well with experimentally derived values. In particular, it predicts and explains why the experimental values are always much smaller than numbers predicted based on a naive harmonic picture. Our work is a first step towards a more complete understanding of the $\tilde{C}$ state potential energy surface near the equilibrium geometry and it is relevant to the question of how vibronic coupling between $\tilde{C}^1B_2$ state and higher lying A_1 state(s) gives rise to unequal S-O bond length.

MJ14 **5:23 – 5:38**

MEASUREMENT AND MODELING OF COLD $^{13}CH_4$ SPECTRA FROM 2.1 TO 2.7 μm

LINDA BROWN, KEEYOON SUNG, TIMOTHY J CRAWFORD, *Jet Propulsion Laboratory, California Institute of Technology, Pasadena, CA, USA*; ANDREI V. NIKITIN, SERGEY TASHKUN, *Atmospheric Spectroscopy Div., Institute of Atmospheric Optics, Tomsk, Russia*; MICHAEL REY, VLADIMIR TYUTEREV, *Groupe de Spectrométrie Moléculaire et Atmosphérique, UMR CNRS 7331, Université de Reims, Reims Cedex 2, France*; MARY ANN H. SMITH, *Science Directorate, NASA Langley Research Center, Hampton, VA, USA*; ARLAN MANTZ, *Department of Physics, Astronomy and Geophysics, Connecticut College, New London, CT, USA.*

A new study of $^{13}CH_4$line positions and intensities in the Octad region between 3600 and 4800 cm^{-1}will be reported. Nine spectra were recorded with two Fourier transform spectrometers (the McMath-Pierce FTS at Kitt Peak Observatory and the Bruker 125 HR FTS at the Jet Propulsion Laboratory) using ^{13}C-enriched samples at temperatures from 299 K to 80 K. Line positions and intensities were retrieved by non-linear least squares curve-fitting procedures and analyzed using the effective Hamiltonian and the effective Dipole moment expressed in terms of irreducible tensor operators adapted to spherical top molecules. Quantum assignments were found for all the 24 sub-vibrational states of the Octad (some as high as J=10). Over 4750 experimental line positions and 3300 line intensities were fitted with RMS standard deviations of 0.004 $cm^{-1}$and 6.9%, respectively. A new linelist of over 9600 measured positions and intensities from 3607 to 4735 $cm^{-1}$was produced, with known quantum assignments given for 45% of the features.[a]

[a] Part of the research described in this paper was performed at the Jet Propulsion Laboratory, California Institute of Technology, NASA Langley Research Center, and Connecticut College, under contracts and cooperative agreements with the National Aeronautics and Space Administration. The support of the Groupement de Recherche International SAMIA between CNRS (France) and RFBR (Russia) is acknowledged.

TA. Metal containing

Tuesday, June 23, 2015 – 8:30 AM

Room: 116 Roger Adams Lab

Chair: Jacob Stewart, Emory University, Atlanta, GA, USA

TA01 **8:30 – 8:45**

BONDING AT THE EXTREME. DETECTION AND CHARACTERIZATION OF THORIUM DIMER, Th_2

TIMOTHY STEIMLE[a], SETH MUSCARELLA, DAMIAN L KOKKIN, *Department of Chemistry and Biochemistry, Arizona State University, Tempe, AZ, USA.*

Due to the difficultly of working with actinides (radioactive, short lifetimes) and the number of electrons in these systems our chemical understanding either experimentally or theoretically on these systems is very limited. The electronic spectrum of thorium dimer, Th_2, is expected to be heavily congested due to the predicted twelve electronic states within an energy less then 1 eV of the calculated $^3\Delta_g$ ground state. The chemical bond is predicted to be a quadruple bond in both the ground state and low lying $^1\Sigma_g^+$ state (T_e=400 cm^{-1})[b]. Experimentally Th_2 was been detected in the gas phase by mass spectrometry[c]. Here we report on the detection of the gas fluorescence spectrum of Th_2 in the 495-560 nm range via application of 2D LIF spectroscopy and attempts to record high resolution field free and Zeeman spectra.

[a]NSF CHE-1265885

[b]B.J. Roos, P.-Å. Malmqvist and L. Gagliardi, J. Am. Chem. Soc. 128, 17000-17006, 2006

[c]M.C. Heaven, B.J. Barker and I.O. Antonov, J. Phys. Chem A. 118, 10867-10881, 2014

TA02 **8:47 – 9:02**

THE QUINTESSENTIAL BOND OF MODERN SCIENCE. THE DETECTION AND CHARACTERIZATION OF DIATOMIC GOLD SULFIDE, AuS.

DAMIAN L KOKKIN, RUOHAN ZHANG, TIMOTHY STEIMLE, *Department of Chemistry and Biochemistry, Arizona State University, Tempe, AZ, USA*; BRADLEY W PEARLMAN, IAN A WYSE, THOMAS D. VARBERG, *Chemistry Department , Macalester College, Saint Paul, Minnesota, USA.*

The gold sulfur bond is becoming ever more important to a vast range of scientific endeavors. We have recorded the electronic spectrum of gas-phase AuS, at vibrational resolution, over the 440-740 nm wavelength range. By application of a synergy of production techniques, hot hollow-cathode sputtering source and cold laser ablation molecular beam source, excitation from both spin components of the inverted $^2\Pi$ ground state is possible. Excitation into four different excited electronic states involving approximately 100 red-degraded bands has been observed. The four excited states have been characterized as $a^4\Sigma_{1/2}$, $A^2\Sigma^+_{1/2}$, $B^2\Sigma^-_{1/2}$ and $C^2\Delta_i$. The observed red-degraded vibronic bands where then globally analyzed to determine an accurate set of term energies and vibrational constants for the excited and ground electronic states. The electronic configurations from which these states arise will be discussed.

TA03 **9:04 – 9:19**

LASER SPECTROSCOPY OF RUTHENIUM CONTAINING DIATOMIC MOLECULES: RuH/D AND RuP.

ALLAN G. ADAM, RICARDA M. KONDER, NICOLE M. NICKERSON, *Department of Chemistry, University of New Brunswick, Fredericton, NB, Canada*; COLAN LINTON, D. W. TOKARYK, *Department of Physics, University of New Brunswick, Fredericton, NB, Canada.*

In the last few years, the Cheung group in Hong Kong and the Steimle group in Arizona have successfully studied several ruthenium containing diatomic molecules, RuX (X =C[a], O[b], N[c], B[d]), using the laser-ablation molecular jet technique. Based on this success, the UNB spectroscopy group decided to try and find the optical signatures of other RuX molecules. Using CH_3OH and PH_3 as reactant gases, the RuH and RuP diatomic molecules have been detected in surveys of the 420 - 675 nm spectral region. RuD has also been made using fully deuterated methanol as a reactant. Dispersed fluorescence experiments have been performed to determine ground state vibrational frequencies and the presence of any low-lying electronic states. Rotationally resolved spectra for these molecules have also been taken and the analysis is proceeding. The most recent results will be presented.

[a]F. Wang et al., Journal of Chemical Physics 139, 174318 (2013).

[b]N. Wang et al., Journal of Physical Chemistry A 117, 13279 (2013).

[c]T. Steimle et al., Journal of Chemical Physics 119, 12965 (2003).

[d]N. Wang et al., Chemical Physics Letters 547, 21 (2012).

TA04 **9:21 – 9:36**

OPTICAL ZEEMAN SPECTROSCOPY OF CALCIUM FLUORIDE, CaF.

TIMOTHY STEIMLE[a], DAMIAN L KOKKIN, *Department of Chemistry and Biochemistry, Arizona State University, Tempe, AZ, USA*; JACK DELVIN, MICHAEL TARBUTT, *Centre for Cold Matter, Blackett Laboratory, Imperial College London, London, United Kingdom.*

Recently laser cooling has been demonstrated for the diatomic radical calcium fluoride, CaF[b]. The mechanism of magneto-optical trapping for diatomic molecules has been elucidated recently by Tarbutt[c] where a rate model was used to model the interaction of molecules with multiple frequencies of laser light. It was shown that the correct choice of laser polarization depends on the sign of the upper state magnetic g-factor. The magnetic tuning of the low rotational levels in the $X^2\Sigma^+$, $A^2\Pi$ and $B^2\Sigma^+$ electronic states of CaF, have been experimentally investigated using high resolution optical Zeeman spectroscopy of a cold molecular beam sample. The observed Zeeman-induced shifts and splittings were successfully modeled using a traditional effective Hamiltonian approach to account for the interaction between the (ν=0) $A^2\Pi$ and (ν=0) $B^2\Sigma^+$ states. The determined magnetic g-factors for the $X^2\Sigma^+$, $A^2\Pi$ and $B^2\Sigma^+$ states are compared to those predicted by perturbation theory.

[a]NSF CHE-1265885

[b]V. Zhelyazkova, A. Cournol, T.E. Wall, A. Matsushima, J.J. Hudson, E.A. Hinds, M.R. Tarbutt and B.E. Sauer, Phys. Rev. A 89, 053416 (2014)

[c]M. R. Tarbutt, New J. Phys 17, 015007 (2015)

TA05 **9:38 – 9:53**

ELECTRONIC TRANSITIONS OF YTTRIUM MONOPHOSPHIDE

ALLAN S.C. CHEUNG, *Department of Chemistry, The University of Hong Kong, Hong Kong, Hong Kong*; BIU WA LI, *Department of Chemistry, The Chinese University of Hong Kong, Hong Kong, Hong Kong, China*; MAN-CHOR CHAN, *Department of Chemistry, The Chinese University of Hong Kong, Hong Kong, Hong Kong, China.*

Electronic transition spectrum of the yttrium monophosphide (YP) molecule in the visible region between 715 nm and 880 nm has been recorded using laser ablation/reaction free-jet expansion and laser induced fluorescence spectroscopy. The YP molecule was produced by reacting laser - ablated yttrium atoms with PH_3 seeded in argon. Thirteen vibrational bands were analyzed and five electronic transition systems have identified, namely the [12.2] $\Omega = 3$ - $X^3\Pi_2$ transition, [13.3] $\Omega = 3$ - $X^3\Pi_2$ transition, [13.4] $\Omega = 3$ - $X^3\Pi_2$ transition, [13.5] $\Omega = 3$ - $X^3\Pi_2$ transition, and [13.4] $\Omega = 2$ - $X^3\Pi_2$ transition. Least squares fits of the measured rotational lines yielded molecular constants for the ground and excited states. The ground state symmetry and the bond length r_0 of the YP molecule have been determined to be a $X^3\Pi_2$ state and 2.4413 Å respectively in this work. A molecular orbital energy level diagram has been used to help the assignment of the observed electronic states. This work represents the first experimental investigation of the spectrum of the YP molecule.

TA06 **9:55 – 10:10**

ROTATIONALLY RESOLVED SPECTROSCOPY OF THE $B^1\Pi \leftarrow X^1\Sigma^+$ AND $C^1\Sigma^+ \leftarrow X^1\Sigma^+$ ELECTRONIC BANDS OF CaO

MICHAEL SULLIVAN, JACOB STEWART, MICHAEL HEAVEN, *Department of Chemistry, Emory University, Atlanta, GA, USA.*

The $B^1\Pi \leftarrow X^1\Sigma^+$ and $C^1\Sigma^+ \leftarrow X^1\Sigma^+$ transitions of CaO, at energies below 30,000 cm^{-1}, were previously investigated by Lagerqvist[a]. The arc source used in that work yielded spectra at energies above 30,000 cm^{-1} that were too congested for analysis. In the present study we have used jet-cooling of CaO to extend the characterization of the $B \leftarrow X$ and $C \leftarrow X$ band systems up to 35,000 cm^{-1}. Analyses of these data and spectroscopic constants will be reported. This work is being carried out in support of two-color photoionization studies of the cation, where the higher energy vibronic levels of the B and C states are used as the first excitation step.

[a]A. Lagerqvist, *Arkiv För Fysik* **8**, 83, 1954

Intermission

TA07 **10:29 – 10:44**

HIGH RESOLUTION LASER SPECTROSCOPY OF NICKEL MONOBORIDE, NiB

E. S. GOUDREAU, COLAN LINTON, D. W. TOKARYK, *Department of Physics, University of New Brunswick, Fredericton, NB, Canada*; ALLAN G. ADAM, *Department of Chemistry, University of New Brunswick, Fredericton, NB, Canada.*

Diatomic nickel boride, NiB, has been produced in the UNB laser ablation molecular jet source. Survey spectra, taken at medium resolution with a pulsed dye laser in the 415 – 510 nm region, showed an intense band system which had previously been observed and assigned as a $^2\Pi_{3/2}$ - $^2\Sigma^+$ transition by Zhen et al.[a] Using a single frequency ring dye laser, we have obtained high resolution spectra of the 0-0, 2-0 and 3-0 bands of the most abundant isotopologue, $^{58}Ni^{11}B$, and the 2-0 band of $^{60}Ni^{11}B$. The rotational analysis showed that the transition was from an $\Omega = 0.5$ upper state to the ground $X^2\Sigma^+$ state. The data were found to fit equally well as $^2\Sigma^+$ - $^2\Sigma^+$ or $^2\Pi_{1/2}$ - $^2\Sigma^+$. The fine structure e/f parity splitting was examined for each of the two options in an attempt to determine the identity of the upper state. Partially resolved hyperfine structure due to the ^{11}B nuclear spin, I = 3/2, was observed and analyzed to try and determine the nature of the boron atom contribution to the ground $^2\Sigma^+$ state configuration. The results of the rotational and hyperfine structure analysis will be discussed.

[a] J-f. Zhen, L. Wang, C-b. Qin, Q. Zhang, Y. Chen, Chinese J. Chem. Phys. 23, 626 (2010).

TA08 **10:46 – 11:01**

MOLECULAR LINE LISTS FOR SCANDIUM AND TITANIUM HYDRIDE USING THE DUO PROGRAM

LORENZO LODI, *Department of Physics and Astronomy, University College London, London, IX, United Kingdom*; SERGEI N. YURCHENKO, *Department of Physics and Astronomy, University College London, Gower Street, London WC1E 6BT, United Kingdom*; JONATHAN TENNYSON, *Department of Physics and Astronomy, University College London, London, IX, United Kingdom.*

Transition-metal-containing (TMC) molecules often have very complex electronic spectra because of their large number of low-lying, interacting electronic states, of the large multi-reference character of the electronic states and of the large magnitude of spin-orbit and relativistic effects. As a result, fully ab initio calculations of line positions and intensities of TMC molecules have an accuracy which is considerably worse than the one usually achievable for molecules made up by main-group atoms only. In this presentation we report on new theoretical line lists for scandium hydride ScH and titanium hydride TiH[a]. Scandium and titanium are the lightest transition metal atoms and by virtue of their small number of valence electrons are amenable to high-level electronic-structure treatments and serve as ideal benchmark systems. We report for both systems energy curves, dipole curves and various coupling curves (including spin-orbit) characterising their electronic spectra up to about 20 000 cm-1. Curves were obtained using Internally-Contracted Multi Reference Configuration Interaction (IC-MRCI) as implemented in the quantum chemistry package MOLPRO. The curves where used for the solution of the coupled-surface ro-vibronic problem using the in-house program DUO [b]. DUO is a newly-developed, general program for the spectroscopy of diatomic molecules and its main functionality will be described. The resulting line lists for ScH and TiH are made available as part of the Exomol project [c].

[a] L. Lodi, S. N. Yurchenko and J. Tennyson, Mol. Phys. (Handy special issue) in press.
[b] S. N. Yurchenko, L. Lodi, J. Tennyson and A. V. Stolyarov, Computer Phys. Comms., to be submitted.
[c] J. Tennyson and S. N. Yurchenko, Mon. Not. R. Astr. Soc. 2012, 425, 21. See also www.exomol.com.

TA09 **11:03 – 11:18**

UV SPECTROSCOPY ON GAS PHASE Cu(I)-BIPYRIDYL COMPLEXES

SHUANG XU, *JILA and Department of Physics, University of Colorado at Boulder, Boulder, CO, USA*; CASEY CHRISTOPHER, J. MATHIAS WEBER, *JILA and the Department of Chemistry and Biochemistry, University of Colorado-Boulder, Boulder, CO, USA.*

Transition metal complexes with bipyridine ligands are of great interest in metal-organic chemistry, since they are prototypes for many applications in photochemistry and homogeneous catalysis. Under-coordinated bipyridyl complexes are elusive species in the condensed phase, and the ligand-induced changes in electronic structure are of fundamental interest. We present UV photodissociation spectra of mass-selected monocationic copper(I)-bipyridyl complexes $[bpy\text{-}Cu\text{-}L]^+$ with different ligands (L = H_2O, D_2, N_2, MeOH, Cl). Complexes were prepared via electrospray ionization of copper/bipyridine solutions followed by accumulation and buffer gas cooling in a cryogenic Paul trap. In addition, we show spectra of similar species based on copper oxide, $[bpy\text{-}CuO\text{-}L]^+$.

TA10 **11:20 – 11:35**

ANION PHOTOELECTRON SPECTROSCOPY OF NbW^- and W_2^-

D. ALEX SCHNEPPER, MELISSA A. BAUDHUIN, DOREEN LEOPOLD, *Chemistry Department, University of Minnesota, Minneapolis, MN, USA*; SEAN M. CASEY, *Department of Chemistry, University of Nevada, Reno, Reno, NV, USA.*

The 488 nm vibrationally-resolved photoelectron spectra of NbW^- and W_2^- are reported. The electron affinity of W_2 ($^1\Sigma_g{}^+ \leftarrow {}^2\Sigma_u{}^+$) is found to be 1.118 ± 0.007 eV, which differs from the value reported in a previous anion photoelectron spectroscopic study of W_2^- (1.46 eV)[a], but was accurately predicted by density functional calculations (1.12 eV)[b]. The fundamental vibrational frequency of W_2 is measured to be 345 ± 15 cm^{-1}, in agreement with the value previously reported in matrix resonance Raman studies (337 cm^{-1})[c]. The W_2^- anion is measured to have a fundamental frequency of 320 $\pm$ 15 cm^{-1}. Several weak transitions to excited electronic states are seen and tentatively assigned based on calculated energies. NbW has an electron affinity of 0.856 ± 0.007 eV. Vibrational frequencies are found, by Franck-Condon fitting of overlapping transitions, to be 365 ± 20 cm^{-1} for NbW^- and 410 ± 20 cm^{-1} for NbW. This increase in vibrational frequency upon photodetachment suggests that the extra electron is in an antibonding orbital, leading to ground state assignments of $^3\Delta$ and $^2\Delta$ for the anion and neutral, respectively. These results are compared to those obtained for other Group V and Group VI transition metal dimers and trends are discussed.

[a]H. Weidele et al., Chem. Phys. Lett. 237 (1995) 425-431
[b]Z. J. Wu, X. F. Ma, Chem. Phys. Lett. 371 (2003) 35-39
[c]Z. Hu, J.-G. Dong, J. R. Lombardi, D. M. Lindsay, J. Chem. Phys. 97 (1992) 8811-8812

TB. Mini-symposium: Accelerator-Based Spectroscopy

Tuesday, June 23, 2015 – 8:30 AM

Room: 100 Noyes Laboratory

Chair: Jennifer van Wijngaarden, University of Manitoba, Winnipeg, MB, Canada

TB01 *INVITED TALK* **8:30 – 9:00**

JET-COOLED SPECTROSCOPY ON THE AILES INFRARED BEAMLINE OF THE SYNCHROTRON RADIATION FACILITY SOLEIL

ROBERT GEORGES, *IPR UMR6251, CNRS - Université Rennes 1, Rennes, France.*

The Advanced Infrared Line Exploited for Spectroscopy (AILES) extracts the bright far infrared (FIR) synchrotron continuum of the third generation radiation facility SOLEIL. This beamline is equipped with a high resolution (10^{-3} cm^{-1}) Bruker IFS125 Fourier transform spectrometer which can be operated in the FIR but also in the mid and near infrared by using its internal conventional sources. The jet-AILES consortium (IPR, PhLAM, MONARIS, SOLEIL) has implemented a supersonic-jet apparatus on the beamline to record absorption spectra at very low temperature (5-50 K) and in highly supersaturated gaseous conditions. Heatable slit-nozzles of various lengths and widths are used to set properly the stagnation conditions. A mechanical pumping (roots pumps) was preferred for its ability to evacuate important mass flow rates and therefore to boost the experimental sensitivity of the set-up, the counterpart being a non-negligible consumption of both carrier (argon, helium or nitrogen) and spectroscopic gases. Various molecular systems were investigated up to now using the Jet-AILES apparatus. The very low temperature achieved in the gas expansion was either used to simplify the rotation-vibration structure of monomers, such as SF_6[a], CF_4 or naphthalene [b], or to stabilize the formation of weakly bonded molecular complexes such as the trimer of HF[c] or the dimer of acetic acid[d]. The nucleation of water vapor and the nuclear spin conversion of water were also investigated under free-jet conditions in the mid infrared.

[a]High-resolution spectroscopy and analysis of the $\nu_2 + \nu_3$ combination band of SF_6 in a supersonic jet expansion. V. Boudon, P. Asselin, P. Soulard, M. Goubet, T. R. Huet, R. Georges, O. Pirali, P. Roy, Mol. Phys. 111, 2154–2162 (2013)

[b]The far infrared spectrum of naphthalene characterized by high resolution synchrotron FTIR spectroscopy and anharmonic DFT calculations. O. Pirali, M. Goubet, T.R. Huet, R. Georges, P. Soulard, P. Asselin, J. Courbe, P. Roy and M. Vervloet, Phys. Chem. Chem. Phys. 15, 10141-10150 (2013)

[c]The cyclic ground state structure of the HF trimer revealed by far-infrared jet-cooled Fourier transform spectroscopy. P. Asselin, P. Soulard, B. Madebène, M. Goubet, T. R. Huet, R. Georges, O. Pirali and P. Roy, Phys. Chem. Chem. Phys. 16(10), 4797-806 (2014)

[d]Standard free energy of the equilibrium between the trans-monomer and the cyclic-dimer of acetic acid in the gas phase from infrared spectroscopy. M. Goubet, P. Soulard, O. Pirali, P. Asselin, F. Réal, S. Gruet, T. R. Huet, P. Roy and R. Georges, Phys. Chem. Chem. Phys. DOI: 10.1039/c4cp05684a

TB02 **9:05 – 9:20**

LOWEST VIBRATIONAL STATES OF ACRYLONITRILE

ZBIGNIEW KISIEL, *ON2, Institute of Physics, Polish Academy of Sciences, Warszawa, Poland*; MARIE-ALINE MARTIN-DRUMEL, *Spectroscopy Lab, Harvard-Smithsonian Center for Astrophysics, Cambridge, MA, USA*; OLIVIER PIRALI, *AILES beamline, Synchrotron SOLEIL, Saint Aubin, France.*

Recent studies of the broadband rotational spectrum of acrylonitrile, $H_2C{=}CHC{\equiv}N$, revealed the presence of multiple resonances between rotational levels in different vibrational states. The resonances affect even the ground state transitions and their analysis allowed determination of vibrational term values for the first three excited states above the ground state[a] and of vibrational energy differences in several polyads above these states.[b] At that time there was no infrared data of sufficient resolution to assess the reliability of the resonance based vibrational energy determinations.

We presently report results based on a 40-700 cm^{-1} high-resolution spectrum of acrylonitrile recorded at the AILES beamline of the SOLEIL synchrotron. This spectrum was reduced by using the AABS package[a,c] and allowed assignment of vibration-rotation transitions in four fundamentals, five hot bands, and one overtone band. The infrared data and previous measurements made with microwave techniques have been combined into a single global fit encompassing over 31000 measured transitions. Precise vibrational term values have been determined for the eight lowest excited vibrational states. The new results validate the previous estimates from rotational perturbations and are also compared with results of *ab initio* anharmonic force field calculations.

[a]Z. Kisiel, et al., *J. Mol. Spectrosc.* **280** 134 (2012).

[b]A. López, et al., *Astron. & Astrophys.* **572**, A44 (2014).

[c]Z. Kisiel, et al., *J. Mol. Spectrosc.* **233** 231 (2005).

TB03 **9:22 – 9:37**

FIR SYNCHROTRON SPECTROSCOPY OF HIGH TORSIONAL LEVELS OF CD_3OH: THE TAU OF METHANOL

RONALD M. LEES, LI-HONG XU, *Department of Physics, University of New Brunswick, Saint John, NB, Canada*; BRANT E BILLINGHURST, *EFD, Canadian Light Source Inc., Saskatoon, Saskatchewan, Canada.*

Sub-bands involving high torsional levels of the CD_3OH isotopologue of methanol have been analyzed in Fourier transform spectra recorded at the Far-Infrared beamline of the Canadian Light Source synchrotron in Saskatoon. Energy term values for A and E torsional species of the third excited torsional state, $v_t = 3$, are now almost complete up to rotational levels $K = 15$, and thirteen substates have so far been identified for $v_t = 4$. The spectra show interesting close groupings of high-v_t sub-bands related by Dennison's torsional symmetry label τ, rather than A and E, that can be understood in terms of a simple and universal free-rotor "spectral predictor" chart. Transitions between states on the same free rotor curve have torsional overlap matrix elements close to unity, so give rise to strong sub-bands providing radiative routes for rapid population transfer through the high torsional manifold. Where the energy curves for the $v_t = 3$ and 4 ground-state torsional levels pass through the excited vibrational states, strong resonances can occur and a number of anharmonic and Coriolis interactions have been detected through perturbations to the spectra and appearance of forbidden transitions due to strong mixing and intensity borrowing.

TB04 **9:39 – 9:54**

FAR-INFRARED SYNCHROTRON-BASED SPECTROSCOPY OF PROTON TUNNELLING IN MALONALDEHYDE

E. S. GOUDREAU, D. W. TOKARYK, STEPHEN CARY ROSS, *Department of Physics, University of New Brunswick, Fredericton, NB, Canada.*

Malonaldehyde ($C_3O_2H_4$) is a prototype molecule for the study of intramolecular tunnelling proton transfer. In the case of malonaldehyde, this transfer occurs between the two terminal oxygen atoms in its open-ring structure. Although the ground state tunnelling splitting of 21 cm^{-1} has been accurately determined from microwave studies[a], the splitting has never been obtained with high resolution in any excited vibrational state. The ν_6 vibrational band was investigated in a diode laser jet experiment[b] in 2004, but the researchers were not able to identify the (-) parity tunnelling component and so could not determine the splitting. We have collected high-resolution far-IR Fourier transform spectra from a number of fundamental vibrational bands of malonaldehyde at the CLS (Canadian Light Source) synchrotron in Saskatoon, Saskatchewan, exploiting the considerable gain in signal-to-noise ratio at the highest resolution available afforded by the intense and well-collimated beam. We will report on our tunnelling-rotation analysis of the anti-symmetric out-of-plane bend near 384 cm^{-1} and present its tunnelling splitting value.

[a] T. Baba, T. Tanaka, I. Morinoa, K. M. T. Yamada, K. Tanaka. *Detection of the tunneling-rotation transitions of malonaldehyde in the submillimeter-wave region.* J. Chem. Phys., **110**, 4131-4133 (1999)

[b] C. Duan, D. Luckhaus. *High resolution IR-diode laser jet spectroscopy of malonaldehyde.* Chem. Phys. Lett., **391**, 129-133 (2004)

Intermission

TB05 *INVITED TALK* **10:13 – 10:43**

THE DISCRETE NATURE OF THE COHERENT SYNCHROTRON RADIATION

STEFANO TAMMARO, *AILES beam line, Synchrotron Soleil, Gif-sur-Yvette, France*; OLIVIER PIRALI, P. ROY, *AILES beamline, Synchrotron SOLEIL, Saint Aubin, France*; JEAN FRANÇOIS LAMPIN, GAËL DUCOURNEAU, *Institut d'Electronique de Microélectronique et de Nanotechnologie, Université de Lille 1, Villeneuve d'Ascq, France*; ARNAUD CUISSET, FRANCIS HINDLE, GAËL MOURET, *Laboratoire de Physico-Chimie de l'Atmosphère, Université du Littoral Côte d'Opale, Dunkerque, France.*

Frequency Combs (FC) have radically changed the landscape of frequency metrology and high-resolution spectroscopy investigations extending tremendously the achievable resolution while increasing signal to noise ratio. Initially developed in the visible and near-IR spectral regions [a], the use of FC has been expanded to mid-IR [b], extreme ultra-violet [c] and X-ray [d]. Significant effort is presently dedicated to the generation of FC at THz frequencies. One solution based on converting a stabilized optical frequency comb using a photoconductive terahertz emitter, remains hampered by the low available THz power [e]. Another approach is based on active mode locked THz quantum-cascade-lasers providing intense FC over a relatively limited spectral extension [f]. Alternatively, we show that dense powerful THz FC is generated over one decade of frequency by coherent synchrotron radiation (CSR). In this mode, the entire ring behaves in a similar fashion to a THz resonator wherein electron bunches emit powerful THz pulses quasi-synchronously. The observed FC has been fully characterized and is demonstrated to be offset free. Based on these recorded specifications and a complete review of existing THz frequency comb, a special attention will be paid onto similarities and differences between them.

[a]Udem, Th., Holzwarth, H., Hänsch, T. W., Optical frequency metrology. Nature 416, 233-237 (2002)

[b]Schliesser, A., Picqué, N., Hänsch, T. W., Mid-infrared frequency combs. Nature Photon. 6, 440 (2012)

[c]Zinkstok, R. Th., Witte, S., Ubachs, W., Hogervorst, W., Eikema, K. S. E., Frequency comb laser spectroscopy in the vacuum-ultraviolet region. Physical Review A 73, 061801 (2006)

[d]Cavaletto, S. M. et al. Broadband high-resolution X-ray frequency combs. Nature Photon. 8, 520-523 (2014)

[e]Tani, M., Matsuura, S., Sakai, K., Nakashima, S. I., Emission characteristics of photoconductive antennas based on low-temperature-grown GaAs and semi-insulating GaAs. Applied Optics 36, 7853-7859 (1997)

[f]Burghoff, D. et al. Terahertz laser frequency combs. Nature Photon. 8, 462-467 (2014)

TB06 **10:48 – 11:03**

LOW-TEMPERATURE COLLISIONAL BROADENING IN THE FAR-INFRARED CENTRIFUGAL DISTORTION SPECTRUM OF CH_4

VINCENT BOUDON, *Laboratoire ICB, CNRS/Université de Bourgogne, DIJON, France*; JEAN VANDER AUWERA, *Service de Chimie Quantique et Photophysique, Université Libre de Bruxelles, Brussels, Belgium*; LAURENT MANCERON, *Synchrotron SOLEIL, CNRS-MONARIS UMR 8233 and Beamline AILES, Saint Aubin, France*; F. KWABIA TCHANA, *LISA, CNRS, Universités Paris Est Créteil et Paris Diderot, Créteil, France*; TONY GABARD, BADR AMYAY, *Laboratoire ICB, CNRS/Université de Bourgogne, DIJON, France*; MBAYE FAYE, *AILES beamline, Synchrotron SOLEIL, Saint Aubin, France.*

Previously, we could record on the AILES Beamline at the SOLEIL Synchrotron facility the first resolved centrifugal distorsion spectrum of methane (CH_4) in the THz region, which led to a precise determination of line intensities [a]. Later, we could measure collisional self- and N_2-broadening coefficients at room temperature[b]. This time, we reinvestigated this topic by measuring these broadening coefficients at low temperature (between 120 K and 160 K) for $J = 5$ to 12, thanks to a cryogenic multipass cell[c]. We used a 93 m total optical path length. Five pure methane pressures (from 10 to 100 mbar) and four CH_4/N_2 mixtures (20 % of methane with a total pressure from 100 to 800 mbar) were used. These measurements allow us to obtain data for physical conditions approaching those of Titan's atmosphere and to estimate temperature exponents.

[a]V. Boudon, O. Pirali, P. Roy, J.-B. Brubach, L. Manceron and J. Vander Auwera, *J. Quant. Spectrosc. Radiate. Transfer*, **111**, 1117–1129 (2010).

[b]M. Sanzharov, J. Vander Auwera, O. Pirali, P. Roy, J.-B. Brubach, L. Manceron, T. Gabard and V. Boudon, *J. Quant. Spectrosc. Radiate. Transfer*, **113**, 1874–1886 (2012).

[c]F. Kwabia Tchana, F. Willaert, X. Landshere, J.-M. Flaud, L. Lago, M. Chapuis, C. Herbeaux, P. Roy and L. Manceron, *Rev. Sci. Instrum.*, **84**, 093101 (2013).

TB07 **11:05 – 11:20**

HYDROGEN AND NITROGEN BROADENED ETHANE AND PROPANE ABSORPTION CROSS SECTIONS

ROBERT J. HARGREAVES, *Department of Chemistry and Biochemistry, Old Dominion University, Norfolk, VA, USA*; DOMINIQUE APPADOO, *800 Blackburn Road, Australian Synchrotron, Melbourne, Victoria, Australia*; BRANT E BILLINGHURST, *EFD, Canadian Light Source Inc., Saskatoon, Saskatchewan, Canada*; PETER F. BERNATH, *Department of Chemistry and Biochemistry, Old Dominion University, Norfolk, VA, USA*.

High-resolution infrared absorption cross sections are presented for the ν_9 band of ethane (C_2H_6) at 823 cm^{-1}. These cross sections make use of spectra recorded at the Australian Synchrotron using a Fourier transform infrared spectrometer with maximum resolution of 0.00096 cm^{-1}. The spectra have been recorded at 150, 120 and 90 K for hydrogen and nitrogen broadened C_2H_6. They cover appropriate temperatures, pressures and broadening gases associated with the atmospheres of the Outer Planets and Titan, and will improve atmospheric retrievals. The THz/Far-IR beamline at the Australian Synchrotron is unique in combining a high-resolution Fourier transform spectrometer with an 'enclosive flow cooling' (EFC) cell designed to study molecules at low temperatures. The EFC cell is advantageous at temperatures for which the vapor pressure is very low, such as C_2H_6 at 90 K.

Hydrogen broadened absorption cross sections of propane between 700 and 1200 cm^{-1} will also be presented based on spectra obtained at the Canadian Light Source.

TC. Mini-symposium: Spectroscopy in the Classroom

Tuesday, June 23, 2015 – 8:30 AM

Room: B102 Chemical and Life Sciences

Chair: S. A. Cooke, Purchase College SUNY, Purchase, NY, USA

TC01 *INVITED TALK* **8:30 – 9:00**

PGOPHER IN THE CLASSROOM AND THE LABORATORY

COLIN WESTERN, *School of Chemistry, University of Bristol, Bristol, United Kingdom.*

PGOPHER[ab] is a general purpose program for simulating and fitting rotational, vibrational and electronic spectra. As it uses a graphical user interface the basic operation is sufficiently straightforward to make it suitable for use in undergraduate practicals and computer based classes. This talk will present two experiments that have been in regular use by Bristol undergraduates for some years based on the analysis of infra-red spectra of cigarette smoke and, for more advanced students, visible and near ultra-violet spectra of a nitrogen discharge and a hydrocarbon flame. For all of these the rotational structure is analysed and used to explore ideas of bonding. The talk will discuss the requirements for the apparatus and the support required. Other ideas for other possible experiments and computer based exercises will also be presented, including a group exercise.

The PGOPHER program is open source, and is available for Microsoft Windows, Apple Mac and Linux. It can be freely downloaded from the supporting website http://pgopher.chm.bris.ac.uk. The program does not require any installation process, so can be run on student's own machines or easily setup on classroom or laboratory computers.

[a] PGOPHER, a Program for Simulating Rotational, Vibrational and Electronic Structure, C. M. Western, University of Bristol, http://pgopher.chm.bris.ac.uk

[b] PGOPHER version 8.0, C M Western, 2014, University of Bristol Research Data Repository, doi:10.5523/bris.huflggvpcuc1zvliqed497r2

TC02 **9:05 – 9:20**

SPECTROSCOPY FOR THE MASSES

ROBERT J. LE ROY[a], SCOTT HOPKINS, WILLIAM P. POWER, TONG LEUNG, *Department of Chemistry, University of Waterloo, Waterloo, ON, Canada*; JOHN HEPBURN[b], *Departments of Chemistry, Physics and Astronomy, University of British Columbia, Vancouver, BC, Canada.*

Undergraduate students in all areas of science encounter one or more types of spectroscopy as an essential tool in their discipline, but most never take the advanced physics or chemistry courses in which the subject is normally taught.

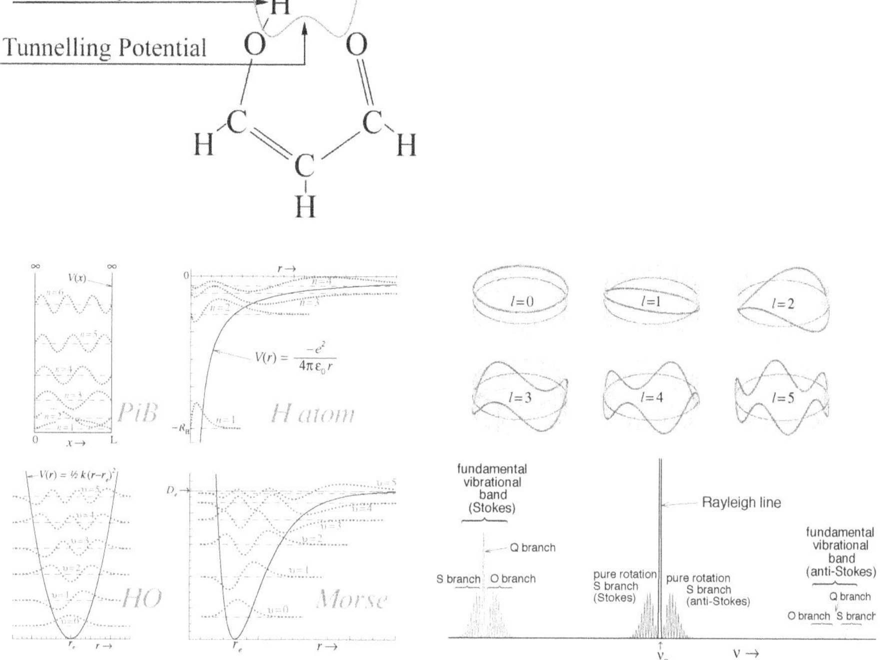

To address this problem, for over 20 years our department has been teaching a popular Introductory Spectroscopy course that assumes as background only a one-term introductory chemistry course containing a unit on atomic theory, and a familiarity with rudimentary calculus. This survey course provides an introduction to microwave, infrared, Raman, electronic, photoelectron and NMR spectroscopy in a manner that allows students to understand many of these phenomena as intuitive generalizations of the problem of a particle in a 1-D box or a particle-on-a-ring, and does not require any high level mathematics.

[a]http://leroy.uwaterloo.ca
[b]present address: Department of Chemistry, University of British Columbia, Vancouver, BC, Canada.

TC03 **9:22 – 9:37**

RESEARCH AT A LIBERAL ARTS COLLEGE: MAKE SURE YOU HAVE A NET FOR YOUR HIGH WIRE ACT

MARK D. MARSHALL, HELEN O. LEUNG, *Chemistry Department, Amherst College, Amherst, MA, USA.*

A career as a spectroscopist at a primarily (or exclusively) undergraduate institution presents both great rewards and significant challenges. Strategies that we have found helpful in meeting some of the challenges are presented along with some of the work we have been able to accomplish with undergraduate students. The most important resource is a network of colleagues who can provide mentoring and collaboration, and the role of the International Symposium on Molecular Spectroscopy in facilitating this support is highlighted.

TC04 **9:39 – 9:54**

A SPECTROSCOPY BASED P-CHEM LAB, INCLUDING A DETAILED TEXT AND LAB MANUAL

JOHN MUENTER, *Department of Chemistry, University of Rochester, Rochester, NY, USA.*

Rochester's second semester physical chemistry lab course is based on spectroscopy experiments and follows a full semester of quantum mechanics lectures. The laboratory course is fully separate from the traditional physical chemistry course and has its own lectures. The lab course is constructed to achieve three major goals: provide a detailed knowledge of the instrumentation that acquires data, establish a good understanding of how that data is analyzed, and give students a familiarity with spectroscopic techniques and quantum mechanical models. Instrumentation is emphasized by using common components to construct different experiments. Microwave, modulation and detection components are used for both OCS pure rotation and ESR experiments. Optical components, a monochromator, and PMT detectors are used in a HeNe laser induced fluorescence experiment on I_2 *(J. Chem. Ed. 73, 576 (1996))* and a photoluminescence experiment on pyrene *(J. Chem. Ed. 73, 580 (1996))*. OCS is studied in both the microwave and infrared regions, and the C=S stretching vibration is identified through microwave intensity measurements. Lecture notes and laboratory instructions are combined in an exhaustive text of more than 400 pages, containing 325 figures, 285 equations and numerous MathCad data analysis programs. This text can be downloaded as a 10 Mbyte pdf file at chem.rochester.edu/~muenter/CHEM232Manual.

Intermission

TC05 **10:13 – 10:28**

HOW WE KNOW: SPECTROSCOPY IN THE FIRST YEAR AND BEYOND

KRISTOPHER J OOMS, *Chemistry, The King's University, Edmonton, Alberta, Canada.*

Chemical educators face the never ending challenge of showing students that the content written in their textbook arises from a rich interplay of experimentation, imagination and a desire to understand and impact the world. We have found that asking three simple questions – What do we know, How do we know it, Why do we care – is an effective strategy to guide the content and pedagogy within our chemistry classes. Of these three questions What we know is the most thoroughly covered and with the growing use of rich context teaching, the Why we care is becoming more central to our chemistry teaching. How are we doing on telling students How we know?

Spectroscopy is at the core of our ability to answer questions about how we know things about the molecular world. Yet the teaching of spectroscopy is not a central part of student's early chemistry learning, often being left to the later stages of degrees and courses. For example, a brief look at common North American general chemistry text books reveals almost no discussion of spectroscopic techniques and their centrality to understanding chemistry.

In this talk I will discuss efforts to bring spectroscopy into the first year course and some of the repercussions this has for the whole chemistry undergraduate curriculum. The goal is to make students better aware of where the ideas in chemistry arise from, the strengths and weaknesses of spectroscopic experiments, and how our models of the molecular world are built on rigorous experimentation.

TC06 **10:30 – 10:40**

EXPANDED CHOICES FOR VIBRATION-ROTATION SPECTROSCOPY IN THE PHYSICAL CHEMISTRY TEACHING LABORATORY

JOEL R SCHMITZ, DAVID A DOLSON, *Department of Chemistry, Wright State University, Dayton, OH, USA.*

Many third-year physical chemistry laboratory students in the US analyze the vibration-rotation spectrum of HCl in support of lecture concepts in quantum theory and molecular spectroscopy. Contemporary students in physical chemistry teaching laboratories increasingly have access to FTIR spectrometers with 1/8th cm^{-1}resolution, which allows for expanded choices of molecules for vibration-rotation spectroscopy. Here we present the case for choosing HBr/DBr for such a study, where the 1/8th cm^{-1}resolution enables the bromine isotopic lines to be resolved. Vibration-rotation lines from the fundamental and first-overtone bands of four hydrogen bromide isotopomers are combined in a global analysis to determine molecular spectroscopic constants. Sample production, spectral appearance, analysis and results will be presented for various resolutions commonly available in teaching laboratories.

TC07 **10:42 – 10:57**

SPECTROSCOPIC CASE-BASED STUDIES IN A FLIPPED QUANTUM MECHANICS COURSE

STEVEN SHIPMAN, *Department of Chemistry, New College of Florida, Sarasota, FL, USA.*

Students in a flipped Quantum Mechanics course were expected to apply their knowledge of spectroscopy to a variety of case studies involving complex mixtures of chemicals. They used simulated data, prepared in advance by the instructor, to determine the major chemical constituents of complex mixtures. Students were required to request the appropriate data in order to ultimately make plausible guesses about the composition of the mixtures, allowing them ownership over the discovery process. This talk will describe how these activities worked in practice, give caveats for instructors who wish to adopt them in the future, and discuss how the results of these exercises can be used for both formative and summative assessment.

TC08 **10:59 – 11:09**

THE H-ATOM SPECTRUM: NOT A CLASSROOM DEMONSTRATION ...

WOLFGANG JÄGER, *Department of Chemistry, University of Alberta, Edmonton, AB, Canada.*

The spectrum of the hydrogen atom is topic of every freshmen chemistry course and at the same time a first brush with quantum mechanics for many students. A picture of the four visible emission lines of the Balmer series is shown in probably every introductory Chemistry textbook, but only few students have likely seen those lines with their own eyes.

I will tell you about a simple in-class activity that allows the students to see those lines and can be done in large classes (I have done it in classes with up to 500 students) at low cost.

TC09 **11:11 – 11:26**

RAMAN INVESTIGATION OF TEMPERATURE PROFILES OF PHOSPHOLIPID DISPERSIONS IN THE BIOCHEMISTRY LABORATORY

NORMAN C. CRAIG, *Department of Chemistry and Biochemistry, Oberlin College, Oberlin, OH, USA.*

The temperature dependence of self-assembled, cell-like dispersions of phospholipids is investigated with Raman spectroscopy in the biochemistry laboratory. Vibrational modes in the hydrocarbon interiors of phospholipid bilayers are strongly Raman active, whereas the vibrations of the polar head groups and the water matrix have little Raman activity. From Raman spectra increases in fluidity of the hydrocarbon chains can be monitored with intensity changes as a function of temperature in the CH-stretching region. The experiment uses detection of scattered 1064-nm laser light (Nicolet NXR module) by a Fourier transform infrared spectrometer (Nicolet 6700). A thermoelectric heater-cooler device (Melcor) gives convenient temperature control from 5 to 95°C for samples in melting point capillaries. Use of deuterium oxide instead of water as the matrix avoids some absorption of the exciting laser light and interference with intensity observations in the CH-stretching region. Phospholipids studied range from dimyristoylphosphotidyl choline (C_{14}, transition T = 24°C) to dibehenoylphosphotidyl choline (C_{22}, transition T = 74°C).

TC10 *Post-Deadline Abstract* **11:28 – 11:43**

ONLINE AND CERTIFIABLE SPECTROSCOPY COURSES USING INFORMATION AND COMMUNICATION TOOLS. A MODEL FOR CLASSROOMS AND BEYOND

MANGALA SUNDER KRISHNAN, *Department of Chemistry, Indian Institute of Technology Madras, Chennai , Tamil Nadu, India.*

Online education tools and flipped (reverse) class models for teaching and learning and pedagogic and andragogic approaches to self-learning have become quite mature in the last few years because of the revolution in video, interactive software and social learning tools. Open Educational resources of dependable quality and variety are also becoming available throughout the world making the current era truly a renaissance period for higher education using Internet. In my presentation, I shall highlight structured course content preparation online in several areas of spectroscopy and also the design and development of virtual lab tools and kits for studying optical spectroscopy.

Both elementary and advanced courses on molecular spectroscopy are currently under development jointly with researchers in other institutions in India. I would like to explore participation from teachers throughout the world in the teaching-learning process using flipped class methods for topics such as experimental and theoretical microwave spectroscopy of semi-rigid and non-rigid molecules, molecular complexes and aggregates. In addition, courses in Raman, Infrared spectroscopy experimentation and advanced electronic spectroscopy courses are also envisaged for free, online access. The National Programme on Technology Enhanced Learning (NPTEL) and the National Mission on Education through Information and Communication Technology (NMEICT) are two large Government of India funded initiatives for producing certified and self-learning courses with financial support for moderated discussion forums. The learning tools and interactive presentations so developed can be used in classrooms throughout the world using flipped mode of teaching. They are very much sought after by learners and researchers who are in other areas of learning but want to contribute to research and development through inter-disciplinary learning. NPTEL is currently is experimenting with Massive Open Online Course (MOOC) strategy, but with proctored and certified examination processes for large numbers in some of the above courses.

I would like to present a summary of developments in these areas to help focus classroom (online and offline) learning of Molecular spectroscopy.

TD. Conformers, isomers, chirality, stereochemistry

Tuesday, June 23, 2015 – 8:30 AM

Room: 274 Medical Sciences Building

Chair: Emilio J. Cocinero, Universidad del País Vasco (UPV-EHU), Leioa, Spain

TD01 **8:30 – 8:45**

A JOINT THEORETICAL AND EXPERIMENTAL STUDY OF THE SiH_2OO ISOMERIC SYSTEM

MICHAEL C McCARTHY, *Atomic and Molecular Physics, Harvard-Smithsonian Center for Astrophysics, Cambridge, MA, USA*; JÜRGEN GAUSS, *Institut für Physikalische Chemie, Universität Mainz, Mainz, Germany*.

In contrast to the CH_2OO isomers, those of SiH_2OO have received relatively little attention, either theoretically or experimentally. High-level coupled cluster calculations predict a much different energy ordering in comparison to that found for CH_2OO, with the three conformers of Si-dihydroxycarbene, HOSiOH, most stable, followed by the Si-analogues of *cis* and *trans* formic acid, and then a cyclic isomer with C_{2v} symmetry, *c*-SiH_2OO. Guided by these theoretical predictions, rotational lines of the *cis, trans* isomer of HOSiOH, as well as *c*-SiH_2OO, have been detected by Fourier transform microwave spectroscopy. The lines of the cyclic form are sufficiently strong that several rare isotopic species have also been found, enabling, in combination with calculated vibrational corrections, a precise semi-experimental structure to be derived. This talk will provide a status report on our joint study of this unusual isomeric system, and an update on searches for still other isomers.

TD02 **8:47 – 9:02**

A MINTY MICROWAVE MENAGERIE: THE ROTATIONAL SPECTRA OF MENTHONE, MENTHOL, CARVACROL, AND THYMOL

DAVID SCHMITZ, V. ALVIN SHUBERT, THOMAS BETZ, *CoCoMol, Max-Planck-Institut für Struktur und Dynamik der Materie, Hamburg, Germany*; BARBARA MICHELA GIULIANO, *Department of Chemistry, University of Bologna, Bologna, Italy*; MELANIE SCHNELL, *CoCoMol, Max-Planck-Institut für Struktur und Dynamik der Materie, Hamburg, Germany*.

Terpenes represent one of the largest classes of secondary metabolites in nature and are derived from adding substituents to their core building block, isoprene. They exhibit a huge assortment of structures and thus a variety of chemical and biological activities. We recently investigated a number of monoterpenoids using broadband rotational spectroscopy in the 2-8.5 GHz frequency range.

We present a comparative study of the aromatic monoterpenoids thymol and carvacrol and aliphatic menthone and menthol. The differences in their electronic and steric structures significantly influence molecular properties such as internal rotation barriers and conformational flexibility. These influences are revealed in the rotational spectra. We report the rotational spectra and the experimentally determined molecular parameters. Results from extensive quantum chemical calculations of the conformational spaces of these molecules are compared with the experimentally determined molecular parameters.

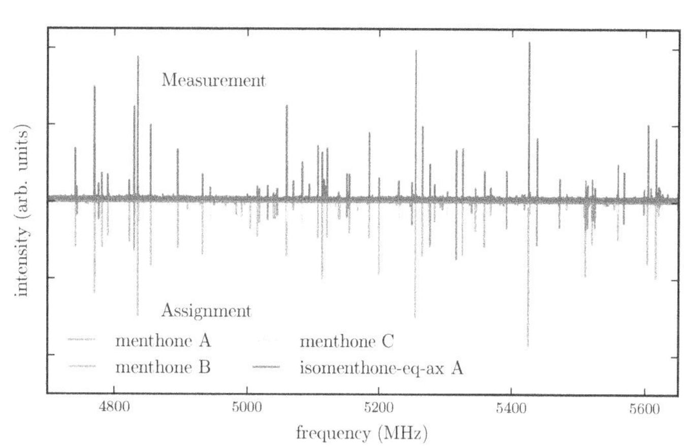

TD03 **9:04 – 9:19**

THE ROTATIONAL SPECTRUM AND CONFORMATIONAL STRUCTURES OF METHYL VALERATE

HA VINH LAM NGUYEN, *CNRS et Universités Paris Est et Paris Diderot, Laboratoire Interuniversitaire des Systèmes Atmosphériques (LISA), Créteil, France*; WOLFGANG STAHL, *Institute for Physical Chemistry, RWTH Aachen University, Aachen, Germany.*

Methyl valerate, $C_4H_9COOCH_3$, belongs to the class of fruit esters, which play an important role in nature as odorants of different fruits, flowers, and wines. A sufficient explanation for the structure–odor relation of is not available. It is known that predicting the odor of a substance is not possible by knowing only its chemical formula. A typical example is the blueberry- or pine apple-like odor of ethyl isovalerate while its isomers ethyl valerate and isoamyl acetate smell like green apple and banana, respectively. Obviously, not only the composition but also the molecular structures are not negligible by determining the odor of a substance. Gas phase structures of fruit esters are thus important for a first step towards the determination of structure–odor relation since the sense of smell starts from gas phase molecules.

For this purpose, a combination of microwave spectroscopy and quantum chemical calculations (QCCs) is an excellent tool. Small esters often have sufficient vapor pressure to be transferred easily in the gas phase for a rotational study but already contain a large number of atoms which makes them too big for classical structure determination by isotopic substitution and requires nowadays a comparison with the structures optimized by QCCs. On the other hand, the results from QCCs have to be validated by the experimental values.

About the internal dynamics, the methoxy methyl group $-COOCH_3$ of methyl acetate shows internal rotation with a barrier of $424.581(56)\ \mathrm{cm}^{-1}$. A similar barrier height of $429.324(23)\ \mathrm{cm}^{-1}$was found in methyl propionate, where the acetyl group is extended to the propionyl group. The investigation on methyl valerate fits well in this series of methyl alkynoates. In this talk, the structure of the most energetic favorable conformer as well as the internal rotation shown by the methoxy methyl group will be reported. It could be confirmed that the internal rotation barrier of the methoxy methyl group remains by longer alkyl chain.

TD04 **9:21 – 9:36**

ROTATIONAL SPECTRUM OF THE METHYL SALICYLATE-WATER COMPLEX: THE MISSING CONFORMER AND THE TUNNELING MOTIONS

SUPRIYA GHOSH, JAVIX THOMAS, YUNJIE XU, WOLFGANG JÄGER, *Department of Chemistry, University of Alberta, Edmonton, AB, Canada.*

Methyl salicylate is a naturally occurring organic ester produced by wintergreen and other plants. It is also found in many over-the-counter remedies, such as muscle ache creams. The rotational spectrum of the methyl salicylate monomer was reported previously, where the most stable, dominant conformer was identified.[a] The methyl salicylate-water complex was first studied using fluorescence-detected infrared spectroscopy; only one monohydrate conformer was found in that work.[b] In the present study, we employed both broadband chirped and cavity based Fourier transform microwave spectroscopy to examine the competition between intra- and intermolecular hydrogen-bonding interactions and possible large amplitude motions associated with the methyl group and the water subunit. In contrast to the previous infrared study, two monohydrate conformers were identified, with carbonyl O or hydroxyl O as the hydrogen bond acceptors. Detailed analyses of the observed hyperfine structures will be presented, as well as our efforts to extend the study to larger methyl salicylate hydration clusters.

[a] S. Melandri, B. M. Giuliano, A. Maris, L. B. Favero, P. Ottaviani, B. Velino, W. Caminati, J. Phys. Chem. A. 2007, 111, 9076.

[b] A. Mitsuzuka, A. Fujii, T. Ebata, N. Mikami, J. Phys. Chem. A 1998, 102, 9779.

TD05 **9:38 – 9:53**

UNRAVELLING THE CONFORMATIONAL LANDSCAPE OF NICOTINOIDS: THE STRUCTURE OF COTININE BY BROADBAND ROTATIONAL SPECTROSCOPY

ICIAR URIARTE, PATRICIA ECIJA, EMILIO J. COCINERO, *Departamento de Química Física, Universidad del País Vasco (UPV-EHU), Bilbao, Spain*; CRISTOBAL PEREZ, *CoCoMol, Max-Planck-Institut für Struktur und Dynamik der Materie, Hamburg, Germany*; ELENA CABALLERO-MANCEBO, ALBERTO LESARRI, *Departamento Química Física y Química Inorgánica , Universidad de Valladolid, Valladolid, Spain.*

Alkaloids such as nicotine, cotinine or anabasine share a common floppy structural motif consisting of a two-ring assembly with a 3-pyridil methylamine skeleton. In order to investigate the structure-activity relationship of these biomolecules, structural studies with rotational resolution have been carried out for nicotine[a] and anabasine[b] in the gas phase, where these molecules can be probed in an "interaction-free" environment (no solvent or crystal-packing interactions).

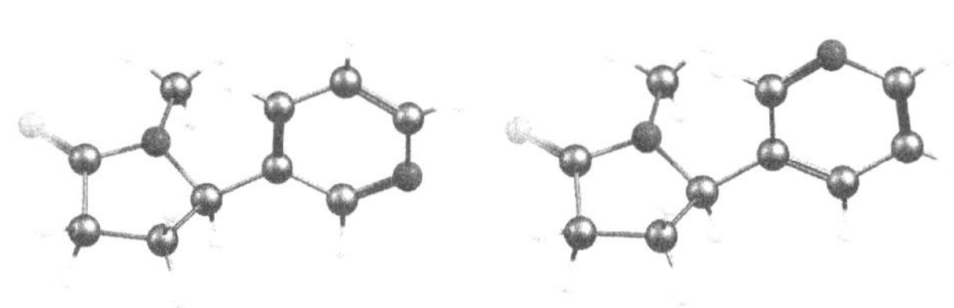

We hereby present a structural investigation of cotinine in a jet expansion using the chirped-pulse Fourier-transform microwave (CP-FTMW) spectrometer recently built at the University of the Basque Country (UPV-EHU). The rotational spectrum (6-18 GHz) reveals the presence of two different conformations. The conformational preferences of cotinine originate from the internal rotation of the two ring moieties, the detected species differing in a near 180° rotation of pyridine. The final structure is modulated by steric effects.

[a]J.-U. Grabow, S. Mata, J. L. Alonso, I. Peña, S. Blanco, J. C. López, C. Cabezas, *Phys. Chem. Chem. Phys.* 2011, **13**, 21063.
[b]A. Lesarri, E. J. Cocinero, L. Evangelisti, R. D. Suenram, W. Caminati, J.-U. Grabow, *Chem. Eur. J.* 2010, **16**, 10214.

TD06 **9:55 – 10:10**

CONFORMATIONALLY RESOLVED STRUCTURES OF JET-COOLED PHENACETIN AND ITS HYDRATED CLUSTERS

CHEOL JOO MOON, AHREUM MIN, AHREUM AHN, MYONG YONG CHOI, *Department of Chemistry, Gyeongsang National University, JinJu, GyeongsangNamDo, Korea.*

Phenacetin (PA) is one of the typical synthetic fever reducers as similar to acetaminophen (AAP), a major ingredient of Tylenol®. PA and AAP are both derivatives of acetanilide (AA), substituted by ethoxyl group and hydroxyl group in the para position of AA, respectively. In this work, we present the conformational investigations and photochemistry of jet-cooled PA and its 1:1 hydrates using resonance enhanced multiphoton ionization (REMPI), UV-UV hole-burning and IR-dip spectroscopy. Moreover, we calculated the optimized structures of PA and its 1:1 hydrates by density functional theory. Here, we report the structural information of PA and its 1:1 hydrates with an aid of the experimental data and the ab initio calculations.

TD07 **10:12 – 10:27**

CONFORMATIONAL STRUCTURES OF JET-COOLED ACETAMINOPHEN-WATER CLUSTERS BY IR-DIP SPECTROSCOPY AND COMPUTATIONAL CALCULATIONS

AHREUM MIN, AHREUM AHN, CHEOL JOO MOON, MYONG YONG CHOI, *Department of Chemistry, Gyeongsang National University, JinJu, GyeongsangNamDo, Korea.*

Acetaminophen (AAP) is a widely used over-the-counter antipyretic and analgesic, a major ingredient in various cold and flu drugs, Tylenol®. In a previous study, we reported the conformational structures of jet-cooled AAP monomer by resonance enhanced multi-photon ionization (REMPI) and UV-UV hole-burning spectroscopy in the gas phase, providing the identification of two almost isoenergetic conformers, cis- and trans-, in the free-jet experiments. [Phys.Chem.Chem.Phys., 2011, 13, 16537-16541]. In this talk, we like to step further on the study of AAP-water clusters via the REMPI, UV-UV hole-burning and IR-dip spectroscopy. The conformational structures of AAP-water clusters will be compared with the spectroscopic results and theoretical calculations.

Intermission

TD08 **10:46 – 11:01**

THE INHERENT CONFORMATIONAL PREFERENCES OF GLUTAMINE-CONTAINING PEPTIDES: THE ROLE FOR SIDE-CHAIN BACKBONE HYDROGEN BONDS

PATRICK S. WALSH, *Department of Chemistry, Purdue University, West Lafayette, IN, USA*; CARL McBURNEY, SAMUEL H. GELLMAN, *Department of Chemistry, University of Wisconsin–Madison, Madison, WI, USA*; TIMOTHY S. ZWIER, *Department of Chemistry, Purdue University, West Lafayette, IN, USA.*

Glutamine is widely known to be found in critical regions of peptides which readily fold into amyloid fibrils, the structures commonly associated with Alzheimer's disease and glutamine repeat diseases such as Huntington's disease. Building on previous single-conformation data on Gln-containing peptides containing an aromatic cap on the N-terminus (Z-Gln-OH and Z-Gln-NHMe), we present here single-conformation UV and IR spectra of Ac-Gln-NHBn and Ac-Ala-Gln-NHBn, with its C-terminal benzyl cap. These results point towards side-chain to backbone hydrogen bonds dominating the structures observed in the cold, isolated environment of a molecular beam. We have identified and assigned three main conformers for Ac-Gln-NHBn all involving primary side-chain to backbone interactions. Ac-Ala-Gln-NHBn extends the peptide chain by one amino acid, but affords an improvement in the conformational flexibility. Despite this increase in the flexibility, only a single conformation is observed in the gas-phase: a structure which makes use of both side-chain-to-backbone and backbone-to-backbone hydrogen bonds.

TD09 **11:03 – 11:18**

APPLICATIONS OF STRUCTURAL MASS SPECTROMETRY TO METABOLOMICS: CLARIFYING BOND SPECIFIC SPECTRAL SIGNATURES WITH ISOTOPE EDITED SPECTROSCOPY

OLGA GORLOVA, CONRAD T. WOLKE, JOSEPH FOURNIER, SEAN COLVIN, MARK JOHNSON, SCOTT MILLER, *Department of Chemistry, Yale University, New Haven, CT, USA.*

Comprehensive FTIR, MS/MS and NMR of pharmaceuticals are generally readily available but characterization of their metabolites has been an obstacle. Atorvastatin is a statin drug responsible for the maintenance of cholesterol in the body. Diovan is an angiostensin receptor antagonist used to treat high blood pressure and congestive heart failure. The field of metabolomics, however, is struggling to obtain the identity of their structures. We implement mass spectrometry with cryogenic ion spectroscopy to study gaseous ions of the desired metabolites which, in combination, not only identify the mass of the metabolite but also elucidate their structures through isotope-specific infrared spectroscopy.

TD10 **11:20 – 11:35**

ALKALI METAL-GLUCOSE INTERACTION PROBED WITH INFRARED PRE-DISSOCIATION SPECTROSCOPY

STEVEN J. KREGEL, BRETT MARSH, JIA ZHOU, ETIENNE GARAND, *Department of Chemistry, The Univeristy of Wisconsin, Madison, WI, USA.*

The efficient extraction of cellulose from biomass and its subsequent conversion to glucose derivatives is an attractive goal in the field of energy science. However, current industrial methods require high ionic strength and harsh conditions. Ionic liquids (IL's) are a class of "green" compounds that have been shown to dissolve cellulose in concentrations of up to 25 wt%. In order to understand IL's extraordinary cellulose dissolving power, a molecular level understanding of the IL-cellulose interaction is needed. Toward that end, we have acquired infrared pre-dissociation spectra of M^+-glucose, where M^+=Li^+, Na^+, or K^+. Through comparisons with density functional theory calculations, we have determined the relative abundances of various M^+-glucose binding motifs in both the thermodynamic and kinetic limits. These results provide insight on the hydrogen bonding dynamics of glucose and are a step towards a fuller understanding of cellulose interactions with ionic liquids.

TD11 **11:37 – 11:52**

PROBING THE CONFORMATIONAL LANDSCAPE OF A POLYETHER BUILDING BLOCK BY RAMAN JET SPECTROSCOPY

SEBASTIAN BOCKLITZ, MARTIN A. SUHM, *Institute of Physical Chemistry, Georg-August-Universität Göttingen, Göttingen, Germany.*

Polyethylene oxides (Polyethylene glycoles) represent a prominent class of water-soluble polymers. Surprisingly, already 1,2-dimethoxyethane as the simplest representative of this polymer family has an undetermined conformational preference in the gas phase. Here, we address this problem by spontaneous Raman scattering in a supersonic jet.

Variation of carrier gas, stagnation pressure, nozzle distance and temperature provides information on the three lowest conformations and their mutual interconversion during collisions in the expansion. The results are compared to quantum chemical calculations of the potential energy landscape and of normal modes.

TE. Instrument/Technique Demonstration

Tuesday, June 23, 2015 – 8:30 AM

Room: 217 Noyes Laboratory

Chair: Ken Leopold, University of Minnesota, Minneapolis, MN, USA

TE01 **8:30 – 8:45**

ELIMINATION OF THE VACUUM PUMP REQUIREMENT FOR HIGH-RESOLUTION ROTATIONAL SPECTROSCOPY.

JENNIFER HOLT, *Department of Physics, The Ohio State University, Columbus, OH, USA*; RYAN W DALY, *Battelle Memorial Institute, Columbus, Ohio, USA*; CHRISTOPHER F. NEESE, FRANK C. DE LUCIA, *Department of Physics, The Ohio State University, Columbus, OH, USA.*

It has been observed that with the advances being driven by the wireless communications industry, the microwave components for submillimeter wave spectrometers and sensors will become almost "free". Moreover, these electronic components will require little power. However, neither of these attributes applies to the vacuum requirements for high-resolution rotational spectroscopy. We will report on the design, construction, and operation of a simple spectroscopic cell that overcomes these problems.

TE02 **8:47 – 8:57**

3-D PRINTED SLIT NOZZLES FOR FOURIER TRANSFORM MICROWAVE SPECTROSCOPY

CHRIS DEWBERRY, BECCA MACKENZIE, *Chemistry Department, University of Minnesota, Minneapolis, MN, USA*; SUSAN GREEN, *Chemistry Department , Macalester College, Saint Paul, Minnesota, USA*; KEN LEOPOLD, *Chemistry Department, University of Minnesota, Minneapolis, MN, USA.*

3-D printing is a new technology whose applications are only beginning to be explored. In this report, we describe the application of 3-D printing to the facile design and construction of supersonic nozzles. The efficacy of a variety of designs is assessed by examining rotational spectra OCS and Ar-OCS using a Fourier transform microwave spectrometer with tandem cavity and chirped-pulse capabilities. This work focuses primarily on the use of slit nozzles but other designs have been tested as well. New nozzles can be created for $0.50 or less each, and the ease and low cost should facilitate the optimization of nozzle performance (e.g., jet temperature or cluster size distribution) for the needs of any particular experiment.

TE03 **8:59 – 9:14**

IMPLEMENTATION OF CMOS MILLIMETER-WAVE DEVICES FOR ROTATIONAL SPECTROSCOPY

BRIAN DROUIN, ADRIAN TANG, ERICH T SCHLECHT, ADAM M DALY, EMILY BRAGEOT, *Jet Propulsion Laboratory, California Institute of Technology, Pasadena, CA, USA*; QUN JANE GU, YU YE, RAN SHU, *Department of Electrical and Computer Engineering, University of California - Davis, Davis, CA, USA*; M.-C. FRANK CHANG, ROD M. KIM, *Electrical Engineering, University of California - Los Angeles, Los Angeles, CA, USA.*

The extension of radio-frequency CMOS circuitry into millimeter wavelengths promises the extension of spectroscopic techniques in compact, power efficient systems. We are now exploring the use of CMOS millimeter devices for low-mass, low-power instrumentation capable of remote or in-situ detection of gas composition during space missions. This effort focuses on the development of a semi-confocal Fabry-Perot cavity with mm-wavelength CMOS transmitter and receiver attached directly to a cavity coupler. Placement of the devices within the cavity structure bypasses problems encountered with signal injection and extraction in traditional cavity designs and simultaneously takes full advantage of the miniaturized form of the CMOS hardware. The presentation will provide an overview of the project and details of the accomplishments thus far, including the development and testing of a pulse modulated 83-98 GHz transmitter.

TE04 **9:16 – 9:31**

FAST SWEEPING DIRECT ABSORPTION (SUB)MILLIMETER SPECTROSCOPY BASED ON CHIRPED-PULSE TECHNOLOGY

BRIAN HAYS, *Department of Chemistry, Emory University, Atlanta, GA, USA*; STEVEN SHIPMAN, *Department of Chemistry, New College of Florida, Sarasota, FL, USA*; SUSANNA L. WIDICUS WEAVER, *Department of Chemistry, Emory University, Atlanta, GA, USA.*

Chriped-pulse Fourier Transform Microwave (CP-FTMW) technology has transformed traditional microwave spectroscopy into a rapid-acquisition, broadband spectral technique. The CP-FT technique has recently been expanded to the millimeter-wave region, but this approach requires costly equipment that is not readily available in most spectroscopy labs. To overcome this challenge, a new experiment has been designed that combines the broadband aspects of CP-FTMW with the high sensitivity of (sub) millimeter absorption spectroscopy. Using the arbitrary waveform generator from a CP-FTMW experiment, and the frequency multipliers and hot electron bolometer detector from a (sub) millimeter wave experiment, we have designed and benchmarked a highly sensitive spectrometer that offers broad spectral coverage and rapid spectral acquisition speeds. This technique is comparable in performance to other rapid-acquisition techniques currently used in the (sub) millimeter range, but offers more sensitivity after averaging. The design of this instrument and the benchmarking results will be presented.

TE05 **9:33 – 9:48**

FAST SWEEPING DOUBLE RESONANCE MICROWAVE-(SUB)MILLIMETER SPECTROSCOPY BASED ON CHIRPED PULSE TECHNOLOGY

BRIAN HAYS, SUSANNA L. WIDICUS WEAVER, *Department of Chemistry, Emory University, Atlanta, GA, USA*; STEVEN SHIPMAN, *Department of Chemistry, New College of Florida, Sarasota, FL, USA.*

Microwave-millimeter double resonance spectroscopy has been commonly applied by driving absorption with the millimeter light and then probing the resonance using a Fourier Transform Microwave (FTMW) spectrometer. We will present data from an inverse scheme, in which millimeter light is used to probe a transition whose intensity is modulated by the application of microwave radiation. This detection scheme is effective in aiding the assignment of millimeter-wave transitions by revealing which energy levels are associated with particular spectral lines. To increase the speed of this detection technique, we incorporated an arbitrary waveform generator into the microwave source to rapidly sweep the microwave radiation through a broad frequency range. We will discuss this approach as applied to pulsed valve experiments and in combination with a laser-induced chemistry experiment. Potential applications to other experimental designs will also be discussed.

Intermission

TE06 **10:07 – 10:22**

ON THE PHASE DEPENDENCE OF DOUBLE-RESONANCE EXPERIMENTS IN ROTATIONAL SPECTROSCOPY

DAVID SCHMITZ, V. ALVIN SHUBERT, ANNA KRIN, *CoCoMol, Max-Planck-Institut für Struktur und Dynamik der Materie, Hamburg, Germany*; DAVID PATTERSON, *Department of Physics, Harvard University, Cambridge, MA, USA*; MELANIE SCHNELL, *CoCoMol, Max-Planck-Institut für Struktur und Dynamik der Materie, Hamburg, Germany.*

We report double-resonance experiments using broadband chirped-pulse Fourier transform microwave spectroscopy that facilitate spectral assignment and yield information about weak transitions with high resolution and sensitivity. Using the diastereomers menthone and isomenthone as examples, we investigate both the amplitude and the phase dependence of the free-induction decay of the microwave signal transition from pumping a radio frequency transition sharing a common level.

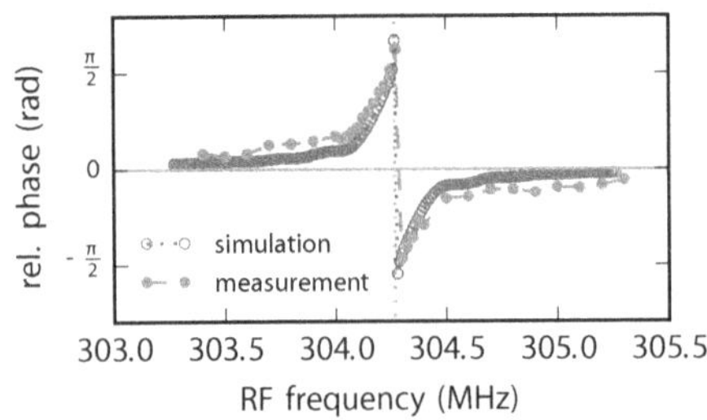

We observe a strong phase change when scanning the radio frequency through molecular resonance. The direction of the phase change depends on the energy level arrangement, i.e., if it is progressive or regressive. The experimental results can be simulated using the density-matrix formalism using the three-level Bloch equations and are best described with the AC Stark effect within the dressed-state picture, resulting in an Autler-Townes splitting. The characteristic phase inversion allows for a) the precise frequency determination of the typically weak radio frequency transitions exploiting the high sensitivity of the connected strong microwave signal transition and b) definitive information about the connectivity of the energy levels involved, i.e., progressive vs. regressive arrangements.

TE07 **10:24 – 10:39**

MICROWAVE THREE-WAVE MIXING EXPERIMENTS FOR CHIRALITY DETERMINATION: CURRENT STATUS

CRISTOBAL PEREZ, V. ALVIN SHUBERT, DAVID SCHMITZ, CHRIS MEDCRAFT, ANNA KRIN, MELANIE SCHNELL, *CoCoMol, Max-Planck-Institut für Struktur und Dynamik der Materie, Hamburg, Germany.*

Microwave three-wave mixing experiments have been shown to provide a novel and sensitive way to generate and measure enantiomer-specific molecular signatures. The handedness of the sample can be obtained from the phase of the molecular free induction decay whereas the enantiomeric excess can be determined by the amplitude of the chiral signal. After the introduction of this technique by Patterson et al.[a] remarkable improvements have been realized and experimental strategies for both absolute phase determination and enantiomeric excess have been presented.[b,c] This technique has been also successfully implemented at higher microwave frequencies.[d] Here we present the current status of this technique as well future directions and perspectives. This will be illustrated through our systematic study of chiral terpenes as well as preliminary results in molecular clusters.

[a]Patterson, D.; Schnell, M.; Doyle, J. M. Enantiomer-Specific Detection of Chiral Molecules via Microwave Spectroscopy. Nature 2013, 497, 475–477.

[b]Patterson, D.; Doyle, J. M. Sensitive Chiral Analysis via Microwave Three-Wave Mixing. Phys. Rev. Lett. 2013, 111, 023008.

[c]Shubert, V. A.; Schmitz, D.; Patterson, D.; Doyle, J. M.; Schnell, M. Identifying Enantiomers in Mixtures of Chiral Molecules with Broadband Microwave Spectroscopy. Angew. Chem. Int. Ed. 2014, 53, 1152–1155.

[d]Lobsiger, S.; Perez, C.; Evangelisti, L.; Lehmann, K. K.; Pate, B. H. Molecular Structure and Chirality Detection by Fourier Transform Microwave Spectroscopy. J. Phys. Chem. Lett. 2014, 6, 196–200.

TE08 **10:41 – 10:56**

A SEMI-AUTOMATED COMBINATION OF CHIRPED-PULSE AND CAVITY FOURIER TRANSFORM MICROWAVE SPECTROSCOPY

KYLE N CRABTREE, *Department of Chemistry, The University of California, Davis, CA, USA*; MARIE-ALINE MARTIN-DRUMEL, *Spectroscopy Lab, Harvard-Smithsonian Center for Astrophysics, Cambridge, MA, USA*; MICHAEL C McCARTHY, *Atomic and Molecular Physics, Harvard-Smithsonian Center for Astrophysics, Cambridge, MA, USA*; SYDNEY A GASTER, TAYLOR M HALL, DEONDRE L PARKS, GORDON G BROWN, *Department of Science and Mathematics, Coker College, Hartsville, SC, USA*.

A combination of chirped-pulse Fourier transform microwave (CP-FTMW) spectroscopy and cavity Fourier transform microwave (c-FTMW) spectroscopy has been used to analyze the spectra of 3,4-difluorobenzaldehyde and two distinct fluoropyridine – carbon dioxide complexes. In all cases, the 8 – 18 GHz CP-FTMW spectrum was measured, and the most intense transitions were chosen for further analysis. The intensities of the identified transitions were measured at multiple polarization powers using the c-FTMW spectrometer. Subsequently, a series of double-resonance experiments were performed on these transitions, again using the c-FTMW spectrometer, in order to discover which transitions shared a common quantum state. Following the double-resonance experiments, the assignments of the spectra were trivial. The results of the spectroscopic analysis, as well as the semi-automated method, will be presented.

TE09 **10:58 – 11:13**

SUBMILLIMETER ABSORPTION SPECTROSCOPY IN SEMICONDUCTOR MANUFACTURING PLASMAS AND COMPARISON TO THEORETICAL MODELS

YASER H. HELAL, CHRISTOPHER F. NEESE, FRANK C. DE LUCIA, *Department of Physics, The Ohio State University, Columbus, OH, USA*; PAUL R. EWING, *Applied Materials, Austin, TX, USA*; ANKUR AGARWAL, BARRY CRAVER, PHILLIP J. STOUT, MICHAEL D. ARMACOST, *Applied Materials, Sunnyvale, CA, USA*.

Plasmas used in the semiconductor manufacturing industry are of a similar nature to the environments often created for submillimeter spectroscopic study of astrophysical species. At the low operating pressures of these plasmas, submillimeter absorption spectroscopy is a method capable of measuring the abundances and temperatures of molecules, radicals, and ions without disturbing any of the properties of the plasma. These measurements provide details and insight into the interactions and reactions occurring within the plasma and their implications for semiconductor manufacturing processes. A continuous wave, 500 to 750 GHz, absorption spectrometer was designed and used to make measurements of species in semiconductor processing plasmas. Comparisons with expectations from theoretical plasma models provide a basis for validating and improving these models, which is a complex and difficult science itself. Furthermore, these comparisons are an evaluation for the use of submillimeter spectroscopy as a diagnostic tool in manufacturing processes.

TE10 **11:15 – 11:30**

CLOUD COMPUTING FOR THE AUTOMATED ASSIGNMENT OF BROADBAND ROTATIONAL SPECTRA: PORTING AUTOFIT TO AMAZON EC2

AARON C OLINGER, STEVEN SHIPMAN, *Department of Chemistry, New College of Florida, Sarasota, FL, USA*.

Recent developments in instrumentation have made it possible to collect broadband rotational spectra far faster than those spectra can be assigned. As such, we have been working to develop automated assignment algorithms so that the analysis can catch up with the data acquisition. The AUTOFIT program[a] has made strides in this direction, but it is still quite slow on spectra with high line densities, such as those collected near room temperature. Given that the AUTOFIT algorithm is highly parallelizable, we have used Amazon's EC2 webservice to run a modified version of AUTOFIT simultaneously across a large number of cores, allowing us to obtain results in a fraction of the time normally required by a typical desktop computer. In this talk, we will describe how AUTOFIT was modified to run on EC2 and present some benchmark results.

[a]Seifert, N.A., Finneran, I.A., Perez, C., Zaleski, D.P., Neill, J.L., Steber, A.L., Suenram, R.D., Lesarri, A., Shipman, S.T., Pate, B.H., J. Mol. Spec., in press

TF. Mini-symposium: High-Precision Spectroscopy

Tuesday, June 23, 2015 – 1:30 PM

Room: 116 Roger Adams Lab

Chair: Trevor Sears, Brookhaven National Laboratory, Upton, NY, USA

TF01 *INVITED TALK* **1:30 – 2:00**

COMB-REFERENCED SUB-DOPPLER RESOLUTION INFRARED SPECTROMETER

HIROYUKI SASADA, *Department of Physics, Faculty of Science and Technology, Keio University, Yokohama, Japan.*

We have developed a sub-Doppler resolution spectrometer. A difference frequency generation source, which consists of a pump source of a Nd:YAG laser, a signal source of an extended-cavity laser diode, and a waveguide-type PPLN, covers from 87 to 93 THz (2900 to 3100 cm^{-1}). An enhanced-cavity absorption cell remarkably improves the sensitivity of Lamb dips. An optical frequency comb controls the central frequency of the source with an uncertainty of a few kilohertz. Because the idler frequency is swept based on absolute frequency through the comb, recorded spectra can be repeatedly accumulated without any frequency drift. We have applied the spectrometer to resolve the hyperfine structure of the fundamental band of HCl with a spectral resolution of about 250 kHz.

To reduce the transit-time broadening, a novel enhanced-cavity absorption cell coupled with an idler wave of 1.9-mm beam radius at the beam waist has been introduced. The A_1-A_2 splitting of the ν_1 and ν_4 bands of CH_3D is resolved for a few tens low-J transitions with the Lamb-dip linewidth of 60 to 100 kHz.

Very recently, the source linewidth has reduced to 3 kHz using a linewidth transfer technique from the Nd:YAG laser to the extended-cavity laser diode through a novel optical frequency comb with a fast servo control. When methane sample is cooled with liquid-nitrogen, and the beam radius is expanded to 3 mm, the observed Lamb dip is 20 kHz wide without any enhanced-cavity absorption cell.

TF02 **2:05 – 2:20**

SUB-DOPPLER RESOLUTION SPECTROSCOPY OF THE FUNDAMENTAL VIBRATION BAND OF HCl WITH A COMB-REFERENCED SPECTROMETER

KANA IWAKUNI, *Department of Physics, Faculty of Science and Technology, Keio University, Yokohama, Japan*; HIDEYUKI SERA, *Department of Physics, Keio University, Yokohama, IX, Japan*; MASASHI ABE, HIROYUKI SASADA, *Department of Physics, Faculty of Science and Technology, Keio University, Yokohama, Japan.*

Sub-Doppler resolution spectroscopy of the fundamental bands of $H^{35}Cl$ and $H^{37}Cl$ has been carried out from 87 to 90 THz using a comb-referenced difference-frequency-generation (DFG) spectrometer. While the frequencies of the pump and signal waves are locked to that of the individual nearest comb mode, the repetition rate of the comb is varied for sweeping the idler frequency. Therefore, the relative uncertainty of the frequency scale is 10^{-11}, and the spectral resolution remains about 250 kHz even when the spectrum is accumulated for a long time. The hyperfine structures caused by chlorine nucleus are resolved for the R(0) to R(4) transitions. The figure depicts wavelength-modulation spectrum of the R(0) transition of $H^{35}Cl$. Three Lamb dips correspond to the F= 0, 1, and −1 components left to right, and the others with arrows are cross-over resonances which are useful for determining the weak F=−1 component frequencies for the R(1) to R(3) transitions. We have determined 49 and 44 transition frequencies of $H^{35}Cl$ and $H^{37}Cl$ with an uncertainty of 10 kHz. Six molecular constants of the vibrational excited state for each isotopomer are determined. They reproduce the determined frequencies with a standard deviation of about 10 kHz.

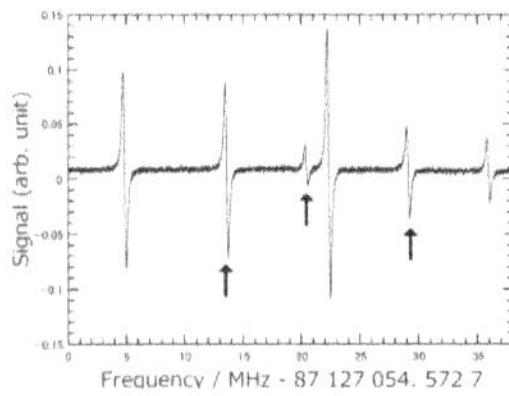

TF03 **2:22 – 2:37**

OBSERVATION AND ANALYSIS OF THE A_1-A_2 SPLITTING OF CH_3D

MASASHI ABE, HIDEYUKI SERA, HIROYUKI SASADA, *Department of Physics, Faculty of Science and Technology, Keio University, Yokohama, Japan.*

Sub-Doppler resolution spectroscopy of CH_3D has been carried out for the ν_1 and ν_4 fundamental bands using a comb-referenced difference-frequency generation spectrometer. Thirty transitions from the low-J'' and $K'' = 3$ levels are observed with a resolution of 60 to 100 kHz, and the A_1-A_2 splitting is resolved for twenty-three of the thirty transitions. Most of them are overlapped in Doppler broadening and resolved for the first time, as far as we know. The absolute transition frequencies are determined with a typical uncertainty of 4 kHz. The A_1-A_2 splitting constant of the $K'' = 3$ levels is yielded as $2h_{3,v=0} = (1.5641 \pm 0.0026)$ Hz for the ground vibrational state. Those of the $K' = 3$ levels for the $v_1 = 1$ states and of the $(K' = 2, l = -1)$ and $(K' = 4, l = 1)$ levels for the $v_4 = 1$ state are also determined including the J'-dependent terms.

TF04 **2:39 – 2:54**

HIGH RESOLUTION SPECTROSCOPY OF NAPHTHALENE CALIBRATED BY AN OPTICAL FREQUENCY COMB

AKIKO NISHIYAMA, KAZUKI NAKASHIMA, AYUMI MATSUBA, MASATOSHI MISONO, *Applied Physics, Fukuoka University, Fukuoka, Japan.*

In high-resolution molecular spectroscopy, the precise measure of the optical frequency is crucial to evaluate minute shifts and splittings of the energy levels. On the other hand, in such spectroscopy, thousands of spectral lines distributed over several wavenumbers have to be measured by a continuously scanning cw laser. Therefore, the continuously changing optical frequency of the scanning laser has to be determined with enough precision.

To satisfy these contradictory requirements, we have been developed two types of high-resolution spectroscopic systems employing an optical frequency comb. One of the systems employs RF band-pass filters to generate equally spaced frequency markers for optical frequency calibration, and is appropriate for wide wavelength-range measurement with relatively high scanning rate.[a] In the other system, the beat frequency between the optical frequency comb and the scanning laser is controlled by an acousto-optic frequency shifter. This system is suitable for more precise measurement, and enables detailed analyses of frequency characteristics of scanning laser.[b]

In the present study, we observe Doppler-free two-photon absorption spectra of $A^1B_{1u}(v_4 = 1) \leftarrow X^1A_g(v = 0)$ transition of naphthalene around 298 nm. The spectral lines are rotationally resolved and the resolution is about 100 kHz. For ${}^qQ$ transition, the rotational lines are assigned, and molecular constants in the excited state are determined. In addition, we analyze the origin of the measured linewidth and Coriolis interactions between energy levels. To determine molecular constants more precisely, we proceed to measure and analyze spectra of other transitions, such as ${}^sS$ transitions.

[a] A. Nishiyama, D. Ishikawa, and M. Misono, J. Opt. Soc. Am. B 30, 2107 (2013).

[b] A. Nishiyama, A. Matsuba, and M. Misono, Opt. Lett. 39, 4923 (2014).

TF05 **2:56 – 3:11**

OPTICAL FREQUENCY COMB FOURIER TRANSFORM SPECTROSCOPY WITH RESOLUTION EXCEEDING THE LIMIT SET BY THE OPTICAL PATH DIFFERENCE

ALEKSANDRA FOLTYNOWICZ, LUCILE RUTKOWSKI, ALEXANDRA C JOHANSSSON, AMIR KHODABAKHSH, *Department of Physics, Umea University, Umea, Sweden*; PIOTR MASLOWSKI, GRZEGORZ KOWZAN, *Institute of Physics, Faculty of Physics, Astronomy and Informatics, Nicolaus Copernicus University, Torun, Poland*; KEVIN LEE, MARTIN FERMANN, *Laser Research, IMRA AMERICA, Inc, Ann Arbor, MI, USA*.

Fourier transform spectrometers (FTS) based on optical frequency combs (OFC) allow detection of broadband molecular spectra with high signal-to-noise ratios within acquisition times orders of magnitude shorter than traditional FTIRs based on thermal sources[a]. Due to the pulsed nature of OFCs the interferogram consists of a series of bursts rather than a single burst at zero optical path difference (OPD). The comb mode structure can be resolved by acquiring multiple bursts, in both mechanical FTS systems[b] and dual-comb spectroscopy[c]. However, in all existing demonstrations the resolution was ultimately limited either by the maximum available OPD between the interferometer arms or by the total acquisition time enabled by the storage memory. We present a method that provides spectral resolution exceeding the limit set by the maximum OPD using an interferogram containing only a single burst. The method allows measurements of absorption lines narrower than the OPD-limited resolution without any influence of the instrumental lineshape function. We demonstrate this by measuring undistorted CO_2 and CO absorption lines with linewidth narrower than the OPD-limited resolution using OFC-based mechanical FTS in the near- and mid-infrared wavelength ranges. The near-infrared system is based on an Er:fiber femtosecond laser locked to a high finesse cavity, while the mid-infrared system is based on a Tm:fiber-laser-pumped optical parametric oscillator coupled to a multi-pass cell. We show that the method allows acquisition of high-resolution molecular spectra with interferometer length orders of magnitude shorter than traditional FTIR.

[a]Mandon, J., G. Guelachvili, and N. Picque, *Nat. Phot.*, 2009. **3**(2): p. 99-102.
[b]Zeitouny, M., et al., *Ann. Phys.*, 2013. **525**(6): p. 437-442.
[c]Zolot, A.M., et al., *Opt. Lett.*, 2012. **37**(4): p. 638-640.

TF06 ***Post-Deadline Abstract*** **3:13 – 3:23**

METROLOGY WITH AN OPTICAL FEEDBACK FREQUENCY STABILIZED CRDS

SAMIR KASSI, JOHANNES BURKART, *UMR5588 LIPhy, Université Grenoble 1/CNRS, Saint Martin D'heres, France*.

We will present a metrological application of our recently developed Optical Feedback Frequency Stabilized - Cavity Ring Down Spectrometer (OFFS-CRDS). This instrument, which ideally fits with an optical frequency comb for absolute frequency calibration, relies on the robust lock of a steady cavity ring down resonator against a highly stable, radiofrequency tuned optical source. At 1.6 μm, over 7 nm, we demonstrate Lamb dip spectroscopy of CO_2 with line frequency retrieval at the kHz level, a dynamic in excess of 700,000 on the absorption scale and a detectivity of $4x10^{-13}cm^{-1}Hz^{-1/2}$. Such an instrument nicely meets the requirements for the most demanding spectroscopy spanning from accurate isotopic ratio determination and very precise lineshape recordings to Boltzmann constant redefinition.

Intermission

TF07 *INVITED TALK* **3:42 – 4:12**

CAVITY ENHANCED ULTRAFAST TRANSIENT ABSORPTION SPECTROSCOPY

THOMAS K ALLISON, *Department of Chemistry, Stony Brook University, Stony Brook, NY, USA*; MELANIE ROBERTS REBER, *Department of Physics and Astronomy, State University of New York, Stony Brook, NY, USA*; YUNING CHEN, *Department of Chemistry, Stony Brook University, Stony Brook, NY, USA*.

Ultrafast spectroscopy on gas phase systems is typically restricted to techniques involving photoionization, whereas solution phase experiments utilize the detection of light. At Stony Brook, we are developing new techniques for performing femtosecond time-resolved spectroscopy using frequency combs and high-finesse optical resonators. A large detection sensitivity enhancement over traditional methods enables the extension of all-optical ultrafast spectroscopies, such as broad-band transient absorption spectroscopy (TAS) and 2D spectroscopy, to dilute gas phase samples produced in molecular beams. Here, gas phase data can be directly compared to solution phase data. Initial demonstration experiments are focusing on the photodissociation of iodine in small neutral argon clusters, where cluster size strongly influences the effects solvent-caging and geminate recombination. I will discuss these initial results, our high power home-built Yb:fiber laser systems, and also extensions of the methods to the mid-IR to study the vibrational dynamics of hydrogen bonded clusters.

TF08 **4:17 – 4:32**

NOISE-IMMUNE CAVITY-ENHANCED OPTICAL FREQUENCY COMB SPECTROSCOPY

LUCILE RUTKOWSKI, AMIR KHODABAKHSH, ALEXANDRA C JOHANSSSON, ALEKSANDRA FOLTYNOWICZ, *Department of Physics, Umea University, Umea, Sweden*.

We present noise-immune cavity-enhanced optical frequency comb spectroscopy (NICE-OFCS), a recently developed technique for sensitive, broadband, and high resolution spectroscopy[a]. In NICE-OFCS an optical frequency comb (OFC) is locked to a high finesse cavity and phase-modulated at a frequency precisely equal to (a multiple of) the cavity free spectral range. Since each comb line and sideband is transmitted through a separate cavity mode in exactly the same way, any residual frequency noise on the OFC relative to the cavity affects each component in an identical manner. The transmitted intensity contains a beat signal at the modulation frequency that is immune to frequency-to-amplitude noise conversion by the cavity, in a way similar to continuous wave noise-immune cavity-enhanced optical heterodyne molecular spectroscopy (NICE-OHMS)[b]. The light transmitted through the cavity is detected with a fast-scanning Fourier-transform spectrometer (FTS) and the NICE-OFCS signal is obtained by fast Fourier transform of the synchronously demodulated interferogram.

Our NICE-OFCS system is based on an Er:fiber femtosecond laser locked to a cavity with a finesse of ~9000 and a fast-scanning FTS equipped with a high-bandwidth commercial detector. We measured NICE-OFCS signals from the $3\nu_1+\nu_3$ overtone band of CO_2 around 1.57 μm and achieved absorption sensitivity 6.4×10^{-11}cm^{-1}Hz$^{-1/2}$ per spectral element, corresponding to a minimum detectable CO_2 concentration of 25 ppb after 330 s integration time[c]. We will describe the principles of the technique and its technical implementation, and discuss the spectral lineshapes of the NICE-OFCS signals.

[a] A. Khodabakhsh, C. Abd Alrahman, and A. Foltynowicz, Opt. Lett. 39, 5034-5037 (2014).

[b] J. Ye, L. S. Ma, and J. L. Hall, J. Opt. Soc. Am. B 15, 6-15 (1998).

[c] A. Khodabakhsh, A. C. Johansson, and A. Foltynowicz, Appl. Phys. B (2015) doi:10.1007/s00340-015-6010-7.

TF09 **4:34 – 4:49**

A NEW BROADBAND CAVITY ENHANCED FREQUENCY COMB SPECTROSCOPY TECHNIQUE USING GHz VERNIER FILTERING.

JÉRÔME MORVILLE, *UMR 5306, ILM University Lyon 1 and CNRS, Villeurbanne, France*; LUCILE RUTKOWSKI, *Department of Physics, Umea University, Umea, Sweden*; GEORGI DOBREV[a], *Department of Physics, Sofia University, Sofia, Bulgaria*; PATRICK CROZET, *UMR 5306, ILM University Lyon 1 and CNRS, Villeurbanne, France*.

We present a new approach to Cavity Enhanced - Direct Frequency Comb Spectroscopy where the full emission bandwidth of a Titanium:Sapphire laser is exploited at GHz resolution. The technique is based on a low-resolution Vernier filtering obtained with an appreciable –actively stabilized– mismatch between the cavity Free Spectral Range and the laser repetition rate, using a diffraction grating and a split-photodiode [b]. This particular approach provides an immunity to frequency-amplitude noise conversion, reaching an absorption baseline noise in the 10^{-9} cm^{-1} range with a cavity finesse of only 3000. Spectra covering 1800 cm^{-1} ($\sim$ 55 THz) are acquired in recording times of about 1 second, providing an absorption figure of merit of a few 10^{-11} cm$^{-1}/\sqrt{Hz}$. Initially tested with ambient air, we report progress in using the Vernier frequency comb method with a discharge source of small radicals.

[a]and ILM University Lyon1
[b]Rutkowski *et al, Opt. Lett., 39(23)2014*

TF10 **4:51 – 5:06**

A DECADE-SPANNING HIGH-RESOLUTION ASYNCHRONOUS OPTICAL SAMPLING BASED TERAHERTZ TIME-DOMAIN SPECTROMETER

JACOB T GOOD, *Division of Chemistry and Chemical Engineering, California Institute of Technology, Pasadena, CA, USA*; DANIEL HOLLAND, *Translational Imaging Center, University of Southern California, Los Angeles, CA, USA*; IAN A FINNERAN, BRANDON CARROLL, MARCO A. ALLODI, GEOFFREY BLAKE, *Division of Chemistry and Chemical Engineering, California Institute of Technology, Pasadena, CA, USA*.

High-resolution ASynchronous OPtical Sampling (ASOPS) is a technique that substantially improves the combined frequency resolution and bandwidth of ASOPS based TeraHertz Time-Domain Spectroscopy (THz-TDS) systems. We employ two mode-locked femtosecond Ti:Sapphire oscillators with repetition frequencies of 80 MHz operating at a fixed repetition frequency offset of 100 Hz. This offset lock is maintained by a Phase-Locked Loop (PLL) operating at the 60th harmonic of the repetition rate of the Ti:Sapphire oscillators. Their respective time delay is scanned across 12.5 ns requiring a scan time of 10 ms, supporting a time delay resolution of up to 15.6 fs. ASOPS-THz-TDS enables high-resolution spectroscopy that is impossible for a THz-TDS system employing a mechanical delay stage. We measure a timing jitter of 1.36 fs for the system using an air-gap etalon and an optical cross-correlator. We report a Root-Mean-Square deviation of 20.7 MHz and a mean deviation of 14.4 MHz for water absorption lines from 0.5 to 2.7. High-resolution ASOPS-THz-TDS enables high resolution spectroscopy of both gas-phase and condensed-phase samples across a decade of THz bandwidth.

TF11 **5:08 – 5:23**

DOPPLER-LIMITED SPECTROSCOPY WITH A DECADE-SPANNING TERAHERTZ FREQUENCY COMB

IAN A FINNERAN, JACOB T GOOD, *Division of Chemistry and Chemical Engineering, California Institute of Technology, Pasadena, CA, USA*; DANIEL HOLLAND, *Translational Imaging Center, University of Southern California, Los Angeles, CA, USA*; BRANDON CARROLL, MARCO A. ALLODI, GEOFFREY BLAKE, *Division of Chemistry and Chemical Engineering, California Institute of Technology, Pasadena, CA, USA.*

We report the generation and detection of a decade-spanning TeraHertz (THz) frequency comb (0.15-2.4 THz) using two Ti:Sapphire femtosecond laser oscillators and ASynchronous OPtical Sampling THz Time-Domain Spectroscopy (ASOPS-THz-TDS). The measured linewidth of the comb at 1.5 THz is 3 kHz over a 60 second acquisition. With time-domain detection of the comb, we measure three transitions of water vapor at 10 mTorr between 1-2 THz with an average Doppler-limited fractional uncertainty of 5.9×10^{-8}. Significant improvements in bandwidth, resolution, and sensitivity are possible with existing technologies and will enable future studies of jet-cooled hydrogen-bonded clusters.

TF12 **5:25 – 5:40**

DUAL COMB RAMAN SPECTROSCOPY ON CESIUM HYPERFINE TRANSITIONS-TOWARD A STIMULATE RAMAN SPECTRUM ON CF_4 MOLECULE

TZE-WEI LIU, *Taiwan International Graduate Program, Academia Sinica, Taipei, Taiwan*; YEN-CHU HSU, *Institute of Atomic and Molecular Sciences, Academia Sinica, Taipei, Taiwan*; WANG-YAU CHENG, *Department of Physics, National Central University, Jhongli, Taiwan.*

Raman spectroscopy is an important spectroscopic technique used in chemistry to provide a fingerprint by which molecules can be identified. It helps us to observe vibration- rotation, and other low-frequency modes in a system. Dual comb Raman spectroscopy allows measuring a wide bandwidth with high resolution in microseconds.

The stimulate Raman spectroscopy had been performed in early days where the nonlinear conversion efficiency depended on laser peak power. Hence we propose an approach for rapidly resolving the Raman spectroscopy of CF_4 molecule by two Ti:sapphire comb lasers. Our progress on this proposal will be presented in the conference.

First, we have realized a compact dual Ti:sapphire comb laser system[a] where the dual Ti:sapphire laser system possesses the specification of 1 GHz repetition rate. In our dual comb system, 1 GHz repetition rate, 100 kHz Δf_{rep} and 2.4 THz optical filter are chosen according to the demands of our future works on spectroscopy. Therefore, the maximum mode number within free spectral range is $5*10^3$, and the widest range of dual-comb based spectra in that each spectrum could be uniquely identified is 5 THz. The actual bandwidth is determined by the employed optical filter and is set to be 2.4 THz here, so that the corresponding data acquisition time is 10 μs.

Secondly, since the identification of the tremendous spectral lines of CF_4 molecule relies on a stable reference and a reliable data-retrieving system, we propose a first-step experiment on atomic system where the direct 6S-8S 822-nm two-photon absorption[b] and 8S-$6P_{3/2}$ (794 nm) enhanced stimulate Raman would be realized directly by using Ti:sapphire laser. We have successfully performed direct comb laser two-photon spectroscopy for both with and without middle-level enhanced. For the level enhanced two-photon spectrum, our experimental setup achieves Doppler-free spectrum and a record narrow linewidth (1 MHz).

[a] T.-W. Liu, C.-M. Wu, Y.–C. Hsu and W.-Y. Cheng, Appl. Phys. B 117, 699 (2014)

[b] P. Fendel, S. D. Bergeson, Th. Udem, and T. W. Hänsch, Opt. Lett. 32, 701 (2007)

TG. Large amplitude motions, internal rotation

Tuesday, June 23, 2015 – 1:30 PM

Room: 100 Noyes Laboratory

Chair: Kaori Kobayashi, University of Toyama, Toyama, Japan

TG01 — 1:30 – 1:45

THE BAND OF CH_3CH_2D FROM 770-880 cm^{-1}

ADAM M DALY, BRIAN DROUIN, JOHN PEARSON, *Jet Propulsion Laboratory, California Institute of Technology, Pasadena, CA, USA*; PETER GRONER, *Department of Chemistry, University of Missouri - Kansas City, Kansas City, MO, USA*; KEEYOON SUNG, LINDA BROWN, *Jet Propulsion Laboratory, California Institute of Technology, Pasadena, CA, USA*; ARLAN MANTZ, *Department of Physics, Astronomy and Geophysics, Connecticut College, New London, CT, USA*; MARY ANN H. SMITH, *Science Directorate, NASA Langley Research Center, Hampton, VA, USA.*

To extend the ethane database we recorded a 0.0028 cm^{-1}resolution spectrum of CH_3CH_2D from 650 to 1500 cm^{-1}using a Bruker IFS-125HR at the Jet Propulsion Laboratory. The 98% deuterium-enriched sample was contained in the 0.2038 m absorption cell; one scan was taken with the sample cryogenically cooled to 130 K and another at room temperature. From the cold data, we retrieved line positions and intensities of 8704 individual absorption features from 770 – 880 cm^{-1}using a least squares curve fitting algorithm. From this set of measurements, we assigned 5041 transitions to the ν_{17} fundamental at 805.3427686(234) cm^{-1}; this band is a c-type vibration, with A and E components arising from internal rotation. The positions were modeled using a 22 term torsional Hamiltonian using SPFIT producing the A and E energy splitting of 5.409(25)x10^{-3} cm^{-1}(162.2(8) MHz) with a standard deviation of 7x10^{-4} cm^{-1}(21 MHz). The calculated line intensities at 130 K agree very well with retrieved intensities. To predict line intensities at different temperatures, the partition function value was determined at eight temperatures between 9.8 and 300 K by summing individual energy levels up to J = 99 and K_a = 99 for the six states up through ν_{17} at 805 cm^{-1}. The resulting prediction of singly-deuterated ethane absorption at 12.5 μm enables its detection in planetary atmospheres, including those of Titan and exoplanets.

TG02 — 1:47 – 2:02

LOW-TEMPERATURE HIGH-RESOLUTION INFRARED SPECTRUM OF ETHANE-1D, C_2H_5D: ROTATIONAL ANALYSIS OF THE ν_{17} BAND NEAR 805 cm^{-1}using ERHAM.

PETER GRONER, *Department of Chemistry, University of Missouri - Kansas City, Kansas City, MO, USA*; ADAM M DALY, BRIAN DROUIN, JOHN PEARSON, KEEYOON SUNG, LINDA BROWN, *Jet Propulsion Laboratory, California Institute of Technology, Pasadena, CA, USA*; ARLAN MANTZ, *Department of Physics, Astronomy and Geophysics, Connecticut College, New London, CT, USA*; MARY ANN H. SMITH, *Science Directorate, NASA Langley Research Center, Hampton, VA, USA.*

The high-resolution infrared spectrum of gaseous ethane-d_1 at 130 K shows transitions that are split into A and E components due to the interaction of overall rotation with the internal rotation of the CH_3 group. An analysis of the spectrum from 680 to 900 $cm^{-1}$with an expanded version of the program ERHAM [a,b] is in progress to assign the bands at E(ν_{17}) = 805 $cm^{-1}$and E(ν_{11}) = 715 cm^{-1}. A discussion of the interactions among the fundamental levels of ν_{17} and ν_{11} with overtone levels of ν_{18} and the(CH_3 torsion) will be given. ERHAM has been and continues to be very successful in the analysis of pure the rotational spectra of molecules containing internal rotation and the vibrational spectrum of C_2H_5D serves as an excellent system to test the extension of the program.

[a] P. Groner, *J. Chem. Phys.* **107** 4483 (1997)
[b] P. Groner, *J. Mol. Spectrosc.* **278** 52 (2012)

TG03 **2:04 – 2:19**

MICROWAVE SPECTROSCOPY OF THE EXCITED VIBRATIONAL STATES OF METHANOL

JOHN PEARSON, ADAM M DALY, *Jet Propulsion Laboratory, California Institute of Technology, Pasadena, CA, USA.*

Methanol is the simplest molecule with a three-fold internal rotation and the observation of its ν_8 band served the primary catalyst for the development of internal rotation theory$^{(a,b)}$. The 75 subsequent years of investigation into the ν_8 band region have yielded a large number assignments, numerous high precision energy levels and a great deal of insight into the coupling of ν_t=3 & 4 with ν_8, ν_7, ν_{11} and other nearby states$^{(c)}$. In spite of this progress numerous assignment mysteries persist, the origin of almost half the far infrared laser lines remain unknown and all attempts to model the region quantum mechanically have had very limited success. The C_{3V} internal rotation Hamiltonian has successfully modeled the ν_t=0,1 & 2 states of methanol and other internal rotors$^{(d)}$. However, successful modeling of the coupling between torsional bath states and excited small amplitude motion remains problematic and coupling of multiple interacting excited small amplitude vibrations featuring large amplitude motions remains almost completely unexplored. Before such modeling can be attempted, identifying the remaining low lying levels of ν_7 and ν_{11} is necessary. We present an investigation into the microwave spectrum of ν_7, ν_8 and ν_{11} along with the underlying torsional bath states in ν_t=3 and ν_t= 4.

$^{(a)}$ A. Borden, E.F. Barker J. Chem. Phys., 6, 553 (1938).
$^{(b)}$ J. S. Koehler and D. M. Dennison, Phys. Rev. 57, 1006 (1940).
$^{(c)}$ R. M. Lees, Li-Hong Xu, J. W. C. Johns, B. P. Winnewisser, and M. Lock, J. Mol. Spectrosc. 243, 168 (2007).
$^{(d)}$ L.-H. Xu, J. Fisher, R.M. Lees, H.Y. Shi, J.T. Hougen, J.C. Pearson, B.J. Drouin, G.A. Blake, R. Braakman J. Mol. Spectrosc., 251, 305 (2008).

TG04 **2:21 – 2:36**

FIRST HIGH RESOLUTION ANALYSIS OF THE ν_{21} BAND OF PROPANE AT 921.4 cm^{-1}: EVIDENCE OF LARGE-AMPLITUDE-MOTION TUNNELLING EFFECTS

AGNES PERRIN, F. KWABIA TCHANA, JEAN-MARIE FLAUD, *LISA, CNRS, Universités Paris Est Créteil et Paris Diderot, Créteil, France*; LAURENT MANCERON, *Synchrotron SOLEIL, CNRS-MONARIS UMR 8233 and Beamline AILES, Saint Aubin, France*; JEAN DEMAISON, NATALJA VOGT, *Section of Chemical Information Systems, Universität Ulm, Ulm, Germany*; PETER GRONER, *Department of Chemistry, University of Missouri - Kansas City, Kansas City, MO, USA*; WALTER LAFFERTY, *Optical Technology Division, National Institute of Standards and Technology, Gaithersburg, MD, USA.*

A high resolution (0.0015 cm^{-1}) IR spectrum of propane, C_3H_8, has been recorded with synchrotron radiation at the French light source facility at SOLEIL coupled to a Bruker IFS-125 Fourier transform spectrometer. A preliminary analysis of the ν_{21} fundamental band (B_1, CH_3 rock) near 921.4 cm^{-1}reveals that the rotational energy levels of 21_1 are split by interactions with the internal rotations of the methyl groups. Conventional analysis of this A-type band yielded band centers at 921.3724(38), 921.3821(33) and 921.3913(44) $\mathrm{cm}^{-1}$for the AA, EE and $AE + EA$ tunneling splitting components, respectively.[a] These torsional splittings most probably are due to anharmonic and/or Coriolis resonance coupling with nearby highly excited states of both internal rotations of the methyl groups. In addition, several vibrational-rotational resonances were observed that affect the torsional components in different ways. The analysis of the B-type band near 870 cm^{-1}(ν_8, sym. C-C stretch) which also contains split rovibrational transitions due to internal rotation is in progress. It is performed by using the effective rotational Hamiltonian method ERHAM[b] with a code that allows prediction and least-squares fitting of such vibration-rotation spectra.

[a] A. Perrin et al., submitted to *J.Mol.Spectrosc.*
[b] P Groner, *J.Chem.Phys.* 107 (1997) 4483; *J.Mol.Spectrosc.* 278 (2012) 52.

TG05 **2:38 – 2:53**

TORSIONAL STRUCTURE IN THE $\tilde{A} - \tilde{X}$ SPECTRUM OF THE CH_3O_2 AND CH_2XO_2 RADICALS

MENG HUANG, ANNE B McCOY, TERRY A. MILLER, *Department of Chemistry and Biochemistry, The Ohio State University, Columbus, OH, USA.*

Large amplitude motions in methyl rotor systems have been well studied, especially the coupling between the CH_3 torsion and the CH stretches. The CH_3OO radical is a example of a system where this coupling is relatively small, but its effects still can be observed in the the infrared spectrum taken by the Lee group.[a] Rotational contour simulations based on an asymmetric rotor model show good agreement with the experimental spectrum except for an unexplained broadening of the Q-branch of one of the CH stretch features. The broadening is likely caused by low frequency torsional modes populated at room temperature resulting in sequence band transitions that are slightly shifted from the origin. A reduced dimension model involving the three CH stretches and the CH_3 torsion is applied to CH_3OO to simulate the observed spectrum. The CH stretches are described by a harmonically coupled anharmonic oscillator model in which the parameters depend on the CH_3 torsion angle. Based on these calculations, the observed broadening of the Q-branch can be qualitatively explained by coupling between two CH stretch/CH_3 torsion combination bands which differ by one quantum in torsional excitation. The $\tilde{A}$-$\tilde{X}$ electronic transitions of halogenated methyl peroxy radicals, CH_2XOO (X-Cl, Br, I), show a complementary structure. At room temperature multiple peaks have been observed in the region of the origin and OO stretch vibronic bands in all three radicals with the spectra for CH_2IO_2 being by far the most complex. This structure may again be the result of hot bands originating from excited torsional levels. Several theoretical models have been investigated to calculate the Franck-Condon factors that govern the structure. A calculation that models the I-C-O-O torsion using curvilinear internal coordinates and molecular geometry and harmonic torsion frequencies predicted by electronic structure calculations shows the best agreement between the CH_2IOO experimental and simulated spectra. The multiple peak structure results from the change in X-C-O-O torsion dihedral between the $\tilde{X}$ state and $\tilde{A}$ states. Interestingly, a similar calculation with Cartesian displacement coordinates fails to explain the torsional structure. This study shows the importance of coordinate system choice if a significant displacement in the torsional coordinate occurs upon electronic excitation.

[a] K.-H. Hsu, Y.-P. Lee, M. Huang, T. A. Miller, TD08, *68th International Symposium of Molecular Spectroscopy* (2013)

TG06 **2:55 – 3:10**

UPDATE OF THE ANALYSIS OF THE PURE ROTATIONAL SPECTRUM OF EXCITED VIBRATIONS OF CH_3CH_2CN

ADAM M DALY, JOHN PEARSON, SHANSHAN YU, BRIAN DROUIN, *Jet Propulsion Laboratory, California Institute of Technology, Pasadena, CA, USA*; CELINA BERMÚDEZ, JOSÉ L. ALONSO, *Grupo de Espectroscopia Molecular, Lab. de Espectroscopia y Bioespectroscopia, Unidad Asociada CSIC, Universidad de Valladolid, Valladolid, Spain.*

The torsion-vibration-rotation analysis of nearly degenerate vibrational states involving both small and large amplitude motion has escaped satisfactory quantum mechanical description. Unfortunately the interstellar medium is filled with many prevalent molecules that feature internal rotation that couples strongly with torsional bath states. Many excited states are observed in emission in hot cores associated with massive star formation and it is likely that absorption in the infrared will be seen by JWST. We present our progress on the analysis of the high resolution pure rotational spectrum of ethyl cyanide, CH_3CH_2CN, which is highly abundant in hot cores with massive star formation and can serve as a sensitive temperature and source size probe[a]. Although the ground state has been assigned to 1.6 THz[b], the two vibrational states ν_{13} and ν_{21}, the C-C-N bend and torsion, have only been assigned up to 400 GHz[c]. It is clear that detailed understanding of excited states will help properly model the temperature dependence of the intensity. We will report the progress on the fit up to 1.5 THz for the states ν_{13} , ν_{21} , ν_{20} and ν_{12}, at 206.5 cm^{-1}, 212 cm^{-1}, 375 $cm^{-1}$and 532 cm^{-1}respectively. In spite of a nearly 1200 cm^{-1}barrier to internal rotation all the vibrational states observed feature A/E splittings inconsistent with such a high barrier suggesting that there is extensive coupling between the torsional bath states and the excited vibrations. The low lying states of ethyl cyanide provide an opportunity to assess all the possible interaction under the C_S group for both A and E symmetry in the high barrier case to serve as a benchmark for developing theory for the analysis of lower barrier cases.

[a] A.M. Daly, C. Bermúdez, A. López, B. Tercero, J.C. Pearson, N. Marcelino, J.L. Alonso, J. Cernicharo *Astrophys. J.*, **768** 81 (2013)
[b] C.S. Brauer, J.C. Pearson, B.J. Drouin, S. Yu *ApJ Suppl. Ser.*, **184** 133 (2009)
[c] D.M. Mehringer, J.C. Pearson, J. Keene, T.G. Phillips *Astrophys. J.*, **608** 306 (2004)

Intermission

TG07 **3:29 – 3:44**

UNUSUAL INTERNAL ROTATION COUPLING IN THE MICROWAVE SPECTRUM OF PINACOLONE

YUEYUE ZHAO, *Institute for Physical Chemistry, RWTH Aachen University, Aachen, Germany*; HA VINH LAM NGUYEN, *CNRS et Universités Paris Est et Paris Diderot, Laboratoire Interuniversitaire des Systèmes Atmosphériques (LISA), Créteil, France*; WOLFGANG STAHL, *Institute for Physical Chemistry, RWTH Aachen University, Aachen, Germany*; JON T. HOUGEN, *Sensor Science Division, National Institute of Standards and Technology, Gaithersburg, MD, USA.*

The molecular-beam Fourier-transform microwave spectrum of pinacolone (methyl *tert*-butyl ketone) has been measured in several regions between 2 and 40 GHz. Assignments of a large number of A and E transitions were confirmed by combination differences, but fits of the assigned spectrum using several torsion-rotation computer programs based on different models led to the unexpected conclusion that no existing program correctly captures the internal dynamics of this molecule. A second puzzle arose when it became clear that roughly half of the spectrum remained unassigned even after all predicted transitions were added to the assignment list. Quantum chemical calculations carried out at the MP2/6-311++G(d,p) level indicate that this molecule does not have a plane of symmetry at equilibrium, and that internal rotation of the light methyl group induces a large oscillatory motion of the heavy *tert*-butyl group from one side of the C_s saddle point to the other. The effect of this non-C_s equilibrium structure was modeled for $J = 0$ levels by a simple two-top torsional Hamiltonian, where magnitudes of the strong top-top coupling terms were determined directly from the *ab initio* two-dimensional potential surface. A plot of the resultant torsional levels on the same scale as a one-dimensional potential curve along the zig-zag path connecting the six (unequally spaced) minima bears a striking resemblance to the 1:2:1 splitting pattern of levels in an internal rotation problem with a six-fold barrier. A plot of the six minima closely resembles the potential surface for methylamine. This talk will focus on implications of these resemblances for future work.

TG08 **3:46 – 4:01**

THE COMPLETE ROTATIONAL SPECTRUM OF CH_3NCO UP TO 376 GHz

ZBIGNIEW KISIEL, *ON2, Institute of Physics, Polish Academy of Sciences, Warszawa, Poland*; LUCIE KOLESNIKOVÁ, JOSÉ L. ALONSO, *Grupo de Espectroscopia Molecular, Lab. de Espectroscopia y Bioespectroscopia, Unidad Asociada CSIC, Universidad de Valladolid, Valladolid, Spain*; IVAN MEDVEDEV, *Department of Physics, Wright State University, Dayton, OH, USA*; SARAH FORTMAN, MANFRED WINNEWISSER, FRANK C. DE LUCIA, *Department of Physics, The Ohio State University, Columbus, OH, USA.*

The methylisocyanate molecule, CH_3NCO, is of interest as a potential astrophysical species and as a model system for the study of quasisymmetric behavior. The rotational spectrum is made very complex by the presence in CH_3NCO of two large-amplitude motions: an almost free internal rotation and a low barrier skeletal bending motion. This challenging spectrum has, nevertheless, been assigned at 8-38 GHz by Stark spectroscopy[a] and has been measured at 117-376 GHz with the broadband FASSST technique.[b]

We presently report the results of measuring this spectrum also in supersonic expansion for the transitions below 40 GHz, and at room-temperature in the region between 40 and 120 GHz. In this way we are finally able to confirm the assignment of the ground state and of the internal rotation m=1 state and to analyse the nitrogen hyperfine splitting structure. It is also possible to confidently transfer the Stark-based assignment to the transition sequences measured in the mm-wave region, and to assign high K_a sequences. Various models for fitting this spectrum are explored but, even without more extensive fits, we are now able to present temperature scalable linelists for astrophysical applications.

[a]J.Koput, *J. Mol. Spectrosc.* **115**, 131 (1986).
[b]Z.Kisiel et al., 65th OSU Symposium on Molecular Spectroscopy, The Ohio State University, Ohio 2010, RC-13.

TG09 **4:03 – 4:18**

GAS PHASE CONFORMATIONS AND METHYL INTERNAL ROTATION FOR 2-PHENYLETHYL METHYL ETHER AND ITS ARGON VAN DER WAALS COMPLEX FROM FOURIER TRANSFORM MICROWAVE SPECTROSCOPY

RANIL M. GURUSINGHE, MICHAEL TUBERGEN, *Department of Chemistry and Biochemistry, Kent State University, Kent, OH, USA.*

A mini-cavity microwave spectrometer was used to record the rotational spectra arising from 2-phenylethyl methyl ether and its weakly bonded argon complex in the frequency range of 10.5 – 22 GHz. Rotational spectra were found for two stable conformations of the monomer: anti-anti and gauche-anti, which are 1.4 kJ mol^{-1} apart in energy at wB97XD/6-311++G(d,p) level. Doubled rotational transitions, arising from internal motion of the methyl group, were observed for both conformers. The program XIAM was used to fit the rotational constants, centrifugal distortion constants, and barrier to internal rotation to the measured transition frequencies of the A and E internal rotation states. The best global fit values of the rotational constants for the anti-anti conformer are A= 3799.066(3) MHz, B= 577.95180(17) MHz, C= 544.7325(3) MHz and the A state rotational constants of the gauche-anti conformer are A= 2676.1202(7) MHz, B= 760.77250(2) MHz, C= 684.78901(2) MHz.

The rotational spectrum of 2-phenylethyl methyl ether – argon complex is consistent with the geometry where argon atom lies above the plane of the benzene moiety of gauche-anti conformer. Tunneling splittings were too small to resolve within experimental accuracy, likely due to an increase in three fold potential barrier when the argon complex is formed. Fitted rotational constants are A= 1061.23373(16) MHz, B= 699.81754(7) MHz, C= 518.33553(7) MHz.

The lowest energy solvated ether - water complex with strong intermolecular hydrogen bonding has been identified theoretically. Progress on the assignment of the water complex will also be presented.

TG10 **4:20 – 4:30**

A COMPARISON OF BARRIER TO METHYL INTERNAL ROTATION OF METHYLSTYRENES: MICROWAVE SPECTROSCOPIC STUDY

RANIL M. GURUSINGHE, MICHAEL TUBERGEN, *Department of Chemistry and Biochemistry, Kent State University, Kent, OH, USA.*

Rotational spectra of α-Methylstyrene, cis-β-Methylstyrene, and trans-β-Methylstyrene were examined to investigate their intrinsic tunneling properties. Theoretical calculations at wB97XD/6-311++G(d,p) level predict only one stable conformer for each molecular system. Spectra were recorded in the frequency range of 10.5 - 22.0 GHz using a cavity based Fourier transform microwave spectrometer.

A relaxed potential scan for the methyl torsion at wB97XD/6-311++G(d,p) level of theory was used to estimate the associated barrier for the hindered internal rotation. The program XIAM was used to fit the rotational constants, distortion constants and barrier to methyl internal rotation to the measured transition frequencies of the A and E internal rotation states.

TG11 **4:32 – 4:47**

MICROWAVE SPECTRA AND AB INITIO STUDIES OF THE NE-ACETONE COMPLEX

JIAO GAO, JAVIX THOMAS, YUNJIE XU, WOLFGANG JÄGER, *Department of Chemistry, University of Alberta, Edmonton, AB, Canada.*

Microwave spectra of the neon-acetone van der Waals complex were measured using a cavity-based molecular beam Fourier-transform microwave spectrometer in the region from 5 to 18 GHz. Both ^{20}Ne and ^{22}Ne containing isotopologues were studied and both *c*- and weaker *a*-type rotational transitions were observed. The transitions are split into multiplets due to the internal rotation of two methyl groups in acetone. Electronic structure calculations were done at the MP2 level of theory with the 6-311++g (2d, p) basis set for all atoms and the internal rotation barrier height of the methyl groups was determined to be about 2.8 kJ/mol. The *ab initio* rotational constants were the basis for our spectroscopic searches, but the multiplet structures and floppiness of the complex made the quantum number assignment very difficult. The assignment was finally achieved with the aid of constructing closed frequency loops and predicting internal rotation splittings using the XIAM code.[a] Analyses of the spectra yielded rotational and centrifugal distortion constants, as well as internal rotation parameters, which were interpreted in terms of structure and internal dynamics of the complex.

[a] H. Hartwig and H. Dreizler, Z. Naturforsch. A **51**, 923 (1996).

TG12 **4:49 – 5:04**

THE EFFECTS OF INTERNAL ROTATION AND ^{14}N QUADRUPOLE COUPLING IN N-METHYLDIACETAMIDE

RAPHAELA KANNENGIEßER, KONRAD EIBL, *Institute for Physical Chemistry, RWTH Aachen University, Aachen, Germany*; HA VINH LAM NGUYEN, *CNRS et Universités Paris Est et Paris Diderot, Laboratoire Interuniversitaire des Systèmes Atmosphériques (LISA), Créteil, France*; WOLFGANG STAHL, *Institute for Physical Chemistry, RWTH Aachen University, Aachen, Germany.*

Acetyl- and nitrogen containing substances play an important role in chemical, physical, and especially biological systems. This applies in particular for acetamides, which are structurally related to peptide bonds. In this work, N-methyldiacetamide, $CH_3N(COCH_3)_2$, was investigated by a combination of molecular beam Fourier transform microwave spectroscopy and quantum chemical calculations.

In N-methyldiacetamide, at least three large amplitude motions are possible: (1) the internal rotation of the methyl group attached to the nitrogen atom and (2, 3) the internal rotations of both acetyl methyl groups. This leads to a rather complicated torsional fine structure of all rotational transitions with additional quadrupole hyperfine splittings caused by the ^{14}N nucleus.

Quantum chemical calculations were carried out at the MP2/6-311++G(d,p) level of theory to support the spectral assignment. Conformational analysis was performed by calculating a full potential energy surface depending on the orientation of the two acetyl groups. This yielded three stable conformers with a maximum energy difference of 35.2 kJ/mol.

The spectrum of the lowest energy conformer was identified in the molecular beam. The quadrupole hyperfine structure as well as the internal rotation of two methyl groups could be assigned. For the N-methyl group and for one of the two acetyl methyl groups, barriers to internal rotation of 147 cm^{-1} and of 680 cm^{-1}, respectively, were determined. The barrier of the last methyl group seems to be so high that no additional splittings could be resolved.

Using the XIAM program, a global fit with a standard deviation on the order of our experimental accuracy could be achieved.

TG13 **5:06 – 5:21**

A NEW HYBRID PROGRAM FOR FITTING ROTATIONALLY RESOLVED SPECTRA OF METHYLAMINE-LIKE MOLECULES: APPLICATION TO 2-METHYLMALONALDEHYDE

ISABELLE KLEINER, *Laboratoire Interuniversitaire des Systèmes Atmosphériques (LISA), CNRS et Universités Paris Est et Paris Diderot, Créteil, France*; JON T. HOUGEN, *Sensor Science Division, National Institute of Standards and Technology, Gaithersburg, MD, USA.*

A new hybrid-model fitting program for methylamine-like molecules has been developed, based on an effective Hamiltonian in which the ammonia-like inversion motion is treated using a tunneling formalism, while the internal-rotation motion is treated using an explicit kinetic energy operator and potential energy function. The Hamiltonian in the computer program is set up as a 2x2 partitioned matrix, where each diagonal block consists of a traditional torsion-rotation Hamiltonian (as in the earlier program BELGI), and the two off-diagonal blocks contain all tunneling terms. This hybrid formulation permits the use of the permutation-inversion group G_6 (isomorphic to C_{3v}) for terms in the two diagonal blocks, but requires G_{12} for terms in the off-diagonal blocks. Our first application of the new program is to 2-methylmalonaldehyde. Microwave data for this molecule were previously fit (essentially to experimental measurement error) using an all-tunneling Hamiltonian formalism to treat both large-amplitude-motions [a]. For 2-methylmalonaldehyde, the hybrid program achieves a fit of nearly the same quality as that obtained by the all-tunneling program, but fits with the hybrid program eliminate a large discrepancy between internal rotation barriers in the OH and OD isotopologues of 2-methylmalonaldehyde that arose in fits with the all-tunneling program. Other molecules for application of the hybrid program will be mentioned.

[a] V.V. Ilyushin, E.A. Alekseev, Yung-Ching Chou, Yen-Chu Hsu, J. T. Hougen, F.J. Lovas, L. Picraux, J. Mol. Spectrosc. 251 (2008) 56-63

TG14 **5:23 – 5:38**

DETERMINATION OF TORSIONAL BARRIERS OF ITACONIC ACID AND N-ACETYLETHANOLAMINE USING CHIRPED-PULSED FTMW SPECTROSCOPY

JOSIAH R BAILEY, TIMOTHY J McMAHON, RYAN G BIRD, *Chemistry, University of Pittsburgh Johnstown, Johnstown, PA, USA*; DAVID PRATT, *Chemistry, University of Vermont, Burlington, VT, USA.*

The ground state rotational spectrum of itaconic acid (methylenesuccinic acid) and N-acetylethanolamine (AEA) have been collected and analyzed over the frequency range of 7-17.5 GHz. Both molecules displayed an unexpected tunneling splitting pattern caused by a V_2 and V_3 barriers, respectively. AEA's methyl rotor is directly connected to a carbonyl and is expected to have too high of a barrier to internal motion. Itaconic acid contains no methyl groups or any symmetry, yet a torsional splitting was observed. The origin of this motion as well their barrier heights and lowest energy conformations will be discussed.

TH. Radicals

Tuesday, June 23, 2015 – 1:30 PM

Room: B102 Chemical and Life Sciences

Chair: Bernadette M. Broderick, University of Georgia, Athens, GA, USA

TH01 **1:30 – 1:45**

AB INITIO SIMULATION OF THE PHOTOELECTRON SPECTRUM FOR METHOXY RADICAL

LAN CHENG, *Department of Chemistry, The University of Texas, Austin, TX, USA*; MARISSA L. WEICHMAN, JONGJIN B. KIM, *Department of Chemistry, The University of California, Berkeley, CA, USA*; TAKATOSHI ICHINO, *Department of Chemistry, The University of Texas, Austin, TX, USA*; DANIEL NEUMARK, *Department of Chemistry, The University of California, Berkeley, CA, USA*; JOHN F. STANTON, *Department of Chemistry, The University of Texas, Austin, TX, USA.*

A theoretical simulation of the photoelectron spectrum for the ground state of methoxy radical is reported based on the quasidiabatic model Hamiltonian originally proposed by Köppel, Domcke, and Cederbaum. The parameters in the model Hamiltonian have been obtained from ab initio coupled-cluster calculations. The linear and quadratic force constants have been calculated using equation-of-motion coupled-cluster ionization potential method with the singles, doubles, and triples (EOMIP-CCSDT) truncation scheme together with atomic natural orbital basis sets of triple-zeta quality (ANO1). The cubic and quartic force constants have been obtained from EOMIP-CCSD calculations with ANO basis sets of double-zeta quality (ANO0), and the spin-orbit coupling constant has been computed at the EOMIP-CCSD/pCVTZ level. The nuclear Schroedinger equation has been solved using the Lanzcos algorithm to obtain vibronic energy levels as well as the corresponding intensities. The simulated spectrum compares favorably with the recent high-resolution slow electron velocity-map imaging experiment for vibronic levels up to 2000 cm^{-1}.

TH02 **1:47 – 2:02**

JAHN-TELLER COUPLING IN THE METHOXY RADICAL: INSIGHTS INTO THE INFRARED SPECTRUM OF MOLECULES WITH VIBRONIC COUPLING

BRITTA JOHNSON, EDWIN SIBERT, *Department of Chemistry, The Univeristy of Wisconsin, Madison, WI, USA.*

The ground $\tilde{X}^2E$ vibrations of the methoxy radical have intrigued both experimentalists and theorists alike due to the presence of a conical intersection at the C_{3v} molecular geometry. This conical intersection causes methoxy's vibrational spectrum to be strongly influenced by Jahn-Teller coupling which leads to large amplitude vibrations and extensive mixing of the two lowest electronic states. The spectrum is further complicated due to spin-orbit and Fermi couplings. The standard diabatic normal mode quantum numbers are poor labels due to this vibronic mixing.

Using the potential energy force field and calculated spectra of the methoxy radical by Nagesh and Sibert[1] as a starting point, we look to develop a method for assigning states to a spectrum with vibronic coupling. We simplify the analysis by considering only the lowest two e modes of methoxy (the rock and the bend). When we include first-order Jahn-Teller coupling between these two modes in a new zero-order Hamiltonian, we are able to use an expanded version of the linear Jahn-Teller quantum numbers to assign the states.[2] This zeroth order representation is nontrivial; therefore, we study the properties of its eigenstates using correlation diagrams with respect to the strength of the Jahn-Teller coupling constant.

[1] Nagesh, J.; Sibert, E. L. *J. Phys. Chem. A* **2012**, *116*, 3846–3855.

[2] Barckholtz, T. A.; Miller, T. A. *Int. Revs. in Phys. Chem.* **1998**, *17*, 435–524.

TH03 **2:04 – 2:19**

RE-EVALUATION OF HO_3 STRUCTURE USING MILLIMETER-SUBMILLIMETER SPECTROSCOPY

LUYAO ZOU, BRIAN HAYS, SUSANNA L. WIDICUS WEAVER, *Department of Chemistry, Emory University, Atlanta, GA, USA.*

The HO_3 radical is of great interest in both atmospheric and astrophysical chemistry. However, its molecular structure has not been fully characterized by previous spectral studies. Microwave spectral studies on the *trans*-HO_3 conformer did not access higher K_a levels due to their limited frequency range. As a result, several centrifugal distortion constants could not be determined. We have therefore conducted spectroscopy of HO_3 in the millimeter and submillimeter ranges, from 70 to 450 GHz, under the guidance of a new fast sweep technique we developed for line searching. Large frequency shifts, primarily due to a large Δ_K centrifugal distortion constant, are observed compared to the spectral extrapolation from previous microwave studies. In addition, new spectral branches have been detected. The measured lines and preliminary spectral analysis will be presented, and the implications of these results will be discussed.

TH04 **2:21 – 2:36**

ON THE STARK EFFECT IN OPEN SHELL COMPLEXES EXHIBITING PARTIALLY QUENCHED ELECTRONIC ANGULAR MOMENTUM

GARY E. DOUBERLY, CHRISTOPHER P. MORADI, *Department of Chemistry, University of Georgia, Athens, GA, USA.*

The Stark effect is considered for polyatomic open shell complexes that exhibit partially quenched electronic angular momentum. Specifically, a zero-field model Hamiltonian is employed that accounts for the partial quenching of electronic orbital angular momentum in hydroxyl radical containing molecular complexes.[a,b] Spherical tensor operator formalism is employed to derive matrix elements of the Stark Hamiltonian in a parity conserving, Hund's case (a) basis for the most general case, in which the permanent dipole moment has projections on all three inertial axes of the system. Ro-vibrational transition intensities are derived, again for the most general case; namely, the laser polarization is projected onto axes parallel and perpendicular to the Stark electric field, and the transition dipole moment vector is projected onto all three inertial axes in the molecular frame. The model discussed here is compared to experimental spectra of OH-(C_2H_2), OH-(C_2H_4), and OH-(H_2O) complexes formed in He nanodroplets.

[a]M. D. Marshall and M. I. Lester, J. Chem. Phys. 121, 3019 (2004).

[b]G. E. Douberly, P. L. Raston, T. Liang, and M. D. Marshall, J. Chem. Phys. in press

TH05 **2:38 – 2:53**

INFRARED LASER SPECTROSCOPY AND AB INITIO COMPUTATIONS OF OH$\cdots$(D_2O)$_N$ COMPLEXES IN HELIUM NANODROPLETS

JOSEPH T. BRICE, CHRISTOPHER M. LEAVITT, CHRISTOPHER P. MORADI, GARY E. DOUBERLY, *Department of Chemistry, University of Georgia, Athens, GA, USA*; FEDERICO J HERNANDEZ, GUSTAVO A PINO, *INFIQC (CONICET – Universidad Nacional de Córdoba) Dpto. de Fisicoquímica – Facultad de Ciencias Quí, Universidad Nacional de Córdoba, Ciudad Universitaria, Córdoba, Argentina.*

OH$\cdots$(D_2O)$_N$ complexes are assembled in He droplets via the sequential pickup of D_2O molecules and the hydroxyl radical, which is formed via the pyrolytic decomposition of tert-butyl hydroperoxide. Bands due to clusters as large as N=4 are observed. Ro-vibrational spectroscopy of the binary complex reveals a vibrationally averaged C_{2v} structure. The effect of partially quenched electronic angular momentum in the complex is partially resolved in the rotational fine structure associated with the ν_1 OH stretch. Stark spectroscopy of this band reveals a permanent electric dipole moment for the binary complex equal to 3.70(5) Debye. OH stretch bands in larger clusters do not exhibit rotational fine structure; however, polarization spectroscopy of the OH$\cdots$(D_2O)$_2$ complex, when compared to predictions from *ab initio* computations, reveals two nearly isoenergetic isomers, both of which resemble the cyclic water trimer. Lower frequency OH stretch bands are assigned to cyclic tetramer and cyclic pentamer clusters on the basis of D_2O pressure dependence and *ab initio* frequency computations.

TH06 **2:55 – 3:10**

VIBRATIONAL-TORSIONAL COUPLING REVEALED IN THE INFRARED SPECTRUM OF HE-SOLVATED *n*-PROPYL RADICAL

CHRISTOPHER P. MORADI, BERNADETTE M. BRODERICK, JAY AGARWAL, HENRY F. SCHAEFER III., GARY E. DOUBERLY, *Department of Chemistry, University of Georgia, Athens, GA, USA.*

The *n*-propyl and *i*-propyl radicals were generated in the gas phase via pyrolysis of *n*-butyl nitrite ($CH_3(CH_2)_3ONO$) and *i*-butyl nitrite ($CH_3CH(CH_3)CH_2ONO$) precursors, respectively. Nascent radicals were promptly solvated by a beam of He nanodroplets, and the infrared spectra of the radicals were recorded in the C-H stretching region. In addition to three vibrations of *n*-propyl previously measured in an Ar matrix,[a] we observe many unreported bands between 2800 and 3150 cm^{-1}, which we attribute to propyl radicals. The C-H stretching modes observed above 2960 cm^{-1} for both radicals are in excellent agreement with anharmonic frequencies computed using VPT2. Between 2800 and 2960 cm^{-1}, however, the spectra of *n*-propyl and *i*-propyl radicals become quite congested and difficult to assign due to the presence of multiple anharmonic resonances. Computations reveal the likely origin of the spectral congestion to be strong coupling between the high frequency C-H stretching modes and a lower frequency torsional motion, which modulates quite substantially a through-space hyperconjugation interaction.

[a]Pacansky, et. al., J. Phys. Chem. 1977, 81, 2149.

TH07 **3:12 – 3:27**

VIBRONIC SPECTROSCOPY OF HETERO DIHALO-BENZYL RADICALS GENERATED BY CORONA DISCHARGE : JET-COOLED CHLOROFLUOROBENZYL RADICALS

YOUNG YOON, SANG LEE, *Department of Chemistry, Pusan National University, Pusan, Korea.*

The technique of corona excited supersonic jet expansion coupled with a pinhole-type glass nozzle was applied to vibronic spectroscopy of jet-cooled chlorofluorobenzyl radicals for the vibronic assignments and measurements of electronic energies of the $D_1 \rightarrow D_0$ transition. The vibronic emission spectra were recorded with a long-path monochromator in the visible region. The 2,3-, 2,4-,[a] and 2,5-[b]chlorofluorobenzyl radicals were generated by corona discharge of corresponding precursor molecules, chlorofluorotoluenes seeded in a large amount of helium carrier gas. The emission spectra show the vibronic bands originating from two benzyl-type radicals, chlorofluorobenzyl and fluorobenzyl benzyl radicals, in which fluorobenzyl radicals were obtained by displacement of Cl by H atom produced by the dissociation of methyl C-H bond. From an analysis of the spectra observed, we could determine the electronic energies in $D_1 \rightarrow D_0$ transition and vibrational mode frequencies at the D_0 state of chlorofluorobenzyl radicals which show the origin band of the electronic transition to be shifted to red region, comparing with the parental benzyl radical. The red-shift is highly sensitive to the number, position, and kind of substituents in chlorofluorobenzyl radicals. From the quantitative analysis of the red-shift, it has been found that the additivity rule, discovered recently by Lee group predicts the observation very well. In addition, the negligible contribution of the substituent at the 4-position, the nodal point of the Hückel's molecular orbital theory, can be well describes by the disconnection of substituent from molecular plane of the benzene ring available for delocalized π electrons. In this presentation, I will discuss the spectroscopic observation of new chlorofluorobenzyl radicals and substituent effect on electronic transition energy which is useful for identification of isomeric substituted benzyl radicals.

[a]C. S. Huh, Y. W. Yoon, and S. K. Lee, *J. Chem. Phys.* **136**, 174306 (2012).
[b]S. Y. Chae, Y. W. Yoon, and S. K. Lee, *Chem. Phys. Lett.* **612**, 134 (2014).

TH08 **3:29 – 3:44**

GROWING UP RADICAL: INVESTIGATION OF BENZYL-LIKE RADICALS WITH INCREASING CHAIN LENGTHS

JOSEPH A. KORN, KHADIJA M. JAWAD, DANIEL M. HEWETT, TIMOTHY S. ZWIER, *Department of Chemistry, Purdue University, West Lafayette, IN, USA.*

Combustion processes involve complex chemistry including pathways leading to polyaromatic hydrocarbons (PAHs) from small molecule precursors. Resonance stabilized radicals (RSRs) likely play an important role in the pathways to PAHs due to their unusual stability. Benzyl radical is a prototypical RSR that is stabilized by conjugation with the phenyl ring. Earlier work on α-methyl benzyl radical showed perturbations to the spectroscopy due to a hindered methyl rotor.[a] If the alkyl chain is lengthened then multiple conformations become possible. This talk will discuss the jet-cooled spectroscopy of α-ethyl benzyl radical and α-propyl benzyl radical produced from the discharge of 1-phenyl propanol and 1-phenyl butanol respectively. Electronic spectra were obtained via resonant two-photon ionization, and IR spectra were obtained by resonant ion-dip infrared spectroscopy.

[a]Kidwell, N. M.; Reilly, N. J.; Nebgen, B.; Mehta-Hurt, D. N.; Hoehn, R. D.; Kokkin, D. L.; McCarthy, M. C.; Slipchenko, L. V.; Zwier, T. S. The Journal of Physical Chemistry A 2013, 117, 13465.

Intermission

TH09 **4:03 – 4:18**

ANALYSIS OF ROTATIONALLY RESOLVED SPECTRA TO NON-DEGENERATE (a_1'') UPPER-STATE VIBRONIC LEVELS IN THE $\tilde{A}^2E'' - \tilde{X}^2A_2'$ ELECTRONIC TRANSITION OF NO_3

MOURAD ROUDJANE, TERRANCE JOSEPH CODD, *Department of Chemistry and Biochemistry, The Ohio State University, Columbus, OH, USA*; MING-WEI CHEN, *Department of Chemistry, University of Illinois at Urbana-Champaign, Urbana, IL, USA*; HENRY TRAN, DMITRY G. MELNIK, TERRY A. MILLER, *Department of Chemistry and Biochemistry, The Ohio State University, Columbus, OH, USA*; JOHN F. STANTON, *Department of Chemistry, The University of Texas, Austin, TX, USA.*

The vibronic structure of the $\tilde{A} - \tilde{X}$ electronic spectrum of NO_3 has been observed using both room-temperature and jet-cooled samples. A recent analysis of this structure is consistent with the Jahn-Teller effect (JTE) in the e' ν_3 vibrational mode (N-O stretch) being quite strong while the JTE in the e' ν_4 mode (O-N-O) bend) is rather weak. Electronic structure calculations qualitatively predict these results but the calculated magnitude of the JTE is quantitatively inconsistent with the spectral analysis.

Rotationally resolved spectra have been obtained for over a dozen vibronic bands of the $\tilde{A} - \tilde{X}$ electronic transition in NO_3. An analysis of these spectra should provide considerably more experimental information about the JTE in the $\tilde{A}$ state of NO_3 as the rotational structure should be quite sensitive to the geometric distortion of the molecule due to the JTE. This talk will focus upon the parallel bands, which terminate on $\tilde{A}$ state levels of a_1'' vibronic symmetry, which were the subject of a preliminary analysis reported at this meeting in 2014. We have now recorded the rotational structure of over a half-dozen parallel bands and have completed analysis on the 3_0^1 and $3_0^1 4_0^1$ transitions with several other bands being reasonably well understood. Two general conclusions emerge from this work. (i) All the spectral bands show evidence of perturbations which can reasonably be assumed to result from interactions of the observed $\tilde{A}$ state levels with high vibrational levels of the $\tilde{X}$ state. The perturbations range from severe in some bands to quite modest in others. (ii) Analyses of observed spectra, insofar as the perturbations permit, have all been performed with an oblate symmetric top model including only additional spin-rotation effects. This result is, of course, consistent with an effective, undistorted geometry for NO_3 of D_{3h} symmetry on the rotational timescale.

TH10 **4:20 – 4:35**

ANALYSIS OF ROTATIONALLY RESOLVED SPECTRA TO DEGENERATE (e′) UPPER-STATE VIBRONIC LEVELS IN THE $\tilde{A}^2E'' - \tilde{X}^2A'_2$ ELECTRONIC TRANSITION OF NO_3

HENRY TRAN, TERRY A. MILLER, *Department of Chemistry and Biochemistry, The Ohio State University, Columbus, OH, USA.*

The vibronic structure of the NO_3 radical in the $\tilde{A}$ state has been the subject of considerable research in our group and others worldwide. Recently we have collected high resolution, rotationally resolved cavity-ringdown spectra of a number of the vibronic bands terminating on levels of the $\tilde{A}^2E''$ state. Parallel bands to non-degenerate levels of a''_1 vibronic symmetry in the $\tilde{A}$ state, can mostly be satisfactorily fit using an oblate symmetric top Hamiltonian including the effects of spin rotation. The perpendicular bands, to levels of e' symmetry, are not as satisfactorily described using this Hamiltonian. In particular, the rotational structure of the e' levels has more transitions than the oblate top model predicts. For this reason we have developed a new rovibronic Hamiltonian capable of analyzing the vibronically degenerate levels. This Hamiltonian is based upon a D_{3h} configuration for NO_3 corresponding to rotation of an oblate symmetric top. Terms corresponding to coriolis, spin-rotation, spin-orbit, and Jahn-Teller distortions are then added. The simulations of the e' bands using this model show generally better agreement with the high resolution spectra. Our preliminary analysis indicates only modest effects on the rotational structure due to Jahn-Teller distortion. Details of the analysis of the e' bands, particularly 2^1_0, will be presented.

TH11 **4:37 – 4:52**

ROVIBRONIC VARIATIONAL CALCULATIONS OF THE NITRATE RADICAL

BRYAN CHANGALA, *JILA, National Institute of Standards and Technology and Univ. of Colorado Department of Physics, University of Colorado, Boulder, CO, USA*; JOSHUA H BARABAN, *Department of Chemistry, University of Colorado, Boulder, CO, USA*; JOHN F. STANTON, *Department of Chemistry, The University of Texas, Austin, TX, USA.*

In recent years, sophisticated diabatic Hamiltonians have been developed in order to understand the low-energy vibronic level structure of the nitrate radical (NO_3), which exhibits strong coupling between the X̃ and doubly degenerate B̃ states. Previous studies have reproduced the observed vibronic level positions up to 2000 cm^{-1} above the zero-point level, yet the rotational structure has remained uninvestigated with ab initio methods. In this talk, we present calculations of the $N \geq 0$ rovibronic structure of low-lying vibronic states of NO_3, in which complicated rovibrational and Coriolis interactions have been observed. Our results include calculations using both adiabatic and diabatic Hamiltonians, enabling a direct comparison between the two. We discuss extensions of our treatment to include spin-orbit and spin-rotation effects.

TH12 **4:54 – 5:09**

VIBRONIC STRUCTURE OF THE $\tilde{X}\ ^2A_2'$ STATE OF NO_3

MASARU FUKUSHIMA[a], *Information Sciences, Hiroshima City University, Hiroshima, Japan.*

We have measured dispersed fluorescence (DF) spectra from the single vibronic levels (SVL's) of the $\tilde{B}\ ^2E'$ state of jet cooled $^{14}NO_3$ and $^{15}NO_3$, and found a new vibronic band around the ν_1 fundamental[b]. This new band has two characteristics; (1) inverse isotope shift, and (2) unexpectedly strong intensity, i.e. comparable with that of the ν_1 fundamental. We concluded on the basis of the isotope effect that the terminated (lower) vibrational level of the new vibronic band should have vibrationally a_1' symmetry, and assigned to the third over-tone of the ν_4 asymmetric (e') mode, $3\nu_4$ (a_1'). We also assigned a weaker band at about 160 cm^{-1} above the new band to one terminated to $3\nu_4$ (a_2'). The $3\nu_4$ (a_1') and (a_2') levels are ones with $l = \pm 3$. Hirota proposed new vibronic coupling mechanism[c] which suggests that degenerate vibrational modes can induce electronic orbital angular momentum (L) even in non-degenerate electronic states. We interpret this as a sort of break-down of the Born-Oppenheimer approximation, and think that $\pm l$ induces $\mp\bar{\Lambda}$, where $\bar{\Lambda}$ expresses the pseudo-L; for the present system, one of the components of the third over-tone level, $|\Lambda = 0; v_4 = 3, l = +3\rangle$, can have contributions of $|\bar{\Lambda} = -1; v_4 = 3, l = +2\rangle$ and $|-2; 3, +1\rangle$. Under this interpretation, it is expected that there is sixth-order vibronic coupling, $(q_+^3 Q_-^3 + q_-^3 Q_+^3)$, between $|0; 3, +3\rangle$ and $|0; 3, -3\rangle$. The sixth-order coupling is weaker than the Renner-Teller term (the fourth-order term, $(q_+^2 Q_-^2 + q_-^2 Q_+^2)$), but stronger than the eighth-order term, $(q_+^4 Q_-^4 + q_-^4 Q_+^4)$. It is well known in linear molecules that the former shows huge separation, comparable with vibrational frequency, among the vibronic levels of Π electronic states, and the latter shows considerable splitting, $\sim$10 cm^{-1}, at Δ electronic states. Consequently, the $\sim$160 cm^{-1} splitting at v_4 = 3 is attributed to the sixth-order interaction. The relatively strong intensity for the band to $3\nu_4$ (a_1') can be interpreted as a part of the huge 0-0 band intensity, because the $3\nu_4$ (a_1') level, $|0; 3, \pm 3\rangle$, can connect with the vibrationless level, $|0; 0, 0\rangle$. $3\nu_4$ (a_1') has two-fold intensity because of the vibrational wavefunction, $|0; 3, +3\rangle + |0; 3, -3\rangle$, while negligible intensity is expected for $3\nu_4$ (a_2') with $|0; 3, +3\rangle - |0; 3, -3\rangle$ due to the cancellation. To confirm these interpretations, experiments on rotationally resolved spectra are underway.

[a]Author thanks T. Ishiwata and E. Hirota for their valuable discussion and support.
[b]M. Fukushima and T. Ishiwata, paper WJ03, ISMS2013, and paper MI17, ISMS2014.
[c]E. Hirota, *J.Mol.Spectrosc.*, in press.

TH13 **5:11 – 5:26**

HIGH-RESOLUTION LASER SPECTROSCOPY OF $^{14}NO_3$ RADICAL: VIBRATIONALLY EXCITED STATES OF THE B^2E' STATE

KOHEI TADA, *Graduate School of Science, Kobe University, Kobe, Japan*; SHUNJI KASAHARA, *Molecular Photoscience Research Center, Kobe University, Kobe, Japan*; TAKASHI ISHIWATA, *Information Sciences, Hiroshima City University, Hiroshima, Japan*; EIZI HIROTA, *The Central Office, The Graduate University for Advanced Studies, Hayama, Kanagawa, Japan.*

High-resolution fluorescence excitation spectra of $^{14}NO_3$ radical were intermittently recorded in the region 15860 cm^{-1} to 16050 cm^{-1} corresponding to the transitions to the vibrationally excited states of the B^2E' state. Well-separated rotational lines were found to disappear as the vibrational energy increases. The 16050 cm^{-1} region is almost unstructured even in the high-resolution measurement, and its rotational analysis is almost impossible. The rotational assignment of the 15870 cm^{-1} region is possible and it has been undertaken by the ground state combination differences and the Zeeman effect observation.

TH14 **5:28 – 5:43**

STRUCTURAL CHARACTERIZATION OF HYDROXYL RADICAL ADDUCTS IN AQUEOUS MEDIA

IRENEUSZ JANIK, G. N. R. TRIPATHI, *Radiation Laboratory, University of Notre Dame, Notre Dame, IN, USA.*

The oxidation by the hydroxyl (OH) radical is one of the most widely studied reactions because of its central role in chemistry, biology, organic synthesis, and photocatalysis in aqueous environments, wastewater treatment, and numerous other chemical processes. Although the redox potential of OH is very high, direct electron transfer (ET) is rarely observed. If it happens, it mostly proceeds through the formation of elusive OH adduct intermediate which facilitates ET and formation of hydroxide anion. Using time resolved resonance Raman technique we structurally characterized variety of OH adducts to sulfur containing organic compounds, halide ions as well as some metal cations. The bond between oxygen of OH radical and the atom of oxidized molecule differs depending on the nature of solute that OH radical reacts with. For most of sulfur containing organics, as well as halide and pseudo-halide ions, our observation suggested that this bond has two-center three-electron character. For several metal aqua ions studied, the nature of the bond depends on type of the cation being oxidized. Discussion on spectral parameters of all studied hydroxyl radical adducts as well as the role solvent plays in their stabilization will be presented.

TI. Dynamics/Kinetics/Ultrafast

Tuesday, June 23, 2015 – 1:30 PM

Room: 274 Medical Sciences Building

Chair: Patrick Vaccaro, Yale University, New Haven, CT, USA

TI01 **1:30 – 1:45**

MULTISCALE SPECTROSCOPY OF DIFFUSING MOLECULES IN CROWDED ENVIRONMENTS

AHMED A HEIKAL, *Chemistry and Biochemistry, University of Minnesota Duluth, Duluth, MN, USA.*

Living cells are known to be crowded with organelles, biomembranes, and macromolecules such as proteins, DNA, RNA, and actin filaments. It is believed that such macromolecular crowding affect biomolecular diffusion, protein-protein and protein-substrate interaction, and protein folding. In this contribution, I will discuss our recent results on rotational and translational diffusion of small and large molecules in crowded environments using time-resolved anisotropy and fluorescence correlation spectroscopy methods. In these studies, rhodamine green and enhanced green fluorescent protein are used as fluorescent probes diffusing in buffers enriched with biomimetic crowding agents such as Ficoll-70, bovine serum albumin (BSA), and ovalbumin. Controlled experiments on pure and glycerol-rich buffers were carried out as environments with variable, homogeneous viscosity. Our results indicate that the microviscosity differs from the corresponding bulk viscosity, depending on the nature of crowding agents (i.e., proteins versus polymers), the concentration of crowding agents and spatio-temporal scaling of our experimental approach. Our findings provide a foundation for fluorescence-based studies of diffusion and binding of biomolecules in the crowded milieu of living cells.

TI02 **1:47 – 1:57**

INVESTIGATING THE ROLE OF HUMAN SERUM ALBUMIN ON THE EXCITED STATE DYNAMICS OF INDOCYANINE GREEN USING SHAPED FEMTOSECOND LASER PULSES[a]

MUATH NAIRAT[b], ARKAPRABHA KONAR, MARIE KANIECKI, VADIM V. LOZOVOY, MARCOS DANTUS[c], *Department of Chemistry, Michigan State University, East Lansing, MI, USA.*

Differences in the excited state dynamics of molecules and photo-activated drugs either in solution or confined inside protein pockets or large biological macromolecules occur within the first few hundred femtoseconds. Shaped femtosecond laser pulses are used to probe the behavior of indocyanine green (ICG), the only Food and Drug administration (FDA) approved near-infrared dye and photodynamic therapy agent, while free in solution and while confined inside the pocket of the human serum albumin (HSA) protein. Experimental findings indicate that the HSA pocket hinders torsional motion and thus mitigates the triplet state formation in ICG. Low frequency vibrational motion of ICG is observed more clearly when it is bound to the HSA protein.

---

[a] Phys. Chem. Chem. Phys 17, 5872-5877 (2015)

[b] nairatmu@chemistry.msu.edu

[c] dantus@chemistry.msu.edu

TI03 **1:59 – 2:14**

ULTRAFAST SPECTROSCOPIC AND *AB INITIO* COMPUTATIONAL INVESTIGATIONS ON SOLVATION DYNAMICS OF NEUTRAL AND DEPROTONATED TYROSINE.

TAKASHIGE FUJIWARA, *Center for Photochemical Sciences, Bowling Green State University, Bowling Green, OH, USA*; MAREK Z. ZGIERSKI, *Steacie Laboratory, National Research Council of Canada, Ottawa, ON, Canada.*

We have studied one of the aromatic amino acids, tyrosine, regarding its photophysical properties in various solvent conditions by using a femtosecond fluorescence up-conversion technique and high-level TDDFT and CC2 computations. In this talk, profound details not only on ultrafast solvation dynamics on a neutral tyrosine in various solvents, but also on the excited-state dynamics for a single- (or doubly-) deprotonated tyrosine under various pH solutions will be presented. In high basicity, a tyrosine shows different absorption/emission spectra, and a total spectrum consists of a combination of these individual spectra that depend on the pH of the solution. The time scale of acid–base equilibrium is essential in solvation dynamics; whereas the protonation is simply controlled by diffusion, the de-protonation is considered to be slow process such that acid–base equilibrium may not be reached in the short-lived excited state after photo-excitation. Experimental and computational approaches taken and insights obtained in this concerted work will be described.

TI04 **2:16 – 2:31**

WHICH ELECTRONIC AND STRUCTURAL FACTORS CONTROL THE PHOTOSTABILITY OF DNA AND RNA PURINE NUCLEOBASES?

MARVIN POLLUM, CHRISTIAN REICHARDT, CARLOS E. CRESPO-HERNÁNDEZ, *Chemistry, Case Western Reserve, Cleveland, OH, USA*; LARA MARTÍNEZ-FERNÁNDEZ, INÉS CORRAL, *Departamento de Quimica, Universidad Autonoma de Madrid, Madrid, Spain*; CLEMENS RAUER, SEBASTIAN MAI, PHILIPP MARQUETAND, LETICIA GONZÁLEZ, *Institute for Theoretical Chemistry, University of Vienna, Vienna, Austria.*

Following ultraviolet excitation, the canonical purine nucleobases, guanine and adenine, are able to efficiently dissipate the absorbed energy within hundreds of femtoseconds. This property affords these nucleobases with great photostability. Conversely, non-canonical purine nucleobases exhibit high fluorescence quantum yields or efficiently populate long-lived triplet excited states from which chemistry can occur. Using femtosecond broadband transient absorption spectroscopy in combination with ab initio static and surface hopping dynamics simulations we have determined the electronic and structural factors that regulate the excited state dynamics of the purine nucleobase derivatives. Importantly, we have uncovered that the photostability of the guanine and adenine nucleobases is not due to the structure of the purine core itself and that the substituent at the C6 position of the purine nucleobase plays a more important role than that at the C2 position in the ultrafast relaxation of deleterious electronic energy. [The authors acknowledge the CAREER program of the National Science Foundation (Grant No. CHE-1255084) for financial support.]

TI05 **2:33 – 2:48**

ULTRAFAST DYNAMICS IN DNA AND RNA DERIVATIVES MONITORED BY BROADBAND TRANSIENT ABSORPTION SPECTRSCOPY

MATTHEW M BRISTER, *Department of Chemistry, Case Western Reserve University, Cleveland, OH, USA*; CARLOS E. CRESPO-HERNÁNDEZ, *Chemistry, Case Western Reserve, Cleveland, OH, USA.*

The ultrafast dynamics of nucleic acids have been under scrutiny for the past couple of decades because of the role that the high-energy electronic states play in mutagenesis and carcinogenesis. Kinetic models have been proposed, based on both experimental and theoretical discoveries. Direct experimental evidence of the intersystem crossing rate and population of the triplet state for most nucleic acid bases has yet to be reported, even though the triplet state is thought to be the most reactive species. Utilizing broadband femtosecond transient absorption spectroscopy, we reveal the time scale at which singlet-to-triplet population transfer occurs in several nucleic acid derivatives in the condensed phase. The implication of these results to the current understanding of the DNA and RNA photochemistry will be discussed. The authors acknowledge the CAREER program of the National Science Foundation (Grant No. CHE-1255084) for financial support.

TI06 **2:50 – 3:05**

CAN FEMTOSECOND TRANSIENT ABSORPTION SPECTROSCOPY PREDICT THE POTENTIAL OF SMALL MOLECULES AS PERSPECTIVE DONORS FOR ORGANIC PHOTOVOLTAICS?

REGINA DISCIPIO, GENEVIEVE SAUVE, *Department of Chemistry, Case Western Reserve University, Cleveland, OH, USA*; CARLOS E. CRESPO-HERNÁNDEZ, *Chemistry, Case Western Reserve, Cleveland, OH, USA.*

The utility of a perspective donor or acceptor molecule for photoelectric applications is difficult to predict a priori. This hinders productive synthetic exploration and necessitates lengthy device optimization procedures for reasonable estimation of said molecule's applicability. Using femtosecond broadband transient absorption spectroscopy, supported by time-dependent density functional theory computations and steady-state-absorption and emission spectroscopies, we have characterized a family of perspective optoelectronic compounds, in an effort to predict their relative performance in organic photovoltaic devices from information accrued from excited-state dynamics and photophysical properties.

A series of tetraphenylazadipyrromethene (ADP) complexes chelated with three different metal centers was investigated. We have determined that the chelating metal has little effect on the ground state properties of this family. However their excited state dynamics are strongly modulated by the metal. Specifically, the zinc-chelated ADP complex remains in the excited state tenfold longer than the cobalt or nickel complexes. We assert that this is key photophysical property that should make the zinc complex outperform the other two complexes in photovoltaic applications. This hypothesis is supported by preliminary power conversion efficiency results in devices.

TI07 **3:07 – 3:22**

MOLECULE-LIKE CdSe NANOCLUSTERS PASSIVATED WITH STRONGLY INTERACTING LIGANDS: ENERGY LEVEL ALIGNMENT AND PHOTOINDUCED ULTRAFAST CHARGE TRANSFER PROCESSES

YIZHOU XIE, *Department of Chemistry, University of Louisville, Louisville, KY, USA*; MEGHAN B TEUNIS, *Department of Chemistry, Indiana University-Purdue University Indianapolis, Indianapolis, IN, USA*; BILL PANDIT[a], *Department of Chemistry, University of Louisville, Louisville, KY, USA*; RAJESH SARDAR, *Department of Chemistry, Indiana University-Purdue University Indianapolis, Indianapolis, IN, USA*; JINJUN LIU, *Department of Chemistry, University of Louisville, Louisville, KY, USA*.

Semiconductor nanoclusters (SCNCs) are promising electronic materials for use in solid-state device fabrication, where device efficiency is strongly controlled by charge generation and transfer from SCNCs to their surroundings. In this paper we report the excited-state dynamics of molecule-like 1.6 nm diameter CdSe SCNCs, which are passivated with the highly conjugated ligand phenyldithiocarbamate (PDTC) or para-substituted PDTCs. Femtosecond transient absorption studies reveal sub-picosecond hole transfer ($\tau \approx 0.9$ ps) from a SCNC to its ligand shell based on strong electronic interaction and hole delocalization, and hot electron transfer ($\tau \approx 0.2$ ps) to interfacial states created by charge separation. A series of control experiments were performed by varying SCNC size (1.6 nm v.s. 2.9 nm) and photon energy of the pump laser (388 nm v.s. 490 nm), as well as addition of electron quencher (benzoquinone) and hole quencher (pyridine), which rules out alternative mechanisms and confirms the critical role of energy level alignment between the SCNC and its passivating ligands.

[a]Current address: Department of Chemistry, Northwestern University

TI08 **3:24 – 3:39**

TOWARD THE ACCURATE SIMULATION OF TWO-DIMENSIONAL ELECTRONIC SPECTRA

ANGELO GIUSSANI, ARTUR NENOV, JAVIER SEGARRA-MARTÍ, VISHAL K. JAISWAL, *Dipartimento di Chimica G. Ciamician, Università di Bologna, Bologna, Italy*; IVAN RIVALTA, ELISE DUMONT, *Laboratoire de Chimie, Ecole Normale Suprieure de Lyon, Lyon, FR*; SHAUL MUKAMEL, *Department of Chemistry, University of California, Irvine, Irvine, CA, USA*; MARCO GARAVELLI, *Dipartimento di Chimica G. Ciamician, Università di Bologna, Bologna, Italy*.

Two-dimensional pump-probe electronic spectroscopy is a powerful technique able to provide both high spectral and temporal resolution, allowing the analysis of ultrafast complex reactions occurring via complementary pathways by the identification of decay-specific fingerprints. [1-2] The understanding of the origin of the experimentally recorded signals in a two-dimensional electronic spectrum requires the characterization of the electronic states involved in the electronic transitions photoinduced by the pump/probe pulses in the experiment. Such a goal constitutes a considerable computational challenge, since up to 100 states need to be described, for which state-of-the-art methods as RASSCF and RASPT2 have to be wisely employed. [3] With the present contribution, the main features and potentialities of two-dimensional electronic spectroscopy are presented, together with the machinery in continuous development in our groups in order to compute two-dimensional electronic spectra. The results obtained using different level of theory and simulations are shown, bringing as examples the computed two-dimensional electronic spectra for some specific cases studied. [2-4]

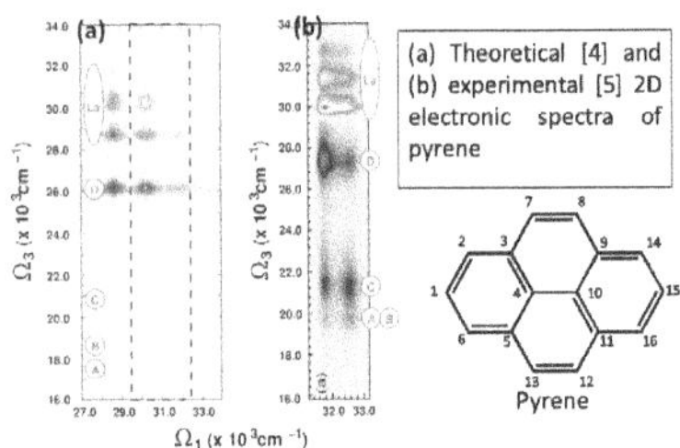

(a) Theoretical [4] and (b) experimental [5] 2D electronic spectra of pyrene

[1] Rivalta I, Nenov A, Cerullo G, Mukamel S, Garavelli M, Int. J. Quantum Chem., 2014, 114, 85 [2] Nenov A, Segarra-Martí J, Giussani A, Conti I, Rivalta I, Dumont E, Jaiswal V K, Altavilla S, Mukamel S, Garavelli M, Faraday Discuss. 2015, DOI: 10.1039/C4FD00175C [3] Nenov A, Giussani A, Segarra-Martí J, Jaiswal V K, Rivalta I, Cerullo G, Mukamel S, Garavelli M, J. Chem. Phys. submitted [4] Nenov A, Giussani A, Fingerhut B P, Rivalta I, Dumont E, Mukamel S, Garavelli M, Phys. Chem. Chem. Phys. Submitted [5] Krebs N, Pugliesi I, Riedle E, New J. Phys., 2013,15, 085016

Intermission

TI09 **3:58 – 4:13**

ULTRAFAST TERAHERTZ KERR EFFECT SPECTROSCOPY OF LIQUIDS AND BINARY MIXTURES

MARCO A. ALLODI, IAN A FINNERAN, GEOFFREY BLAKE, *Division of Chemistry and Chemical Engineering, California Institute of Technology, Pasadena, CA, USA.*

The ultrafast TeraHertz Kerr effect (TKE) has recently been demonstrated as a nonlinear spectroscopic technique capable of measuring the dielectric relaxation of liquids. The true power of this technique lies in its ability to provide complementary information to measurements taken using heterodyne-detected optical Kerr effect (OKE) spectroscopy. The optical pulses in OKE measurements interact with the sample via the molecular polarizability, a rank-two tensor, in contrast with THz pulses that interact with the molecules via the dipole moment, a rank-one tensor. Given the different light-matter interactions in the two techniques, TKE measurements help complete the physical picture of intermolecular interactions at short timescales.

We report here our implementation of heterodyne-detected TKE spectroscopy, along with measurements of pure liquids, and binary mixtures. Some of the liquids presented here were previously believed to be TKE inactive, thus showing that we have achieved a greater sensitivity than the previous implementation in the literature. In addition, we will discuss a variety of binary mixtures and show how the TKE data can be compared with OKE data to deepen our physical understanding of intermolecular interactions in liquids.

TI10 **4:15 – 4:30**

ULTRAFAST TERAHERTZ KERR EFFECT SPECTROSCOPY OF AROMATIC LIQUIDS

IAN A FINNERAN, MARCO A. ALLODI, GEOFFREY BLAKE, *Division of Chemistry and Chemical Engineering, California Institute of Technology, Pasadena, CA, USA.*

Ultrafast Terahertz Kerr Effect (TKE) spectroscopy is a relatively new nonlinear THz technique that is sensitive to the orientational dynamics of anisotropic, condensed-phase samples. The sample is excited by a single high field strength ~1 picosecond THz pulse, and the resulting transient birefringence is measured by a ~40 femtosecond 800 nm probe pulse. We have measured the TKE response of several aromatic liquids at room temperature, including benzene, benzene-d6, hexafluorobenzene, pyridine, and toluene. The measured decay constants range from ~1-10 ps, and, along with previous optical Kerr effect results in the literature[a], give insights into intermolecular interactions in these liquids.

[a] Loughnane et al. JPCB 110.11 (2006): 5708-5720.

TI11 **4:32 – 4:47**

VIBRATIONALLY-RESOLVED KINETIC ISOTOPE EFFECTS IN THE PROTON-TRANSFER DYNAMICS OF GROUND-STATE TROPOLONE

KATHRYN CHEW, ZACHARY VEALEY, PATRICK VACCARO, *Department of Chemistry, Yale University, New Haven, CT, USA.*

The vibrational and isotopic dependence of the hindered (tunneling-mediated) proton-transfer reaction taking place in the ground electronic state ($\tilde{X}^1A_1$) of monodeuterated tropolone (TrOD) has been explored under ambient (bulk-gas) conditions by applying two-color variants of resonant four-wave mixing (RFWM) spectroscopy in conjunction with polarization-resolved detection schemes designed to alleviate spectral complexity and facilitate rovibrational assignments. Full rotation-tunneling analyses of high-resolution spectral profiles acquired for the fundamental and first-overtone bands of a reaction-promoting O–D· · · O deformation/ring-breathing mode, $\nu_{36}(a_1)$, were performed, thereby extracting refined structural and dynamical information that affords benchmarks for the quantitative interpretation of tunneling-induced signatures found in long-range scans of $\tilde{X}^1A_1$ vibrational levels residing below $E_{vib}^{\tilde{X}} = 1700\ \mathrm{cm}^{-1}$. Observed kinetic isotope effects, which reflect changes in both reaction kinematics and vibrational displacements, will be discussed, with high-level quantum-chemical calculations serving to elucidate state-resolved propensities for proton transfer in TrOH and TrOD.

TI12 **4:49 – 5:04**

CHARACTERIZATION OF $CHBrCl_2$ PHOTOLYSIS BY VELOCITY MAP IMAGING

W G MERRILL, AMANDA CASE, *Department of Chemistry, The Univeristy of Wisconsin, Madison, WI, USA*; BENJAMIN C. HAENNI, *Department of Chemistry, University of Wisconsin–Madison, Madison, WI, USA*; ROBERT J. McMAHON, FLEMING CRIM, *Department of Chemistry, The Univeristy of Wisconsin, Madison, WI, USA.*

Halomethanes have attracted extensive research efforts of considerable variety, owing to their relative simplicity and ubiquitous presence in synthetic and environmental settings as well as their amenability to benchmark problems in physical chemistry. Their role in atmospheric processes is well known, most famously as the source of atomic halogens which catalyze the depletion of stratospheric ozone. Indeed, the photolytic cleavage of the carbon-halogen bond is the primary fate of halomethanes in the atmosphere. We utilize laser-induced photolysis to study the C-Br bond cleavage in $CHBrCl_2$ in a molecular beam. Atomic bromine fragments are probed with resonance enhanced multiphoton ionization (REMPI), which allows ground state and spin-excited products to be independently detected. Action spectroscopy in conjunction with velocity map imaging is used to determine the internal energy of the $CHCl_2$ partner fragment. Product state distributions as a function of photolysis energy may be discerned with these techniques. Current results will be presented.

TI13 **5:06 – 5:21**

REVERSIBILITY OF INTERSYSTEM CROSSING IN THE $\tilde{a}^1A_1(000)$ and $\tilde{a}^1A_1(010)$ STATES OF METHYLENE, CH_2

ANH T. LE, TREVOR SEARS[a], GREGORY HALL, *Chemistry Department, Brookhaven National Laboratory, Upton, NY, USA.*

The lowest energy singlet ($\tilde{a}^1A_1$) and triplet ($\tilde{X}^3B_1$) electronic states of methylene, CH_2, are only separated by 3150 cm^{-1}, but differ greatly in chemical reactivity. Overall methylene reaction rates and chemical behavior are therefore strongly dependent on collisionally-mediated singlet-triplet interconversion. Collisions with inert partners tend to depopulate the excited singlet state and populate vibrationally excited triplet levels in CH_2. This process is generally considered as irreversible for large molecules, however, this is not the case for small molecules such as CH_2. An investigation of the decay kinetics of CH_2 in the presence of argon and various amounts of oxygen has been carried out using transient frequency modulation (FM) absorption spectroscopy, to monitor *ortho* and *para* rotational levels in both the $\tilde{a}^1A_1(000)$ and $\tilde{a}^1A_1(010)$ states. In the $\tilde{a}^1A_1(000)$ state, all observed rotational levels follow double exponential decay kinetics, a direct consequence of reversible intersystem crossing. The relative amplitude of the slower decay component is an indicator of how quickly the reverse crossing from excited triplet levels becomes significant during the reaction and relaxation of singlet methylene. The *para* rotational levels show more obvious signs of reversibility than *ortho* rotational levels. Adding oxygen enhances the visibility of reversibility for both *ortho* and *para* levels. However, in the $\tilde{a}^1A_1(010)$ state where the FM signal is 5-10 times smaller than the $\tilde{a}^1A_1(000)$ state, there is no evidence of double exponential decay kinetics.

Acknowledgments: Work at Brookhaven National Laboratory was carried out under Contract No. DE-AC02-98CH10886 and DE-SC0012704 with the U.S. Department of Energy and supported by its Office of Basic Energy Sciences, Division of Chemical Sciences, Geosciences and Biosciences.

[a]Also, Chemistry Department, Stony Brook University, Stony Brook, New York 11794

TI14 **5:23 – 5:38**

EFFICIENT SUPER ENERGY TRANSFER COLLISIONS THROUGH REACTIVE-COMPLEX FORMATION: H + SO_2

JONATHAN M. SMITH, MICHAEL J. WILHELM, *Department of Chemistry, Temple University, Philadelphia, PA, USA*; JIANQIANG MA, *Chemistry, Columbia University, New York, New York, USA*; HAI-LUNG DAI, *Department of Chemistry, Temple University, Philadelphia, PA, USA.*

Translational-to-vibrational energy transfer (ET) from a hyperthermal H atom to ambient SO_2 was characterized using time-resolved Fourier transform infrared emission spectroscopy. Vibrational excitation of SO_2, following collisions with H atoms containing 59 kcal/mol of kinetic energy, generated from the 193 nm photolysis of HBr, is detected in two distinct energy distributions: one with excitation predominantly at the fundamental vibrational levels is attributable to classical impulsive collisions, while the other, accounting for 80% of the excited SO_2 with vibrational energy as high as 14,000 cm^{-1}, is proposed to arise from the formation of a transient reactive-complex during the collision. The cross-section for this super ET collision is determined to be 0.53±0.05 Å^2, or roughly 2% of all hard sphere collisions. This observation reveals that in collisions between a hyperthermal atom and an ambient molecule, for which a reactive-complex exists on the potential energy surface, a large quantity of translational energy can be transferred to the molecule with high efficiency.

TI15 **5:40 – 5:55**

FOURTH-ORDER VIBRATIONAL TRANSITION STATE THEORY AND CHEMICAL KINETICS

JOHN F. STANTON, *Department of Chemistry, The University of Texas, Austin, TX, USA*; DEVIN A. MATTHEWS, JUSTIN Z GONG, *Department of Chemistry and Biochemistry, The University of Texas, Austin, TX, USA.*

Second-order vibrational perturbation theory (VPT2) is an enormously successful and well-established theory for treating anharmonic effects on the vibrational levels of semi-rigid molecules. Partially as a consequence of the fact that the theory is exact for the Morse potential (which provides an appropriate qualitative model for stretching anharmonicity), VPT2 calculations for such systems with appropriate *ab initio* potential functions tend to give fundamental and overtone levels that fall within a handful of wavenumbers of experimentally measured positions. As a consequence, the next non-vanishing level of perturbation theory – VPT4 – offers only slight improvements over VPT2 and is not practical for most calculations since it requires information about force constants up through sextic. However, VPT4 (as well as VPT2) can be used for other applications such as the next vibrational correction to rotational constants (the "gammas") and other spectroscopic parameters. In addition, the marriage of VPT with the semi-classical transition state theory of Miller (SCTST) has recently proven to be a powerful and accurate treatment for chemical kinetics. In this talk, VPT4-based SCTST tunneling probabilities and cumulative reaction probabilities are give for the first time for selected low-dimensional model systems. The prospects for VPT4, both practical and intrinsic, will also be discussed.

TJ. Rydberg Atoms and Molecules

Tuesday, June 23, 2015 – 1:30 PM

Room: 217 Noyes Laboratory

Chair: Brian DeMarco, University of Illinois, Urbana, IL, USA

TJ01 **1:30 – 1:45**

PRECISION SPECTROSCOPY IN COLD MOLECULES: THE FIRST ROTATIONAL INTERVALS OF He_2^+ BY HIGH-RESOLUTION SPECTROSCOPY AND RYDBERG-SERIES EXTRAPOLATION

PAUL JANSEN, LUCA SEMERIA, SIMON SCHEIDEGGER, FREDERIC MERKT, *Laboratorium für Physikalische Chemie, ETH Zurich, Zurich, Switzerland.*

Having only three electrons, He_2^+ represents a system for which highly accurate *ab initio* calculations are possible. The latest calculation of rovibrational energies in He_2^+ do not include relativistic or QED corrections but claim an accuracy of about 120 MHz[a]. The available experimental data on He_2^+, though accurate to 300 MHz, are not precise enough to rigorously test these calculations or reveal the magnitude of the relativistic and QED corrections. We have performed high-resolution Rydberg spectroscopy of metastable He_2 molecules and employed multichannel-quantum-defect-theory extrapolation techniques[b] to determine the rotational energy-level structure in the He_2^+ ion. To this end we have produced samples of helium molecules in the $a^3\Sigma_u^+$ state in supersonic beams with velocities tunable down to 100 m/s by combining a cryogenic supersonic-beam source with a multistage Zeeman decelerator[c]. The metastable He_2 molecules are excited to np Rydberg states using the frequency doubled output of a pulse-amplified ring dye laser. Although the bandwidth of the laser systems is too large to observe the reduction of the Doppler width resulting from deceleration, the deceleration greatly simplifies the spectral assignments because of its spin-rotational state selectivity. Our approach enabled us to determine the rotational structure of He_2^+ with unprecedented accuracy, to determine the size of the relativistic and QED corrections by comparison with the results of Ref. *a* and to precisely measure the rotational structure of the metastable state for comparison with the results of Focsa *et al.*[d].

[a] W.-C. Tung, M. Pavanello, L. Adamowicz, *J. Chem. Phys.* **136**, 104309 (2012).
[b] D. Sprecher, J. Liu, T. Krähenmann, M. Schäfer, and F. Merkt, *J. Chem. Phys.* **140**, 064304 (2014).
[c] M. Motsch, P. Jansen, J. A. Agner, H. Schmutz, and F. Merkt, *Phys. Rev. A* **89**, 043420 (2014).
[d] C. Focsa, P. F. Bernath, and R. Colin, *J. Mol. Spectrosc.* **191**, 209 (1998).

TJ02 **1:47 – 2:02**

MICROWAVE SPECTROSCOPY OF THE CALCIUM $4snf \rightarrow 4s(n+1)d$, $4sng$, $4snh$, $4sni$, AND $4snk$ TRANSITIONS

JIRAKAN NUNKAEW, TOM GALLAGHER, *Department of Physics, The University of Virginia, Charlottesville, VA, USA.*

We use a delayed field ionization technique to observe the microwave transitions of calcium Rydberg states, from the $4snf$ states to the $4s(n+1)d$, $4sng$, $4snh$, $4sni$, and $4snk$ states for $18 \leq n \leq 23$. We analyze the observed intervals between the ℓ and $(\ell+1)$, $\ell \geq 5$, states of the same n to determine the Ca^+ $4s$ dipole and quadrupole polarizabilities. We show that the adiabatic core polarization model is not adequate to extract the Ca^+ $4s$ dipole and quadrupole polarizabilities and a non adiabatic treatment is required. We use the non adiabatic core polarization model to determine the ionic dipole and quadrupole polarizabilities to be $\alpha_d = 76.9(3)\ a_0^3$ and $\alpha_q = 206(9)\ a_0^5$, respectively.

TJ03 **2:04 – 2:14**

PHASE DEPENDENCE IN ABOVE THRESHOLD IONIZATION IN THE PRESENCE OF A MICROWAVE FIELD

VINCENT CARRAT, ERIC MAGNUSON, TOM GALLAGHER, *Department of Physics, The University of Virginia, Charlottesville, VA, USA.*

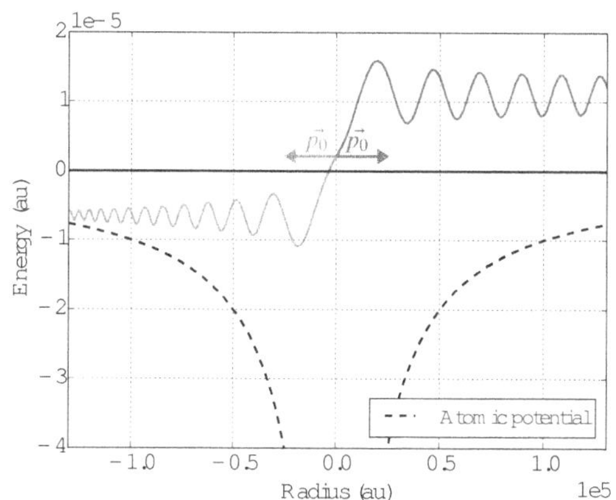

Figure 1: Electrons gaining or losing energy during the first microwave cycle depending of the initial launching direction. Here the phase of the microwave field is adjusted to provide a maximum energy transfer. The oscillations in energy are synchronized with the 14GHz microwave field.

Exciting an atom with high-frequency radiation in the presence of a low frequency field can result in energy transfer between the photoelectron and the low frequency field, depending on the phase of the low frequency field when the excitation occurs. We excite Li atoms with IR lasers in the presence of a microwave field. In a previous experiment, detection of highly excited states with excitation by a ps laser tuned above the limit clearly showed a phase dependence. The variation of the signal due to a phase change reach 0.1% of the total excitation in that case. We are using a new excitation scheme with a CW amplitude modulated laser, the modulation being phase locked to the microwaves. We now observe a signal variation of 10% of the total excitation. The ps pulses spreads the population over a broad energy spectrum while the modulated excitation keeps it in narrow bands. The modulated laser frequency can be tuned to couple one band to the highly excited states, enhancing the collection efficiency, additionally it is closer to the limit. Furthermore, the modulated laser allows the observation of phase dependent transfer to both higher and lower energies. The observations can be described with relatively simple models.

TJ04 **2:16 – 2:26**

MICROWAVE TRANSITIONS BETWEEN PAIR STATES COMPOSED OF TWO Rb RYDBERG ATOMS

JEONGHUN LEE, TOM GALLAGHER, *Department of Physics, The University of Virginia, Charlottesville, VA, USA.*

Microwave transitions between pair states composed of two Rb Rydberg atoms in a magneto-optical trap are investigated. Our current interest is the transition from ndnd to (n+1)d(n-2)f states. This transition is allowed because the dipole-dipole induced configuration interaction between the ndnd state and the energetically close (n+2)p(n-2)f state admixes some of the latter state into the former. The resonance frequencies of the ndnd-(n+1)d(n-2)f transitions for n=35 to 42 have been measured and found to agree well with the calculated values. In addition, the power shifts of the resonance frequencies have been measured for n=35 to 42. The dependence of the fractional population transfer from the ndnd to (n+1)d(n-2)f states on the microwave field strength and atomic density has been measured and can be compared to a simple theoretical model. This work has been supported by the Air Force Office of Scientific Research.

TJ05 **2:28 – 2:43**

HIGH-RESOLUTION SPECTROSCOPY OF LONG-RANGE MOLECULAR STATES OF $^{85}Rb_2$

RYAN CAROLLO, EDWARD E. EYLER, YOANN BRUNEAU, PHILLIP GOULD, W.C. STWALLEY, *Department of Physics, University of Connecticut, Storrs, CT, USA.*

We present analysis of low-n long-range molecular Rydberg states in $^{85}Rb_2$, based on high-resolution spectra. The weakly bound states are accessed by bound-bound transitions from high-v levels of the $a\,^3\Sigma_u^+$ state, which are prepared by photoassociation of laser-cooled atoms. Single-photon transitions to target states near the $5s + 7p$ asymptote are excited by a frequency-doubled pulse-amplified CW laser with a narrow linewidth, under 200 MHz. The long-range portion of the bonding potential is dominated by the elastic scattering interaction of the Rydberg electron of a perturbed $7p$ atom and a nearby ground-state atom, in much the same manner as trilobite states. We use time of flight to selectively measure molecular ions, which are formed via autoionization. This technique gives a two orders-of-magnitude improvement in linewidth over our previous work, reported in Ref. [1]. We also present calculations of a proposed scheme for STIRAP transfer from the current $v'' = 35$ level of the $a\,^3\Sigma_u^+$ state to the $v'' = 39$ level. The long-range states accessible to us are defined in large part by the Franck-Condon factors, which are dominated by the outer lobe of the wavefunction. Thus, choosing a v'' sets R, and determines the Franck-Condon window. The proposed $v'' = 39$ level has a classical outer turning point at $\sim 72\,a_0$, and will provide access to higher-n states with longer-range wells. This work is supported by the NSF and AFOSR.

[1] M. A. Bellos *et al.*, Phys. Rev. Lett. **111**, 053001 (2013)

TJ06 **2:45 – 3:00**

DOUBLE RESONANCE SPECTROSCOPY OF BaF AUTOIONIZING RYDBERG STATES

TIMOTHY J BARNUM, DAVID GRIMES, *Department of Chemistry, MIT, Cambridge, MA, USA*; YAN ZHOU, *JILA, National Institute of Standards and Technology and Univ. of Colorado Department of Physics, University of Colorado, Boulder, Boulder, CO, USA*; ROBERT W FIELD, *Department of Chemistry, MIT, Cambridge, MA, USA.*

We have studied the ν=1 Rydberg states of BaF in the energy region E=38800-39100 cm^{-1}(n*=15-25) via optical-optical double resonance spectroscopy. Rydberg states excited above the first ionization potential spontaneously autoionize and $^{138}Ba^{19}F^+$ ions are detected by TOF-MS. In addition, BaF possesses a particularly low ionization potential, which allows for the study of autoionization dynamics in the absence of predissociative decay. This work extends the assignments of core-penetrating Rydberg states of BaF (Jakubek and Field, 2000) for applications to state-selective ion production schemes. Polarization and Stark spectroscopy techniques will be discussed in the context of accurate and efficient assignment of spectra.

TJ07 **3:02 – 3:17**

MILLIMETER WAVE SPECTROSCOPY OF RYDBERG STATES OF MOLECULES IN THE REGION OF 260-295 GHz

DAVID GRIMES, *Department of Chemistry, MIT, Cambridge, MA, USA*; YAN ZHOU, *JILA, National Institute of Standards and Technology and Univ. of Colorado Department of Physics, University of Colorado, Boulder, Boulder, CO, USA*; TIMOTHY J BARNUM, ROBERT W FIELD, *Department of Chemistry, MIT, Cambridge, MA, USA.*

Free induction decay detected chirped pulse millimeter wave spectroscopy of Rydberg-Rydberg transitions in atoms and molecules is a powerful and flexible method for characterizing the electronic structure of Rydberg states and determining the structure and dynamics of the ion-core. Complicating the use of this technique are the difficulties in reliably and repeatedly accessing not just the most information rich core-nonpenetrating states, but also the low-ℓ core-penetrating Rydberg states in the area of principal quantum number $n^* > 35$. Small transition moments and narrow linewidths for transitions between valence electronic states and high Rydberg states are the primary limiting factor. We demonstrate a simple method to avoid the problem entirely by using chirped pulse technology operating in the frequency range of 260-295 GHz, which allows us to sample a lower range of n^* values than before with comparable frequency resolution and accuracy as our previous W-band experiments. Further improvements to our experiment in order to accurately capture details of Stark demolition, a technique that provides rapid differentiation between core-penetrating and core-nonpenetrating states, will also be discussed.

Intermission

TJ08 **3:36 – 3:51**

EFFECTIVE ION-IN-MOLECULE POTENTIALS FOR NON-PENETRATING RYDBERG STATES OF POLAR MOLECULES

STEPHEN COY, DAVID GRIMES, YAN ZHOU, ROBERT W FIELD, *Department of Chemistry, MIT, Cambridge, MA, USA*; BRYAN M. WONG, *Department of Chemistry, University of California, Riverside, Riverside, Ca, USA*.

Rydberg states of atoms or molecules for which the inner turning point of the Rydberg electron on the radial plus centrifugal potential lies outside the bulk of the ion core electron density are known as core-non-penetrating states. Interpretation of Rydberg spectroscopic data for polar molecules makes use of effective potentials that include ionic bonding and polarizability in order to represent electric properties of the ion core. We examine the accuracy and convergence properties of single-center polarization potentials and show that the center of charge representation, for which the core dipole moment is zero so that first-order l-mixing can be neglected, is excluded by the convergence sphere for use with l-states that can be treated by an expansion about the center or mass, the center of dipole or a newly-defined center of polarizability. The potential expansion converges only outside a sphere enclosing the charge distribution, and the sphere is much larger when the center of charge is used. For higher l-states of the rotating molecule (turning points defined in center of mass), the sphere required for convergence is much smaller for an origin within the charge distribution, so that lower l states are modeled correctly.

TJ09 **3:53 – 4:08**

ELECTRONIC STRUCTURE OF THE X $^1\Sigma^+$ ION CORE OF CaF RYDBERG STATES

STEPHEN COY, JOSHUA H BARABAN, DAVID GRIMES, TIMOTHY J BARNUM, ROBERT W FIELD, *Department of Chemistry, MIT, Cambridge, MA, USA*; BRYAN M. WONG, *Department of Chemistry, University of California, Riverside, Riverside, Ca, USA*.

We use ab-initio calculations to examine the electronic structure of CaF^+,making comparison to the available experimental data and effective potential models. An electron-density-difference plot comparing isolated Ca^{+2} and F^- ions with the CaF^+ ab-initio density shows s-d mixing at Ca, and maintenance of near spherical symmetry at F. This unexpected result is interpreted in terms of the electronic states of Ca^+. Calculation of the effective charge on F spanning the region of the transition from ionic to dissociating Ca^+ F^0 locates the transition very near the crossing of the Ca^{+2} F^- and Ca^+ F^0 curves and additionally determines the width of the ionic-bonding transition region. An accurate non-relativistic long or intermediate range effective potential for the CaF Rydberg electron is obtained by choice of origin at the center of polarizability, with inclusion of multipoles through octopole and the use of anisotropic polarizability. The estimates of CaF^+ polarizability from ab-initio and effective potential models predict high anisotropy, with the parallel dipole polarizability, where the atomic dipoles are mutually enhancing, predicted to be about double the perpendicular polarizability, where the atomic dipoles are mutually antagonistic.

TJ10 **4:10 – 4:25**

SYSTEMATICS OF RYDBERG SERIES OF DIATOMIC MOLECULES AND CORRELATION DIAGRAMS

CHUN-WOO LEE, *Chemistry, Ajou University, Suwon, Republic of Korea*.

Rydberg states are studied for H_2, Li_2, HeH, LiH and BeH using the multi-reference configuration interaction (MRCI) method. The systematics and regularities of the physical properties such as potential energies curves (PECs), quantum defect curves, permanent dipole moment and transition dipole moment curves of the Rydberg series are studied. They are explained using united atom perturbation theory by Bingel and Byers-Brown, Fermi model, Stark theory, and Mulliken's theory. Interesting mirror relationships of the dipole moments are observed between l-mixed Rydberg series, indicating that the members of the l-mixed Rydberg series have dipole moments with opposite directions, which are related to the reversal of the polarity of a dipole moment at the avoided crossing points. The assignment of highly excited states is difficult because of the usual absence of the knowledge on the behaviors of potential energy curves at small internuclear separation whereby the correlation between the united atom limit and separated atoms limit cannot be given. All electron MRCI calculations of PECs are performed to obtain the correlation diagrams between Rydberg orbitals at the united-atom and separated atoms limits.

TJ11 *INVITED TALK* **4:27 – 4:42**

OBSERVATION OF CS TRILOBITE MOLECULES WITH KILO-DEBYE MOLECULAR FRAME PERMANENT ELECTRIC DIPOLE MOMENTS

JAMES P SHAFFER, *Homer L Dodge Department of Physics and Astronomy, University of Oklahoma, Norman, OK, USA.*

We present results on Cs ultracold Rydberg atom experiments involving trilobite and butterfly molecules. Trilobite molecules are predicted to have giant, body-fixed permanent dipole moments, on the order of 1000 Debye. We present spectra for $nS_{1/2}+6S_{1/2}$ $^3\Sigma^+$ molecules, where n=37, 39 and 40, and measurements of the Stark broadenings of selected trilobite states in Cs due to the application of a constant external electric field. These results show that for Cs, because of its near integer s-state quantum defect, it is possible to photoassociate molecules whose wavefunction is predominantly of trilobite character yielding molecular frame dipole moments of around 2000 Debye. In addition, we have also recently observed states whose spectra show characteristics of p-wave dominated butterfly states. The work on what we believe to be the butterfly states will be compared and contrasted to the measurements of the trilobite states.

TJ12 **4:44 – 4:59**

MOLECULE FORMATION AND STATE-CHANGING COLLISIONS OF SINGLE RYDBERG ATOMS IN A BEC

KATHRIN SOPHIE KLEINBACH, MICHAEL SCHLAGMÜLLER, TARA CUBEL LIEBISCH, KARL MAGNUS WESTPHAL, FABIAN BÖTTCHER, ROBERT LÖW, SEBASTIAN HOFFERBERTH, TILMAN PFAU, *5. Physikalisches Institut, Universität Stuttgart, Stuttgart, Germany*; JESÚS PÉREZ-RÍOS, C. H. GREENE, *Department of Physics, Purdue University, West Lafayette, IN, USA.*

A single Rydberg excitation in the high-density and low-temperature environment of a Bose-Einstein condensate (BEC) leads to a fascinating testbed of low-energy electron-neutral and ion-neutral scattering. In particular the small interparticle spacing in a BEC makes it possible to study the role of ion-neutral interactions in l-changing collisions on time scales much shorter than the Rydberg lifetime. We take advantage of the mean field density shift, caused by elastic electron-neutral collisions, to probe density dependent shells of the ^{87}Rb BEC and thereby measure the l-changing collision time versus density and principal quantum number. We report on l-changing collisions due to inelastic scattering of the Rydberg electron with a neutral atom located near the Rydberg ionic core. We measure timescales of both the l-changing collision and the Rb_2 molecule formation of less than one microsecond for $n < 100$ at the highest BEC densities. We extract a change in kinetic energy of the Rydberg atoms that matches well with the energy gap to the next-lowest manifold. We measure Rb_2 signal that decreases with increasing principal quantum number. The mechanism and timescales of the l-changing collision are compared with simulations including the motion of the ionic core and neutral atoms, as well as the Rydberg electron.

TJ13 ***Post-Deadline Abstract*** **5:01 – 5:16**

RYDBERG, VALENCE AND ION-PAIR QUINTET STATES OF O_2

GABRIEL J. VAZQUEZ, *Instituto de Ciencias Fisicas, Universidad Nacional Autonoma de Mexico (UNAM), Cuernavaca, Morelos, Mexico*; HANS P. LIEBERMANN, *Fachbereich C-Physikalische und Theoretische Chemie, Bergische Universität Wuppertal, Wuppertal, Germany*; H. LEFEBVRE-BRION, *Institut des Sciences Moléculaires d'Orsay, Université Paris-Sud, Orsay, France.*

We carried out a relatively comprehensive ab–initio study of the electronic structure of O_2 and O_2^+. We employed the MRD–CI package together with the cc–pV4Z basis set augmented with seven diffuse functions of s, p and d character on each atom. In this contribution we focus on the quintet states. Potential energy curves of about 50 quintet states were computed. The spectroscopic constants of the six valence quintet states ($^5\Sigma_g^+, ^5\Sigma_g^-, ^5\Sigma_u^-, ^5\Pi_u, ^5\Pi_g, ^5\Delta_g$) dissociating to the first dissociation limit O(^{3}P)+O(^{3}P) are reported. The four ion–pair quintet states ($^5\Sigma_g^-, ^5\Sigma_u^-, ^5\Pi_g, ^5\Pi_u$) dissociating into $O^+(^4S)+O^-(^2P)$ at 17.28 eV were also computed and their spectroscopic constants will be presented. A number of bound quintet Rydberg states belonging to series converging to the $a^4\Pi_u$, $b^4\Sigma_g^-$, $f^4\Pi_g$ and $^6\Sigma_u^+$ states of the O_2^+ cation were identified and attributed. Long–range interactions involving the ion–pair states as they slowly approach their dissociation limit will be shown.

TJ14 *Post-Deadline Abstract* **5:18 – 5:33**

AB INITIO STUDY OF THE H, J, I, I′ AND I″ $^3\Pi_u$ SUPEREXCITED STATES OF O_2

GABRIEL J. VAZQUEZ, *Instituto de Ciencias Fisicas, Universidad Nacional Autonoma de Mexico (UNAM), Cuernavaca, Morelos, Mexico*; HANS P. LIEBERMANN, *Fachbereich C-Physikalische und Theoretische Chemie, Bergische Universität Wuppertal, Wuppertal, Germany*; H. LEFEBVRE-BRION, *Institut des Sciences Moléculaires d'Orsay, Université Paris-Sud, Orsay, France.*

In this presentation we report progress in the computation of superexcited states of O_2, namely, of bound $^3\Pi_u$ Rydberg states of the neutral molecule converging to the $a^4\Pi_u$ state of O_2^+. Up to twenty $^3\Pi_u$ potential energy curves were computed. The MRD–CI package together with the cc–pV4Z basis set augmented with seven diffuse functions of s, p and d type on each atom were employed. This study was prompted by the demand of potential curves to try to understand the mechanism of the neutral dissociation of O_2 above the first ionization limit (IP= 12.07 eV) where there exists a competition between autoionization and predissociation. This undertaking focuses on the computation of the I, I′ and I″ $^3\Pi_u$ states that have been postulated as involved in the neutral dissociation of O_2 in the 865–790 Å (14.33–15.69 eV) energy region.

WA. Plenary

Wednesday, June 24, 2015 – 8:30 AM

Room: Foellinger Auditorium

Chair: Leslie Looney, University of Illinois, Urbana, IL, USA

RAO AWARDS **8:30**
Presentation of Awards by Yunjie Xu, University of Alberta

2014 Rao Award Winners
Grant Buckingham, University of Colorado
Kathryn Chew, Yale University
Yu-Hsuan Huang, National Chiao Tung University

MILLER PRIZE **8:40**
Introduction by Mike Heaven, Emory University

WA01 **8:45 – 9:00**

INFRARED LASER STARK SPECTROSCOPY OF THE OH···CH_3OH COMPLEX ISOLATED IN SUPERFLUID HELIUM DROPLETS

CHRISTOPHER M. LEAVITT, JOSEPH T. BRICE, GARY E. DOUBERLY, *Department of Chemistry, University of Georgia, Athens, GA, USA*; FEDERICO J HERNANDEZ, GUSTAVO A PINO, *INFIQC (CONICET – Universidad Nacional de Córdoba) Dpto. de Fisicoquímica – Facultad de Ciencias Quí, Universidad Nacional de Córdoba, Ciudad Universitaria, Córdoba, Argentina.*

The elimination of volatile organic compounds (VOCs) from the atmosphere is initiated by reactions with OH, NO_3 and O_3.[a,b] For oxygenated VOCs, such as alcohols, ketones, ethers, etc., reactions occur nearly exclusively with the hydroxyl radical. Furthermore, the potential energy surfaces associated with reactions between OH and oxygenated VOCs generally feature a pre-reactive complex, stabilized by hydrogen bonding, which results in rate constants that exhibit large negative temperature dependencies.[c] This was explicitly demonstrated recently for the OH + methanol (MeOH) reaction, where the rate constant increased by nearly two orders of magnitude when the temperature decreased from 200 K to below 70 K, highlighting the potential impact of this reaction in the interstellar medium (ISM).[d,e] In this study, we trap this postulated pre-reactive complex formed between OH and MeOH using He nanodroplet isolation (HENDI) techniques, and probe this species using a combination of mass spectrometry and infrared laser Stark spectroscopy.

[a] Atkinson, R.; Arey, J., Chem. Rev. 2003, 103, 4605-4638.
[b] Mellouki, A.; Le Bras, G.; Sidebottom, H., Chem. Rev. 2003, 103, 5077-5096.
[c] Smith, I. W. M.; Ravishankara, A. R., J. Phys. Chem. A 2002, 106, 4798-4807
[d] Shannon, R. J.; Blitz, M. A.; Goddard, A.; Heard, D. E., Nat. Chem. 2013, 5, 745-749.
[e] Martin, J. C. G.; Caravan, R. L.; Blitz, M. A.; Heard, D. E.; Plane, J. M. C., J. Phys. Chem. A 2014, 118, 2693-2701.

FLYGARE AWARDS **9:05**
Introduction by Trevor Sears, Brookhaven National Laboratory

WA02 **9:10 – 9:25**

WHAT CAN WE EXPECT OF HIGH-RESOLUTION SPECTROSCOPIES ON CARBOHYDRATES?

EMILIO J. COCINERO, PATRICIA ECIJA, ICIAR URIARTE, IMANOL USABIAGA, JOSÉ A. FERNÁNDEZ, FRANCISCO J. BASTERRETXEA, *Departamento de Química Física, Universidad del País Vasco (UPV-EHU), Bilbao, Spain*; ALBERTO LESARRI, *Departamento Química Física y Química Inorgánica , Universidad de Valladolid, Valladolid, Spain*; BENJAMIN G. DAVIS, *Department of Chemistry, Oxford University, Oxford, United Kingdom.*

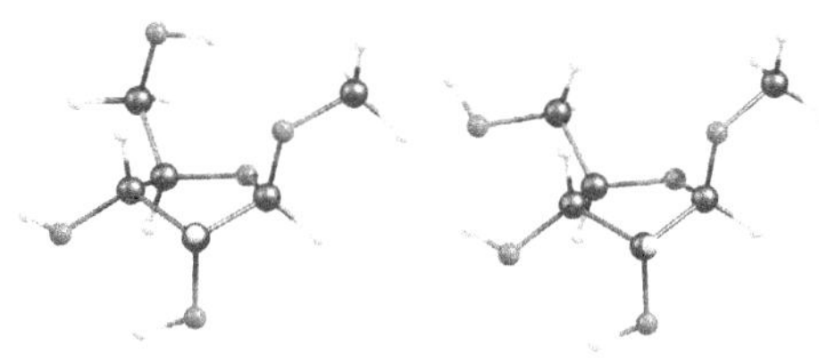

Carbohydrates are one of the most multifaceted building blocks, performing numerous roles in living organisms. We present several structural investigations on carbohydrates exploiting an experimental strategy which combines microwave (MW) and laser spectroscopies in high-resolution. Laser spectroscopy offers high sensitivity coupled to mass and conformer selectivity, making it ideal for polysaccharides studies. On the other hand, microwave spectroscopy provides much higher resolution and direct access to molecular structure of monosaccharides. This combined approach provides not only accurate chemical insight on conformation, structure and molecular properties, but also benchmarking standards guiding the development of theoretical calculations.

In order to illustrate the possibilities of a combined MW-laser approach we present results on the conformational landscape and structural properties of several monosaccharides[a,b] and oligosaccharides including microsolvation and molecular recognition processes of carbohydrates.[c,d]

---

[a] E.J. Cocinero, A. Lesarri, P. Écija, F.J. Basterretxea, J.-U. Grabow, J.A. Fernández and F. Castaño *Angew. Chem. Int. Ed.* **51**, 3119-3124, 2012.

[b] E.J. Cocinero, A. Lesarri, P. Écija, Á. Cimas, B.G. Davis, F.J. Basterretxea, J.A. Fernández and F. Castaño *J. Am. Chem. Soc.* **135**, 2845-2852, 2013.

[c] E.J. Cocinero, P. Çarçabal, T.D. Vaden, J.P. Simons and B.G. Davis *Nature* **469**, 76-80, 2011.

[d] C.S. Barry, E.J. Cocinero, P. Çarçabal, D.P. Gamblin, E.C. Stanca-Kaposta, S. M. Fernández-Alonso, S. Rudić, J.P. Simons and B.G. Davis *J. Am. Chem. Soc.* **135**, 16895-16903, 2013.

WA03 **9:30 – 9:45**

CONSTRUCTION OF POTENTIAL ENERGY SURFACES FOR THEORETICAL STUDIES OF SPECTROSCOPY AND DYNAMICS

RICHARD DAWES, *Department of Chemistry, Missouri University of Science and Technology, Rolla, MO, USA.*

Accurate potential energy surfaces (PESs) combined with methods to solve the Schrödinger equation for the nuclei permit the prediction and interpretation of various types of molecular spectra and/or dynamics.

Part of this talk describes the development of a PES generator (software code) which uses parallel processing on High-Performance Computing (HPC) clusters to construct PESs automatically. Thousands of ab initio data are computed at geometries chosen by an algorithm and fit to a functional form. This strategy is particularly successful when the electronic structure is robustly convergent (such as vdWs systems composed of two closed-shell monomers). Results for a few of such systems [e.g., $(CO)_2$, $(NNO)_2$, CO_2-CS_2, $(OCS)_2$] will be presented.

The electronic structure of molecules is difficult to describe continuously across global reactive PESs since it changes qualitatively as bonds are formed and broken along reaction coordinates. I will discuss a high-level ab initio method (GDW-SA-CASSCF/MRCI) designed to allow the electronic wavefunction to smoothly evolve across the PES and provide an accurate and balanced description of the various regions. These methods are combined to study a number of small gas-phased molecules from the areas of atmospheric, combustion and interstellar chemistry including a large variational calculation of all the bound vibrational states of ozone and the photodissociation dynamics of the simplest Criegee intermediate (CH_2OO).

Intermission

WA04 **10:35 – 10:50**

MILLIMETER AND SUBMILLIMETER STUDIES OF O(^{1}D) INSERTION REACTIONS TO FORM MOLECULES OF ASTROPHYSICAL INTEREST

BRIAN HAYS, NADINE WEHRES, BRIDGET ALLIGOOD DEPRINCE, ALTHEA A. M. ROY, JACOB LAAS, SUSANNA L. WIDICUS WEAVER, *Department of Chemistry, Emory University, Atlanta, GA, USA.*

While both the number of detected interstellar molecules and their chemical complexity continue to increase, understanding of the processes leading to their formation is lacking. Our research group combines laboratory spectroscopy, observational astronomy, and astrochemical modeling for an interdisciplinary examination of the chemistry of star and planet formation. This talk will focus on our laboratory studies of O(^{1}D) insertion reactions with organic molecules to produce molecules of astrophysical interest. By employing these reactions in a supersonic expansion, we are able to produce interstellar organic reaction intermediates that are unstable under terrestrial conditions; we then probe the products using millimeter and submillimeter spectroscopy. We benchmarked this setup using the well-studied O(^{1}D) + methane reaction to form methanol. After optimizing methanol production, we moved on to study the O(^{1}D) + ethylene reaction to form vinyl alcohol (CH_2CHOH), and the O(^{1}D) + methyl amine reaction to form aminomethanol (NH_2CH_2OH). Vinyl alcohol measurements have now been extended up to 450 GHz, and the associated spectral analysis is complete. A possible detection of aminomethanol has also been made, and continued spectral studies and analysis are underway. We will present the results from these experiments and discuss future applications of these molecular and spectroscopic techniques.

WA05 **10:55 – 11:10**

TERAHERTZ AND INFRARED LABORATORY SPECTROSCOPY IN SUPPORT OF NASA MISSIONS

SHANSHAN YU, *Jet Propulsion Laboratory, California Institute of Technology, Pasadena, CA, USA.*

The JPL molecular spectroscopy group supports NASA programs encompassing Astrophysics, Atmospheric Science, and Planetary Science. Ongoing activities include measurement and analysis of molecular spectra in the terahertz and infrared regions under conditions akin to the remote environments under study in NASA missions. This presentation will show the implementation of state-of-the-art spectroscopic techniques to fulfill spectroscopic demands of the Herschel Space Observatory[a] and the Orbiting Carbon Observatory re-flight (OCO-2)[b].

A demonstrative example of the significantly improved frequency predictions for the H_3O^+ ground state high-J transitions will be given. This work was critical to Herschel's successful identification of highly excited metastable H_3O^+ Terahertz lines with $J = K$ up to 11, one of the Herschel mission's many surprising observational results. The observation and subsequent laboratory work revealed that (1) these highly excited H_3O^+ lines had already been observed by European Southern Observatory's Atacama Pathfinder Experiment telescope a few years before but had been classified as U-lines; (2) the H_3O^+ number density was previously underestimated by an order of magnitude, due to ignorance of the population in the metastable states.

A second example focuses on O_2, an important absorber from the microwave through the deep UV. This work is motivated by the challenge of developing an accurate and complete spectroscopic characterization of molecular oxygen across a wide frequency range for current and planned Earth atmospheric observations. Especially, OCO-2 utilizes the O_2 A-band for air mass calibration; extremely accurate O_2 molecular data, i.e., line positions with uncertainty on the order of MHz for the A-band around 13000 cm^{-1}, are required to fulfill the demand of the proposed 0.25% precision for the carbon dioxide concentration retrievals.

[a]G. Pilbratt, J. Riedinger, T. Passvogel, G. Crone, D. Doyle, U. Gageur et al. A&A, 518, L1 (2010).

[b]D. Crisp, B.M Fisher, C. O'Dell, C. Frankenberg, R. Basilio, H. Boesch et al., Atmos. Meas. Tech. 5, 687-707 (2012).

COBLENTZ AWARD **11:15**
Presentation of Award by Mark Druy, Coblentz Society

WA06 *Coblentz Society Award Lecture* **11:20 – 12:00**

LASER SPECTROSCOPY OF RADICALS, CARBENES, AND IONS IN SUPERFLUID HELIUM DROPLETS

GARY E. DOUBERLY, *Department of Chemistry, University of Georgia, Athens, GA, USA.*

The first beam of helium droplets was reported in the 1961 paper *Strahlen aus kondensiertem Helium im Hochvakuum* by Von E. W. Becker and co-workers.[a] However, molecular spectroscopy of helium-solvated dopants wasn't realized until 30 years later in the laboratories of Scoles and Toennies.[b,c] It has now been two decades since this early, seminal work on doped helium droplets, yet the field of helium droplet spectroscopy is still fresh with vast potential. Analogous in many ways to cryogenic matrix isolation spectroscopy, the helium droplet is an ideal environment to spectroscopically probe difficult to prepare molecular species, such as radicals, carbenes and ions. The quantum nature of helium at 0.35 K often results in molecular spectra that are sufficiently resolved to evoke an analysis of line shapes and fine-structure that is worthy of the International Symposium on Molecular Spectroscopy. The present talk will focus on our recent successful attempts to efficiently dope the title molecular species into helium droplets and probe their properties with infrared laser Stark and Zeeman spectroscopies.

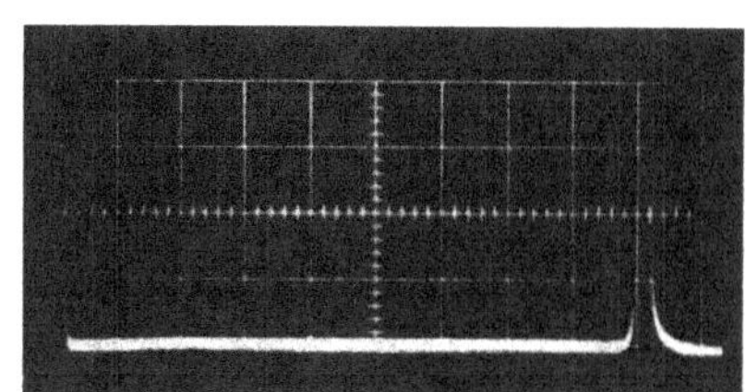

Abb. 1. Laufzeit-Oszillogramm eines im Hochvakuum laufenden kondensierten Heliumstrahls. Die Breite eines Rasterfeldes entspricht 200 μs.

[a] E. W. Becker, R. Klingelhöfer, P. Lohse, Z. *Naturforsch. A* **16A**, 1259 (1961).

[b] S. Goyal, D. L. Schutt, G. Scoles, *Phys. Rev. Lett.* **69**, 933 (1992).

[c] M. Hartmann, R. E. Miller, J. P. Toennies, A. F. Vilesov, *Phys. Rev. Lett.* **75**, 1566 (1995).

WF. Mini-symposium: High-Precision Spectroscopy

Wednesday, June 24, 2015 – 1:30 PM

Room: 116 Roger Adams Lab

Chair: Kevin Cossel, JILA - University of Colorado, Boulder, CO, USA

WF01 *INVITED TALK* **1:30 – 2:00**

ULTRASENSITIVE, HIGH ACCURACY MEASUREMENTS OF TRACE GAS SPECIES

DAVID A. LONG, ADAM J. FLEISHER, *Material Measurement Laboratory, National Institute of Standards and Technology, Gaithersburg, MD, USA*; DAVID F. PLUSQUELLIC, *Physical Measurement Laboratory, National Institute of Standards and Technology, Boulder, CO, USA*; JOSEPH HODGES, *Material Measurement Laboratory, National Institute of Standards and Technology, Gaithersburg, MD, USA.*

Our laboratory seeks to apply novel cavity-enhanced spectroscopic techniques to present problems in atmospheric and physical chemistry. Primarily we use cavity ring-down spectroscopy in which the passive decay of optical power within a Fabry-Pérot resonator is utilized to extract an absorption signal. With this technique we have demonstrated quantum (shot) noise limited sensitivities in both the near-infrared and mid-infrared spectral regions. Both commercial and home-built optical frequency combs are employed either to serve as absolute frequency references for molecular spectra or in a multiheterodyne approach for multiplexed sensing. I will discuss this novel instrumentation as well as measurements we have made of atmospherically relevant species such as CO_2, H_2O, O_2, CH_4, and CO with implications for *in situ* and remote (i.e. satellite-based) sensing. I will conclude by discussing future directions and plans for challenging measurements in the mid-infrared.

WF02 **2:05 – 2:20**

PROBING BUFFER-GAS COOLED MOLECULES WITH DIRECT FREQUENCY COMB SPECTROSCOPY IN THE MID-INFRRARED

BEN SPAUN, BRYAN CHANGALA, BRYCE J BJORK, OLIVER H HECKL, *JILA, National Institute of Standards and Technology and Univ. of Colorado Department of Physics, University of Colorado, Boulder, CO, USA*; DAVID PATTERSON, JOHN M. DOYLE, *Department of Physics, Harvard University, Cambridge, MA, USA*; JUN YE, *JILA, National Institute of Standards and Technology and Univ. of Colorado Department of Physics, University of Colorado, Boulder, CO, USA.*

We present the first demonstration of cavity-enhanced direct frequency comb spectroscopy[a] on buffer-gas cooled molecules[b]. By coupling a mid-infrared frequency comb to a high-finesse cavity surrounding a helium buffer-gas chamber, we can gather rotationally resolved absorption spectra with high sensitivity over a broad wavelength region. The measured $\sim$10 K rotational and translational temperatures of buffer-gas cooled molecules drastically simplify the observed spectra, compared to those of room temperature molecules, and allow for high spectral resolution limited only by Doppler broadening (10-100 MHz). Our system allows for the extension of high-resolution spectroscopy to larger molecules, enabling detailed analysis of molecular structure and dynamics, while taking full advantage of the powerful optical properties of frequency combs.

[a] A. Foltynowicz *et al.* Cavity-enhanced optical frequency comb spectroscopy in the mid-infrared application to trace detection of hydrogen peroxide. Applied Physics B, vol. 110, pp. 163–175, 2013.

[b] D. Patterson and J. M. Doyle. Cooling molecules in a cell for FTMW spectroscopy. Molecular Physics 110, 1757–1766, 2012.

WF03 **2:22 – 2:37**

FREQUENCY-AGILE DIFFERENTIAL CAVITY RING-DOWN SPECTROSCOPY

ZACHARY REED, JOSEPH HODGES, *Chemical Sciences Division, National Institute of Standards and Technology, Gaithersburg, MD, USA.*

The ultimate precision of highly sensitive cavity-enhanced spectroscopic measurements is often limited by interferences (etalons) caused by weak coupled-cavity effects. Differential measurements of ring-down decay constants have previously been demonstrated to largely cancel these effects, but the measurement acquisition rates were relatively low [1,2]. We have previously demonstrated the use of frequency agile rapid scanning cavity ring-down spectroscopy (FARS-CRDS) for acquisition of absorption spectra [3]. Here, the method of rapidly scanned, frequency-agile differential cavity ring-down spectroscopy (FADS-CRDS) is presented for reducing the effect of these interferences and other shot-to-shot statistical variations in measured decay times. To this end, an electro-optic phase modulator (EOM) with a bandwidth of 20 GHz is driven by a microwave source, generating pairs of sidebands on the probe laser. The optical resonator acts as a highly selective optical filter to all laser frequencies except for one tunable sideband. This sideband may be stepped arbitrarily from mode-to-mode of the ring-down cavity, at a rate limited only by the cavity buildup/decay time. The ability to probe any cavity mode across the EOM bandwidth enables a variety of methods for generating differential spectra. The differential mode spacing may be changed, and the effect of this method on suppressing the various coupled-cavity interactions present in the system is discussed. Alternatively, each mode may also be differentially referenced to a single point, providing immunity to temporal variations in the base losses of the cavity while allowing for conventional spectral fitting approaches. Differential measurements of absorption are acquired at 3.3 kHz and a minimum detectable absorption coefficient of 5 $\times 10^{-12}$ cm^{-1} in 1 s averaging time is achieved.

1. J. Courtois, K. Bielska, and J.T Hodges J. Opt. Soc. Am. B, 30, 1486-1495, 2013
2. H.F. Huang and K.K. Lehmann App. Optics 49, 1378-1387, 2010
3. G.-W. Truong, K.O. Douglass, S.E. Maxwell, R.D. van Zee, D.F. Plusquellic, J.T. Hodges, and D.A. Long Nature Photonics, 7, 532-534, 2013

WF04 **2:39 – 2:54**

QUANTUM-NOISE-LIMITED CAVITY RING-DOWN SPECTROSCOPY IN THE MID-INFRARED

ADAM J. FLEISHER, DAVID A. LONG, QINGNAN LIU, JOSEPH HODGES, *Material Measurement Laboratory, National Institute of Standards and Technology, Gaithersburg, MD, USA.*

We report a highly sensitive mid-infrared spectrometer capable of recording cavity ring-down events in the quantum (shot) noise limit. A linear optical cavity of finesse 31,000 was pumped by a distributed feedback quantum cascade laser (DFB-QCL) operating at 4.5 μm until a cavity transmission threshold was reached. A fast optical switch then extinguished optical pumping and initiated a cavity decay which exhibited root-mean-square noise proportional to the square root of optical power (quantum noise) for several cavity time constants until a detector noise floor was reached. This spectrometer has achieved a noise-equivalent absorption of NEA = 2.6×10^{-11} $\mathrm{cm^{-1}Hz^{-1/2}}$ and a minimum absorption coefficient of $\alpha = 2.3 \times 10^{-11}$ $\mathrm{cm^{-1}}$in 3 seconds. Applications for such a highly sensitive spectrometer operating in the mid-infrared region, including ultra-trace molecular spectroscopy of CO_2 isotopologues and the direct interrogation of weak mirror birefringence and polarization-dependent losses, will be discussed.

WF05 **2:56 – 3:11**

MOLECULAR LINE PARAMETERS PRECISELY DETERMINED BY A CAVITY RING-DOWN SPECTROMETER

SHUI-MING HU, YAN TAN, JIN WANG, YAN LU, CUNFENG CHENG, YU ROBERT SUN, AN-WEN LIU, *Hefei National Laboratory for Physical Science at Microscale, University of Science and Technology of China, Hefei, China.*

A cavity ring-down spectrometer calibrated with a set of precise atomic lines was built to retrieve precise line parameters in the near infrared. [1,2] The spectrometer allows us to detect absorptions with a sensitivity of 10^{-11} cm^{-1} and a spectral precision up to 10^{-6} cm^{-1}. Ro-vibrational lines in the second overtone of H_2 have been observed, including the extremely weak $S_3(5)$ line with a line intensity less than 1×10^{-30}cm/molecule, which is among the weakest molecular lines detected by absorption in the gas phase. The absolute line positions of H_2 agree well with the high-level quantum chemical calculations including relativistic and QED corrections, with the deviation being less than 5×10^{-4} cm^{-1}. [3,4] A quantitative study has also been carried out on the $\nu_1 + 5\nu_3$ band of CO_2. [5] It was the first CO_2 band observed 80 years ago in the spectrum of Venus. We determined the line positions with an accuracy of 3×10^{-5} cm^{-1}, two orders of magnitude better than previous studies. Similar studies have been carried out to determine the line parameters of H_2O [6] and CO [7] in the spectral regions near 0.8 μm. The spectroscopic parameters can be used in varies studies, from the atmospheres of the earth-like planets to the test of fundamental physics.

References

[1] H. Pan, *et al.* Rev. Sci. Instrum. **82**, 103110 (2011).
[2] C.-F. Cheng, Opt. Expr. **20**, 9956 (2012).
[3] C.-F. Cheng, *et al.* Phys. Rev. A **85**, 024501 (2012).
[4] y. Tan, *et al.* J. Mol. Spectrosc. **300**, 60 (2014).
[5] Y. Lu, *et al.* Astrophys. J. **775**, 71 (2013).
[6] Y. Lu, *et al.* JQSRT **118**, 96 (2013).
[7] Y. Tan, *et al.* "Ro-vibrational analysis of the fifth overtone of CO at 802 nm", under preparation.

WF06 **3:13 – 3:23**

BROADBAND COMB-RESOLVED CAVITY ENHANCED SPECTROMETER WITH GRAPHENE MODULATOR

KEVIN LEE, CHRISTIAN MOHR, JIE JIANG, MARTIN FERMANN, *Laser Research, IMRA AMERICA, Inc, Ann Arbor, MI, USA*; CHIEN-CHUNG LEE, THOMAS R SCHIBLI, *Department of Physics, University of Colorado, Boulder, CO, USA*; GRZEGORZ KOWZAN, PIOTR MASLOWSKI, *Institute of Physics, Faculty of Physics, Astronomy and Informatics, Nicolaus Copernicus University, Torun, Poland.*

Optical cavities enhance sensitivity in absorption spectroscopy. While this is commonly done with single wavelengths, broad bandwidths can be coupled into the cavity using frequency combs. The combination of cavity enhancement and broad bandwidth allows simultaneous measurement of tens of transitions with high signal-to-noise for even weak near-infrared transitions. This removes the need for time-consuming sequencing acquisition or long-term averaging, so any systematic errors from long-term drifts of the experimental setup or slow changes of sample composition are minimized. Resolving comb lines provides a high accuracy, absolute frequency axis. This is of great importance for gas metrology and data acquisition for future molecular lines databases, and can be applied to simultaneous trace-gas detection of gas mixtures.

Coupling of a frequency comb into a cavity can be complex, so we introduce and demonstrate a simplification. The Pound-Drever-Hall method for locking a cavity and a frequency comb together requires a phase modulation of the laser output. We use the graphene modulator that is already in the Tm fiber laser cavity for controlling the carrier envelope offset of the frequency comb, rather than adding a lossy external modulator. The graphene modulator can operate at frequencies of over 1 MHz, which is sufficient for controlling the laser cavity length actuator which operates below 100 kHz.

We match the laser cavity length to fast variations of the enhancement cavity length. Slow variations are stabilized by comparison of the pulse repetition rate to a GPS reference. The carrier envelope offset is locked to a constant value chosen to optimize the transmitted spectrum. The transmitted pulse train is a stable frequency comb suitable for long measurements, including the acquisition of comb-resolved Fourier transform spectra with a minimum absorption coefficient of about 2×10^{-7} cm^{-1}. For our 38 cm long enhancement cavity, the comb spacing is 394 MHz. With our 300 cm^{-1} bandwidth at 2 μm, we simultaneously measure the full comb line resolved CO_2 vibrational manifold at 4850 cm^{-1}. Other spectral ranges can be accessed by using graphene with different gain fibers or nonlinear frequency conversion.

Intermission

WF07 *INVITED TALK* **3:42 – 4:12**

DUAL-COMB SPECTROSCOPY IN THE OPEN AIR

GREG B RIEKER, *Department of Mechanical Engineering, University of Colorado Boulder, Boulder, CO, USA*; ANDREW KLOSE, SCOTT DIDDAMS, *Time and Frequency Division, National Institute of Standards and Technology, Boulder, CO, USA*; IAN CODDINGTON, FABRIZIO GIORGETTA, LAURA SINCLAIR, ESTHER BAUMANN, GAR-WING TRUONG, GABRIEL YCAS, WILLIAM C SWANN, NATHAN R. NEWBURY, *Quantum Electronics and Photonics Division, National Institute of Standards and Technology, Boulder, CO, USA*.

Dual-comb spectroscopy is arguably the natural successor to FTIR. Based on the interference between two frequency combs, this technique can record broadband spectra with a resolution better than $0.0003\ \mathrm{cm}^{-1}$. Like FTIR, dual-comb spectroscopy measures an entire spectrum simultaneously, allowing for suppression of systematic errors related to temporal dynamics of the sample. Unlike FTIR it records the entire spectrum with virtually no instrument lineshape or error in the frequency axis. The lack of moving parts in dual-comb spectroscopy means that spectra can be recorded in milliseconds to microseconds with the desired signal-to-noise being the only real constrain on the minimum recording time. Finally the high spacial beam quality of the frequency combs allows for increased sensitivity through long interaction paths either in free-space, multi-pass cells or enhancement cavities.

This talk will explore the recent use of dual-comb spectroscopy in the near-infrared to measure atmospheric carbon dioxide, methane and water concentrations over a 2-km outdoor open-air path. Due to many of the strengths just mentioned, precisions of <1 ppm for CO_2 and <3 ppb for CH_4 in 5 min are achieved making this system very attractive for carbon monitoring at length scales relevant to carbon transport models.

Additionally this presentation will address recent work on robust, compact, and portable dual-comb spectrometers as well as dual-comb spectroscopy further into the IR.

WF08 **4:17 – 4:32**

FREQUENCY-COMB REFERENCED SPECTROSCOPY OF ν_4 AND ν_5 HOT BANDS IN THE ν_1+ν_3 COMBINATION BAND OF C_2H_2

SYLVESTRE TWAGIRAYEZU, *Department of Chemistry, Brookhaven National Laboratory, Upton, NY, USA*; MATTHEW CICH, *Department of Chemistry, Stony Brook University, Stony Brook, NY, USA*; TREVOR SEARS[a], *Chemistry Department, Brookhaven National Laboratory, Upton, NY, USA*; C. McRAVEN, *Am Klopfersptiz 19a, Menlo Systems, GmbH, 82152 Martinsried, Germany*; GREGORY HALL, *Chemistry Department, Brookhaven National Laboratory, Upton, NY, USA*.

Doppler-free transition frequencies for ν_4 and ν_5 hot bands in the band of C_2H_2 have been measured using saturation dip spectroscopy with an extended cavity diode laser referenced to a frequency comb. The frequency accuracy of the measured transitions, as judged from line shape model fits and the spectrometer stability, is better than 30 kHz. This is some 2-3 orders of magnitude improvement on the accuracy and precision of previous measurements of the line positions derived from the analysis of high-resolution Fourier transform infrared absorption spectra. The data were analyzed by determining the upper state energies, using known lower state level positions, and fitting them to a $J(J+1)$ polynomial expansion to identify perturbations. The results reveal that the upper rotational energy level structure is mostly regular but suffers $J-$localized perturbations causing level shifts between one and several hundred MHz. These perturbations are due to accidental near degeneracies with energy levels of the same J and larger bending vibrational excitation.

Acknowledgements: We are most grateful to Prof. D.S Perry (U. of Akron) and Prof. M. Herman (U. Libre de Bruxelles) for providing us with detailed results from their work and helpful discussions. Work at Brookhaven National Laboratory is funded by the Division of Chemical Sciences, Geosciences and Biosciences within the Offices of Basic Energy Sciences, Office of Sciences, U.S. Department of Energy under Contract Nos. DE-AC02-98CH10886 and DE-SC0012704.

[a]Also: *Department of Chemistry, Stony Brook University, Stony Brook, NY 11794.*

WF09 **4:34 – 4:49**

LOCAL PERTURBATIONS IN THE (10110) AND (10101) LEVELS OF C_2H_2 FROM FREQUENCY COMB-REFERENCED SPECTROSCOPY

TREVOR SEARS[a], *Chemistry Department, Brookhaven National Laboratory, Upton, NY, USA*; SYLVESTRE TWAGIRAYEZU, DAMIEN FORTHOMME, *Department of Chemistry, Brookhaven National Laboratory, Upton, NY, USA*; GREGORY HALL, *Chemistry Department, Brookhaven National Laboratory, Upton, NY, USA*; MATTHEW CICH, *Department of Chemistry, Stony Brook University, Stony Brook, NY, USA.*

In work reported by Twagirayezu *et al.* at this meeting, the rest frequencies of more than 100 lines in the ν_4 and ν_5 hot bands in the $\nu_1 + \nu_3$ combination band of acetylene have been measured by saturation dip spectroscopy using an extended cavity diode laser locked to a frequency comb. This work was orginally directed towards providing a set of accurate frequencies for the hot band line positions to aid in modeling the lineshapes of the main lines in the band. In analyzing the results, we find that many of the upper levels in the hot band transitions suffer small, and in some cases not so small, local perturbations. These arise because of J-dependent near degeneracies between the title levels and background levels of the same symmetry, mostly derived from zero order states involving multiple quanta of bending excitation. The vibration-rotation levels at the energies in question have previously been modeled using a polyad-based Hamiltonian[b] and the present data can be interpreted on the basis of this model, but they also provide information which can be used to refine the model, and point to terms that may have previously been neglected. The most important result is that the high precision of the measurements gives the opportunity to calibrate the effects of background levels associated with high bending quantum numbers and angular momentun states that are otherwise very difficult to access.

Acknowledgments: We are most grateful to D. S Perry (U. Akron) and M. Herman (U. Libre de Bruxelles) for helpful discussions. Work at Brookhaven National Laboratory is funded by the Division of Chemical Sciences, Geosciences and Biosciences within the Offices of Basic Energy Sciences, Office of Sciences, U.S. Department of Energy under Contract Nos. DE-AC02-98CH10886 and DE-SC0012704.

[a]also: *Chemistry Department, Stony Brook University, Stony Brook, New York 11794*

[b]M. Herman and D. S. Perry, *Phys. Chem. Chem. Phys.*, **15**, 9970-9993 (2013)

WF10 ***Post-Deadline Abstract*** **4:51 – 5:06**

NOISE-IMMUNE CAVITY-ENHANCED OPTICAL HETERODYNE MOLECULAR SPECTROMETRY MODELLING UNDER SATURATED ABSORPTION

PATRICK DUPRÉ, *Laboratoire de Physico-Chimie de l'Atmosphère, Université du Littoral Côte d'Opale, Dunkerque, France.*

The Noise-Immune Cavity-Enhanced Optical Heterodyne Molecular Spectrometry (NICE-OHMS) is a modern technique renowned for its ultimate sensitivity, because it combines long equivalent absorption length provided by a high finesse cavity, and a detection theoretically limited by the sole photon-shot-noise. One fallout of the high finesse is the possibility to accumulating strong intracavity electromagnetic fields (EMF). Under this condition, molecular transitions can be easy saturated giving rise to the usual Lamb dips (or hole burning). However, the unusual shape of the basically trichromatic EMF (due to the RF lateral sidebands) induces nonlinear couplings, i.e., new crossover transitions. An analytical methodology will be presented to calculate spectra provided by NICE-OHMS experiments. It is based on the solutions of the equations of motion of an open two-blocked-level system performed in the frequency-domain (optically thin medium). Knowing the transition dipole moment, the NICE-OHMS signals ("absorption-like" and "dispersion-like") can be simulated by integration over the Doppler shifts and by paying attention to the molecular Zeeman sublevels and to the EMF polarization[a]. The approach has been validated by discussion experimental data obtained on two transitions of C_2H_2 in the near-infrared under moderated saturation[b]. One of the applications of the saturated absorption is to be able to simultaneously determine the transition intensity and the density number while only one these 2 quantities can only be assessed in nonlinear absorption.

[a]J. Opt. Soc. Am. B 32, 838 (2015)

[b]Optics Express 16, 14689 (2008)

WF11 **5:08 – 5:23**

MAGNETIC SPIN-TORSION COUPLING IN METHANOL

L. H. COUDERT, C. GUTLE, *LISA, CNRS, Universités Paris Est Créteil et Paris Diderot, Créteil, France*; T. R. HUET, *Laboratoire PhLAM, UMR 8523 CNRS - Université de Lille 1, Villeneuve d'Ascq, France*; JENS-UWE GRABOW, *Institut für Physikalische Chemie und Elektrochemie, Gottfried-Wilhelm-Leibniz-Universität, Hannover, Germany.*

The hyperfine structure of non-rigid molecules in which hyperfine coupling arises from equivalent nuclei that can be exchanged by large amplitude motions is of great interest and lead to unexpected results. In the non-rigid $(C_2D_2)_2$ and $(D_2O)_2$ dimers, the hyperfine structure arising for nondegenerate tunneling sublevels can be accounted for using an effective quadrupole coupling Hamiltonian with the same coupling constant for all four deuterium atoms.[a] In the non-rigid species CD_3COH and $HCOOCH_3$, the large amplitude torsional motion leads to hyperfine patterns which are qualitatively dependent on the torsional symmetry of the levels.[b] The interaction between a large amplitude torsional motion and the hyperfine coupling may also lead to a less known hyperfine effect, the so-called magnetic spin-torsion coupling, which was first studied by Heuvel and Dymanus[c] and which has not yet been conclusively evidenced.

In this talk, the magnetic hyperfine structure of the non-rigid methanol molecule will be investigated experimentally and theoretically. 13 hyperfine patterns were recorded using two molecular beam microwave spectrometers. These patterns, along with previously recorded ones,[c] were analyzed in an attempt to evidence the effects of the magnetic spin-torsion coupling. The theoretical approach setup to analyze the observed data accounts for the spin-torsion coupling, in addition to the familiar magnetic spin-rotation and spin-spin couplings, and relies on symmetry considerations to build a hyperfine coupling Hamiltonian and a spin-rotation-torsion wavefunction compatible with the Pauli exclusion principle.

In the talk, the results of the analysis will be presented. The hyperfine coupling parameters retrieved will be discussed and we hope to be able to conclusively evidence the effects of the magnetic spin-torsion.

---

[a]Bhattacharjee, Muenter, and Coudert, *J. Chem. Phys.* **97** (1992) 8850; and Stahl and Coudert, *J. Mol. Spectrosc.* **157** (1993) 161.

[b]Coudert and Lopez, *J. Mol. Spectrosc.* **239** (2006) 135; and Tudorie, Coudert, Huet, Jegouso, and Sedes, *J. Chem. Phys.* **134** (2011) 074314.

[c]Heuvel and Dymanus, *J. Mol. Spectrosc.* **45** (1973) 282 and *ibid* **47** (1973) 363.

WF12 **5:25 – 5:40**

SPIN-ROTATION HYPERFINE SPLITTINGS AT MODERATE TO HIGH J VALUES IN METHANOL

LI-HONG XU, *Department of Physics, University of New Brunswick, Saint John, NB, Canada*; JON T. HOUGEN, *Sensor Science Division, National Institute of Standards and Technology, Gaithersburg, MD, USA*; SERGEY BELOV, G YU GOLUBIATNIKOV, ALEXANDER LAPINOV, *Microwave Spectroscopy, Institute of Applied Physics, Nizhny Novgorod, Russia*; V. ILYUSHIN, E. A. ALEKSEEV, A. A. MESCHERYAKOV, *Radiospectrometry Department, Institute of Radio Astronomy of NASU, Kharkov, Ukraine.*

In this talk we present a possible explanation, based on torsionally mediated proton-spin-overall-rotation interaction operators, for the surprising observation in Nizhny Novgorod several years ago[a] of doublets in some Lamb-dip sub-millimeter-wave transitions between torsion-rotation states of E symmetry in methanol. These observed doublet splittings, some as large as 70 kHz, were later confirmed by independent Lamb-dip measurements in Kharkov. In this talk we first show the observed J-dependence of the doublet splittings for two b-type Q branches (one from each laboratory), and then focus on our theoretical explanation. The latter involves three topics: (i) group theoretically allowed terms in the spin-rotation Hamiltonian, (ii) matrix elements of these terms between the degenerate components of torsion-rotation E states, calculated using wavefunctions from an earlier global fit of torsion-rotation transitions of methanol in the $v_t = 0$, 1, and 2 states[b], and (iii) least-squares fits of coefficients of these terms to about 35 experimentally resolved doublet splittings in the quantum number ranges of K = -2 to +2, J = 13 to 34, and $v_t = 0$. Rather pleasing residuals are obtained for these doublet splittings, and a number of narrow transitions, in which no doublet splitting could be detected, are also in agreement with predictions from the theory. Some remaining disagreements between experiment and the present theoretical explanation will be mentioned.

---

[a]G. Yu. Golubiatnikov, S. P. Belov, A. V. Lapinov, "CH_3OH Sub-Doppler Spectroscopy," (Paper MF04) and S.P. Belov, A.V. Burenin, G.Yu. Golubiatnikov, A.V. Lapinov, "What is the Nature of the Doublets in the E-Methanol Lamb-dip Spectra?" (Paper FB07), 68th International Symposium on Molecular Spectroscopy, Columbus, Ohio, June 2013.

[b]Li-Hong Xu, J. Fisher, R.M. Lees, H.Y. Shi, J.T. Hougen, J.C. Pearson, B.J. Drouin, G.A. Blake, R. Braakman, "Torsion-Rotation Global Analysis of the First Three Torsional States (v_t = 0, 1, 2) and Terahertz Database for Methanol," J. Mol. Spectrosc., **251**, 305-313, (2008).

WG. Mini-symposium: Accelerator-Based Spectroscopy

Wednesday, June 24, 2015 – 1:30 PM

Room: 100 Noyes Laboratory

Chair: J. Oomens, Radboud University, Nijmegen, The Netherlands

WG01 *INVITED TALK* **1:30 – 2:00**

IR SPECTROSCOPY ON PEPTIDES AND PROTEINS AFTER ION MOBILITY SELECTION AND IN LIQUID HELIUM DROPLETS

GERT VON HELDEN, *Department of Molecular Physics, Fritz-Haber-Institut der Max-Planck-Gesellschaft, Berlin, Germany.*

IR spectroscopy has become a frequently used tool to characterize gas-phase peptides and proteins. In many experiments, ions are m/z selected, irradiated by intense and tunable IR light and fragmentation is monitored as a function of IR wavelength. The presence of different conformers can, however, complicate the interpretation, as the resulting spectra represent the sum of the spectra of the individual components. We constructed a setup, in which ion mobility methods are used to obtain m/z selected ions of defined shape on which are then further investigated by IR spectroscopy. First results on peptide aggregates are presented and for some of those, the IR spectra show a transition from helical or random coil to beta sheet structures.
In a different experiment, peptide or protein ions are captures in liquid helium droplets prior to IR spectroscopic investigation. The conditions inside a helium droplet are isothermal at 0.38 K and the interaction between the helium matrix and the molecules are weak so that only small perturbations on the molecule are expected. IR spectra for m/z selected peptides with up to 10 aminoacids and proteins containing more than 100 aminoacids have been measured. The spectra of the smaller species show resolved bands of individual oscillators, which can be used for structure assignment. For the larger species, band envelopes are obtained and for the case of highly charged proteins, a transition form helical to extended structures is observed.

WG02 **2:05 – 2:20**

COMBINING THE POWER OF IRMPD WITH ION-MOLECULE REACTIONS: THE STRUCTURE AND REACTIVITY OF RADICAL IONS OF CYSTEINE AND ITS DERIVATIVES

MICHAEL LESSLIE, *Department of Chemistry and Biochemistry, Northern Illinois University, Dekalb, IL, USA*; SANDRA OSBURN, *Department of Chemistry and Biochemistry, Duquesne University, Pittsburgh, PA, USA*; GIEL BERDEN, J. OOMENS, *Institute for Molecules and Materials (IMM), Radboud University Nijmegen, Nijmegen, Netherlands*; VICTOR RYZHOV, *Department of Chemistry and Biochemistry, Northern Illinois University, Dekalb, IL, USA.*

Most of the work on peptide radical cations has involved protons as the source of charge. Nonetheless, using metal ions as charge sources often offers advantages like stabilization of the structure via multidentate coordination and the elimination of the "mobile proton". Moreover, characterization of metal-bound amino acids is of general interest as the interaction of peptide side chains with metal ions in biological systems is known to occur extensively. In the current study, we generate thiyl radicals of cysteine and homocysteine in the gas phase complexed to alkali metal ions. Subsequently, we utilize infrared multiple-photon dissociation (IRMPD) and ion-molecule reactions (IMR) to characterize the structure and reactivity of these radical ions.

Our group has worked extensively with the cysteine-based radical cations and anions, characterizing the gas-phase reactivity and rearrangement of the amino acid and several of its derivatives. In a continuation of this work, we are perusing the effects of metal ions as the charge bearing species on the reactivity of the sulfur radical. Our S-nitroso chemistry can easily be used in conjunction with metal ion coordination to produce initial S-based radicals in peptide radical-metal ion complexes. In all cases we have been able to achieve radical formation with significant yield to study reactivity. Ion-molecule reactions of metallated radicals with allyl iodide, dimethyl disulfide, and allyl bromide have all shown decreasing reactivity going down group 1A.

Recently, we determined the experimental IR spectra for the homocysteine radical cation with Li+, Na+, and K+ as the charge bearing species at the FELIX facility. For comparison, the protonated IR spectrum of homocysteine has previously been obtained by our group. A preliminary match of the IR spectra has been confirmed. Finally, calculations are underway to determine the bond distances of all the metal adduct structures.

WG03 **2:22 – 2:37**

FAR-IR ACTION SPECTROSCOPY OF AMINOPHENOL AND ETHYLVANILLIN: EXPERIMENT AND THEORY

VASYL YATSYNA, RAIMUND FEIFEL, VITALI ZHAUNERCHYK, *Department of Physics, Faculty of Science, University of Gothenburg, Gothenburg, Sweden*; DANIËL BAKKER, ANOUK RIJS, *FELIX Laboratory, Radboud University Nijmegen, Nijmegen, The Netherlands.*

Investigations of molecular structure and conformational isomerism are at the forefront of today's biophysics and biochemistry. In particular, vibrations excited by far-IR radiation can be highly sensitive to the molecular 3D structure as they are delocalized over large parts of the molecule. Current theoretical predictions of vibrational frequencies in the far-IR range are not accurate enough because of the non-local character and anharmonicity of these vibrations. Therefore experimental studies in the far-IR are vital to guide theory towards improved methodology.

In this work we present the conformer-specific far-IR spectra of aminophenol and ethylvanillin molecules in the range of 220-800 cm^{-1}utilizing ion-dip action spectroscopy carried out at the free electron laser FELIX in Nijmegen, Netherlands. The systems studied are aromatic molecules with important functional groups such as the hydroxyl (OH) and amino (NH_2) groups in aminophenol, and the hydroxyl, ethoxy (OCH_2CH_3) and formyl (CHO) groups in ethylvanillin. The experimental spectra show well resolved conformer-specific vibrational bands. In the case of ethylvanillin only two planar conformers have been observed under supersonic jet expansion conditions. Despite the fact that these conformers differ only in the position of oxygen of the formyl group with respect to ethoxy group, they are well distinguishable in far-IR spectra.

The capability of numerical methods based on density functional theory (DFT) for predicting vibrational frequencies in this spectral region within the harmonic approximation has been investigated by using several hybrid-functionals such as B3LYP, PBE0, B2PLYP and CAM-B3LYP. An anharmonic correction based on vibrational second order perturbation theory approach[a] was also applied. We have found that the methods we considered are well suited for the assignment of far-IR vibrational features except the modes which are strongly anharmonic, like the NH_2 wagging mode in aminophenol which is likely to be due to double well potential governing this motion.

[a] V. Barone, Anharmonic vibrational properties by a fully automated second-order perturbative approach, J. Chem. Phys. 122 (2005) 014108.

WG04 **2:39 – 2:54**

OPPORTUNITIES FOR GAS-PHAS MOLECULAR SPECTROSCOPY ON THE VLS-PGM BEAMLINE AT THE CANADIAN LIGHT SOURCE

MICHAEL A MacDONALD, *EFD, Canadian Light Source Inc., Saskatoon, Saskatchewan, Canada.*

The VLS-PGM beamline at the Canadian Light Source cover the energy range from 12eV to 250eV with a resolving power better than 10^4 throughout this range. Associated with this beamline are two endstations designed for gas phase spectroscopy.

The first is a dual toroid electrostatic particle energy analyser. Each toroid can (independently) measure the energy and angular distribution of charged particles emitted from the interaction region and can be set for either positive ions or electrons. This allows both photoelectron and ion kinetic energy spectra to be recorded. Recent results from this instrument will be presented including both high resolution photoelectron spectra and photoelectron asymmetry parameter (β) spectra. Coincidence circuitry exists to allow, in favourable circumstances,the measurement of molecular frame photoelectron angular distributions (MFPADs) where the detection of an ion fragment allows orientation of the parent molecule to be deduced.

The second is a Wiley-McLaren Time-of-Flight mass spectrometer equipped with multihit electronics. This allows partial ion yield (PIY) spectra to be recorded as well as multi-ion coincidence spectra (PePIPICO). Again recent results will be presented looking at double ionisation in benzene like molecules.

WG05 **2:56 – 3:11**

THERMAL DECOMPOSITION OF C_7H_7 RADICALS; BENZYL, TROPYL, AND NORBORNADIENYL

GRANT BUCKINGHAM, BARNEY ELLISON, JOHN W DAILY, *Department of Chemistry and Biochemistry, University of Colorado, Boulder, CO, USA*; MUSAHID AHMED, *UXSL, Chemical Sciences Division, Lawrence Berkeley National Laboratory, Berkeley, CA, USA.*

Benzyl radical ($C_6H_5CH_2$) and two other C_7H_7 radicals are commonly encountered in the combustion of substituted aromatic compounds found in biofuels and gasoline. High temperature pyrolysis of benzyl radical requires isomerization to other C_7H_7 radicals that may include cycloheptatrienyl (tropyl) radical (*cyc*-C_7H_7) and norbornadienyl radical. The thermal decomposition of all three radicals has now been investigated using a micro-reactor that heats dilute gas-phase samples up to 1600 K and has a residence time of about 100 μ-sec. The pyrolysis products exit the reactor into a supersonic expansion and are detected using synchrotron-based photoionization mass spectrometry and matrix-isolation IR spectroscopy. The products of the pyrolysis of benzyl radical ($C_6H_5CH_2$) along with three isotopomers ($C_6H_5^{13}CH_2$, $C_6D_5CH_2$, and $C_6H_5CD_2$) were detected and identified[a]. The distribution of ^{13}C atoms and D atoms indicate that multiple different decomposition pathways are active.

[a]Buckingham, G. T., Ormond, T. K., Porterfield, J. P., Hemberger, P., Kostko, O., Ahmed, M., Robichaud, D. J., Nimlos, M. R., Daily, J. W., Ellison, G. B. 2015, Journal of Chemical Physics **142** 044307

Intermission

WG06 **3:30 – 3:40**

NONDIPOLE EFFECTS IN CHIRAL SYSTEMS MEASURED WITH LINEARLY POLARIZED LIGHT

K P BOWEN, O HEMMERS, *Chemistry, University of Nevada Las Vegas, Las Vegas, NV, USA*; R GUILLEMIN, *Laboratoire de Chimie Physique – Matière et Rayonnement, Université Pierre et Marie Curie, Paris, France*; W C STOLTE, *National Security Technologies, LLC, Livermore, CA, USA*; M N PIANCASTELLI, *Department of Physics and Astronomy, Uppsala Universitet, Uppsala, Sweden*; D W LINDLE, *Chemistry, University of Nevada Las Vegas, Las Vegas, NV, USA.*

With the advent of third-generation synchrotron light sources, it has been demonstrated that higher-order corrections to the dipole approximation are necessary for the description of light-matter interactions in the soft x-ray range. These effects, known as ‘nondipole effects’, present themselves as asymmetries in the angular distributions of photoelectrons. Chiral molecules, known to have asymmetries in photoelectron angular distributions when exposed to circularly polarized light, have been proposed to demonstrate a chiral-specific nondipole effect when exposed to linearly polarized light. We present the first-ever measurement of nondipole chiral angular distributions for the case of each enantiomer of camphor in the photon energy range 296-343eV.

WG07 **3:42 – 3:57**

APPLICATIONS OF THE VUV FOURIER TRANSFORM SPECTROMETER AT SYNCHROTRON SOLEIL

NELSON DE OLIVEIRA, DENIS JOYEUX, *DESIRS Beamline, Synchrotron SOLEIL, Saint Aubin, France*; KENJI ITO, *DESIRS beamline, Synchrotron Soleil, Saint-Aubin, France*; BERENGER GANS, *ISMO, CNRS, Orsay, 91405, France*; LAURENT NAHON, *DESIRS Beamline, Synchrotron SOLEIL, Saint Aubin, France*.

Fourier transform spectrometers (FTS) are usually based upon amplitude division interferometers through beamsplitters (BS) as in the Michelson interferometer geometry. However, the manufacture of broadband BS is difficult and even impossible in the far VUV (below λ = 140 nm). We therefore conceived an instrument based upon an original design involving only reflective plane surfaces, giving access to the whole VUV range without the restrictions associated with BS. The VUV– FTS is a permanent endstation connected to one of the three experimental branches of the DESIRS beamline[a] and devoted to high resolution photoabsorption in the UV-VUV spectral range, typically between λ = 300 and 40 nm[b]. Since 2008, a large international community of users interested in laboratory measurements with applications in astrophysics, molecular physics or planetary atmospheres has been attracted by the VUV - FTS capabilities including its efficiency in terms of signal to noise ratio, even when high spectral resolution was not an issue. A large number of dedicated gas phase sample environments have been developed including a windowless cell that can be cooled down, a heated windowless cell, a free molecular jet set-up and various windowed cells. Besides, a new discharge gas cell for production and study of transient species gave recently its first results. As an illustration, the VUV absorption spectrum of the CH_3 radical down to 140 nm will be shown in this presentation.

[a]Nahon et al., J. Synchrotron Radiat., 19, 508(2012)
[b]De Oliveira et al., Nat. Photonics, 5, 149(2011)

WG08 **3:59 – 4:14**

FORBIDDEN TRANSITIONS IN THE VUV SPECTRUM OF N_2

ALAN HEAYS, *Leiden Observatory, University of Leiden, Leiden, Netherlands*; MING LI NIU, *LaserLaB Amsterdam, Vrije Universiteit Amsterdam, Amsterdam, Netherlands*; NELSON DE OLIVEIRA, *DESIRS Beamline, Synchrotron SOLEIL, Saint Aubin, France*; EDCEL JOHN SALUMBIDES, *Department of Physics and Astronomy, VU University , Amsterdam, Netherlands*; BRENTON R LEWIS, *Research School of Physics and Engineering, Australian National University, Canberra, ACT, Australia*; WIM UBACHS, *Department of Physics and Astronomy, VU University , Amsterdam, Netherlands*; EWINE VAN DISHOECK, *Leiden Observatory, University of Leiden, Leiden, Netherlands*.

The predissociation of N_2 excited levels is enabled by the presence of optically-inaccessible triplet states. We have recorded vacuum ultraviolet (VUV) spectra at the SOLEIL synchrotron which reveal these states through their perturbation of allowed transitions or their direct appearance due to intensity borrowing.

Some of these measurements were recorded at 900 K in order to access high-rotational levels, other measurements investigated weak forbidden transitions at high column density. Following careful analysis, significant new information has been obtained elucidating the states responsible for the astrochemically and atmospherically signficant N_2 predissociation mechanism, and allowing for improvements in its quantitiative modelling.

WG09 **4:16 – 4:31**

SYNCHROTRON INFRARED SPECTROSCOPY OF ν_4, ν_{10}, ν_{11} AND ν_{14} STATES OF THIIRANE

COREY EVANS, JASON P CARTER, *Department of Chemistry, University of Leicester, Leicester, United Kingdom*; DON McNAUGHTON, ANDY WONG, *School of Chemistry, Monash University, Melbourne, Victoria, Australia*; DOMINIQUE APPADOO, *800 Blackburn Road, Australian Synchrotron, Melbourne, Victoria, Australia*.

The high-resolution (0.001 cm^{-1}) spectrum of thiirane has been recorded using the infrared beamline at the Australian synchrotron facility. Spectra have been recorded between 750 cm^{-1} to 1120 cm^{-1} and ro-vibrational transitions associated with four bands have been observed and assigned. Coriolis coupling was observed between the ν_4 (1024 cm^{-1}) and the ν_{14}(1050 cm^{-1}) fundamentals as well as between ν_{11} (825 cm^{-1}) and the ν_8 (895 cm^{-1}) fundamentals. The ν_{10} (945 cm^{-1}) fundamental was also observed and was found to have no significant perturbations associated with it. For each of the observed bands rotational, centrifugal distortion and Coriolis interaction parameters have been determined. The ground state constants have also been further refined.

WG10 **4:33 – 4:48**

FINGERPRINTS OF INTRAMOLECULAR HYDROGEN BONDS: SYNCHROTRON-BASED FAR IR STUDY OF THE CIS AND TRANS CONFORMERS OF 2-FLUOROPHENOL

AIMEE BELL, JAMES SINGER, JENNIFER VAN WIJNGAARDEN, *Department of Chemistry, University of Manitoba, Winnipeg, MB, Canada.*

Rotationally-resolved vibrational spectra of two planar conformers of 2-fluorophenol have been collected from 100-1000 cm^{-1} using the Bruker IFS125HR FTIR spectrometer at the Canadian Light Source with a resolution of 0.000959 cm^{-1}. The cis conformer is lower in energy by 2.9 kcal/mol (MP2/aug-cc-pvDZ) and is thought to be stabilized by an intramolecular hydrogen bond between the hydroxyl group and neighbouring fluorine atom on the ring. The OH out-of-plane torsion bands below 400 cm^{-1} provide the best fingerprint to distinguish between the two conformers in the gas phase spectrum as the *c*-type band origin of the cis conformer is blue-shifted by 36 cm^{-1} from that of the trans conformer as result of the intramolecular interaction. In this talk, we will discuss the progress of the analysis of this complex far infrared spectrum of 2-fluorophenol.

WG11 **4:50 – 5:05**

INFRARED CROSS-SECTIONS OF NITRO-DERIVATIVE VAPORS: NEW SPECTROSCOPIC SIGNATURES OF EXPLOSIVE TAGGANTS AND DEGRADATION PRODUCTS

ARNAUD CUISSET, GAËL MOURET, *Laboratoire de Physico-Chimie de l'Atmosphère, Université du Littoral Côte d'Opale, Dunkerque, France*; OLIVIER PIRALI, SÉBASTIEN GRUET, *AILES beamline, Synchrotron SOLEIL, Saint Aubin, France*; GÉRARD PASCAL PIAU, GILLES FOURNIER, *Airbus Group Innovations, Airbus, Suresnes, France.*

Classical explosives such as RDX or TNT exhibit a very low vapor pressure at room temperature and their detection in air requires very sensitive techniques with levels usually better than 1 ppb. To overcome this difficulty, it is not the explosive itself which is detected, but another compound more volatile present in the explosive. [a] This volatile compound can exist naturally in the explosive due to the manufacturing process. For example, in the case of DiNitroToluene (DNT), the molecule is a degradation product of TNT and is required for its manufacture. Ortho-Mononitrotoluene (2-NT) and para-mononitrotoluene (4-NT) can be also used as detection taggants for explosive detection.

In this study, using the exceptional properties of the SOLEIL synchrotron source, and adapted multipass-cells, gas phase Far-IR rovibrational spectra of different isomers of mononitrotoluene and dinitrotoluene have been investigated. Room temperature Far-IR cross-sections of the 3 isomer forms of mononitrotoluene have been determined for the lowest frequency vibrational bands located below 700 cm^{-1}.[b] Cross sections and their temperature dependences have been also measured in the Mid-IR using conventional FTIR spectroscopy probing the nitro-derivatives vapors in a heated multipass-cell.

[a] J. C. Oxley, J. L. Smith, W. Luo, J. Brady, Prop. Explos. Pyrotec. 34 (2009) 539–543

[b] A. Cuisset, S. Gruet, O. Pirali, G. Mouret, Spectrochimica Acta Part A, 132 (2014) 838-845.

WG12 ***Post-Deadline Abstract*** **5:07 – 5:22**

CHARACTERIZATION OF REACTION PATHWAYS IN LOW TEMPERATURE OXIDATION OF TETRAHYDROFURAN WITH MULTIPLEXED PHOTOIONIZATION MASS SPECTROMETRY TECHNIQUE

IVAN ANTONOV, LEONID SHEPS, *Combustion Research Facility, Sandia National Laboratories, Livermore, CA, USA.*

Tetrahydrofuran (THF) is a prototype biofuel and a common intermediate in combustion of alkanes and alkenes. Photolytic Cl atom-initiated oxidation of THF was studied with multiplexed photoionization mass spectrometry (MPIMS) technique at temperatures 400-650 K and pressures 0.005-2 bar. Photoionization spectra and kinetic time traces were recorded simultaneously for all mass channels. Photoionization spectra, recorded with tunable VUV synchrotron radiation, were used to separate and identify isomers with the same nominal molecular formula, providing mechanistic insight into the the underlying kinetics. Our study suggests that formation of alkylperoxy radicals and their subsequent isomerization to hydroperoxyalkyl radicals plays an important role in low temperature oxidation of THF, while ring opening of THF_{-H} radical (which dominates THF oxidation at T>800 K) is less important at our conditions.

WH. Clusters/Complexes

Wednesday, June 24, 2015 – 1:30 PM

Room: B102 Chemical and Life Sciences

Chair: Elangannan Arunan, Indian Institute of Science, Bangalore, India

WH01 — 1:30 – 1:45

A STRANGE COMBINATION BAND OF THE CROSS-SHAPED COMPLEX CO_2-CS_2

NASSER MOAZZEN-AHMADI, *Department of Physics and Astronomy, University of Calgary, Calgary, AB, Canada*; BOB McKELLAR, *Steacie Laboratory, National Research Council of Canada, Ottawa, ON, Canada.*

The spectrum of the weakly-bound CO_2-CS_2 complex was originally studied by the USC group,[a] using a pulsed supersonic expansion and a tunable diode laser in the CO_2 ν_3 region. Their derived structure was nonplanar X-shaped (C_{2v} symmetry), a relatively unusual geometry among linear molecule dimers. Very recently, there has been a detailed theoretical study of this complex based on a high-level *ab initio* potential surface.[b] The theoretical ground state is X-shaped, in good agreement with experiment, and a very low-lying (3 cm^{-1} at equilibrium, or 8 cm^{-1} zero-point) slipped-parallel isomer is also found.

We report here two new combination bands of X-shaped CO_2-CS_2 which involve the same ν_3 fundamental (2346.546 cm^{-1}) plus a low-frequency intermolecular vibration. The first band has b-type rotational selection rules (the fundamental is c-type). This, and its location (2361.838 cm^{-1}), clearly identify it as being due to the intermolecular torsional mode. The second band (2388.426 cm^{-1}) is a-type and can be assigned to the CO_2 rocking mode. Both observed intermolecular frequencies (15.29 and 41.88 cm^{-1}) are in extremely good agreement with theory (15.26 and 41.92 cm^{-1}).[b] The torsional band is well-behaved, but the 2388 cm^{-1} band is bizarre, with its $K_a = 2 \leftarrow 2$ and $4 \leftarrow 4$ components displaced upward by 2.03 and 2.62 cm^{-1} relative to the $K_a = 0 \leftarrow 0$ origin (odd K_a values are nuclear spin forbidden). A qualitatively similar shift (+2.4 cm^{-1}) was noted for the (forbidden) $K_a = 1$ level of this mode by Brown et al.,[b] but the calculation was limited to $J = 0$ and 1. These huge shifts are presumably due to hindered internal rotation effects.

[a] C.C. Dutton, D.A. Dows, R. Eikey. S. Evans, R.A. Beaudet, J. Phys. Chem. A **102**, 6904 (1998).

[b] J. Brown, X.-G. Wang, T. Carrington, Jr., G.S. Grubbs II, and R. Dawes, J. Chem. Phys. **140**, 114303 (2014).

WH02 — 1:47 – 2:02

RE-ANALYSIS OF THE DISPERSED FLUORESCENCE SPECTRA OF THE C_3-RARE GAS ATOM COMPLEXES

YI-JEN WANG, *Institute of Atomic and Molecular Sciences, Academia Sinica, Taipei, Taiwan*; ANTHONY MERER, *Department of Chemistry, University of British Columbia, Vancouver, BC, Canada*; YEN-CHU HSU, *Institute of Atomic and Molecular Sciences, Academia Sinica, Taipei, Taiwan.*

The dispersed fluorescence (DF) spectra of the C_3Ne, C_3Ar, C_3Kr, and C_3Xe complexes near the 0 2^- 0- 000, 0 4^- 0-000, 0 2^+ 0- 000 and 100-000 bands of the $\tilde{A}$-$\tilde{X}$ system of C_3[a] have been revisited. Some of the DF spectra of the Ne and Ar complexes have been recently obtained with a slightly improved resolution of 6-10 cm^{-1}. All the DF spectra have been reassigned as emission from van der Waals (vdW) complexes and C_3 fragments. The optically excited C_3-Rg (Rg = rare-gas atom) complexes fluorescence and/or decay down to slightly lower (about 2-30 cm^{-1}) vibrational levels without changing the internal energy of C_3 and then predissociate via the continua of the nearby vibronic states of C_3. The available dissociation channels depend on the binding energy of the ground electronic state complex. Exceptions have been found at the vdW bands near the 0 4^- 0- 000 band of C_3. The binding energies of the ground electronic states of these four complexes will be discussed.

[a] G. Zhang, B.-G. Lin, S.-M. Wen, and Y.-C. Hsu, J. Chem. Phys. **120**, 3189(2004); J.-M. Chao, K. S. Tham, G. Zhang, A. J. Merer, Y.-C. Hsu, and W.-P. Hu, J. Chem. Phys. **134**, 074313(2011)

WH03 — 2:04 – 2:19

INFARED SPECTROSCOPY OF $Mn(CO_2)_n^-$ CLUSTER ANIONS

MICHAEL C THOMPSON, J. MATHIAS WEBER, *JILA and the Department of Chemistry and Biochemistry, University of Colorado-Boulder, Boulder, CO, USA.*

We present infrared photodissociation spectra of $Mn(CO_2)_n^-$ ($n = 2 - 10$) cluster ions. The spectra are interpreted in the framework of density functional theory and compared to other first-row transition metals in anionic clusters with CO_2, allowing to draw conclusions to the structure and spin state of the charge carrier.

WH04 **2:21 – 2:36**

INFRARED SPECTROSCOPY OF $(N_2O)_n^-$ AND $(N_2O)_mO^-$ CLUSTER ANIONS

MICHAEL C THOMPSON, J. MATHIAS WEBER, *JILA and the Department of Chemistry and Biochemistry, University of Colorado-Boulder, Boulder, CO, USA.*

We report infrared photodissociation spectra of nitrous oxide cluster anions, $(N_2O)_n^-$ $(n = 7 - 11)$ and $(N_2O)_mO^-$ $(m = 1 - 13)$. Structural changes of the charge carrier in the clusters are driven by increasing levels of solvation. The spectra are interpreted by comparison with quantum chemical calculations.

WH05 **2:38 – 2:53**

INFRARED SPECTROSCOPY OF PHENOL$^+$-TRIETHYLSILANE DIHYDROGEN-BONDED CLUSTER: INTRINSIC STRENGTH OF THE Si-H$\cdots$H-O DYHYDROGEN BOND

HARUKI ISHIKAWA, TAKAYUKI KAWASAKI, RISA INOMATA, *Department of Chemistry, School of Science, Kitasato University, Sagamihara, Japan.*

Dihydrogen bond is known to be one of the unconventional hydrogen bonds. When a hydrogen atom is bonded to an electropositive atom, such as B, Al, and so on, the hydrogen atom has a partial negative charge. Then, a hydrogen-bond type interaction are formed between the oppositely charged two hydrogen atoms. This interaction is called a dihydrogen bond. In the previous study, we reported the infrared spectroscopy of neutral phenol (PhOH)-triethylsilane (TES) cluster[a]. It was suggested that the Si-H$\cdots$H-O dihydrogen bond should be as strong as the π-type hydrogen bond. In the present study, to investigate the intrinsic strength of the Si-H$\cdots$H-O dihydrogen bond, infrared photodissociation spectroscopy on the PhOH$^+$-TES and PhOH$^+$-diethylmethysilane (DEMS) cationic clusteris was carried out[b].

Both of the clusters exhibit a very broad and intense band centered at about 2860 cm^{-1}. This band is assigned as the OH stretching band of the PhOH$^+$ moiety. Based on the amount of the red-shift of the OH stretching band and the results of the theoretical calculations, the intrinsic strength of the Si-H$\cdots$H-O dihydrogen bond is evaluated to be stronger than that of the π-type hydrogen bond. The proton affinities of TES and DEMS estimated by the theoretical calculation are larger than those of benzene and ethylene. These results are consistent with our experimental observations.

[a] H. Ishikawa, T. Kawasaki, RJ02, the 68th International Symposium on Molecular Spectroscopy (2013)
[b] H. Ishikawa, T. Kawasaki, R. Inomata, *J. Phys. Chem. A* **119**, 601 (2015).

WH06 **2:55 – 3:10**

INFRARED SPECTROSCOPY OF HYDROGEN-BONDED CLUSTERS OF PROTONATED HISTIDINE

MAKOTO KONDO, YASUTOSHI KASAHARA, HARUKI ISHIKAWA, *Department of Chemistry, School of Science, Kitasato University, Sagamihara, Japan.*

Histidine(His), one of the essential amino acids, is involved in active sites in many enzyme proteins, and known to play fundamental roles in human body. Thus, to gain detailed information about intermolecular interactions of His as well as its structure is very important. In the present study, we have recorded IR spectra of hydrogen-bonded clusters of protonated His (HisH$^+$) in the gas phase to discuss the relation between the molecular structure and intermolecular interaction of HisH$^+$. Clusters of HisH$^+$-(MeOH)$_n$ $(n = 1, 2)$ were generated by an electrospray ionization of the MeOH solution of L-His hydrochloride monohydrate. IR photodissociation spectra of HisH$^+$-(MeOH)$_{1,2}$ were recorded. By comparing with the results of the DFT calculations, we determined the structures of these clusters. In the case of $n = 1$ cluster, MeOH is bonded to the imidazole ring as a proton acceptor. The most of vibrational bands observed were well explained by this isomer. However, a free NH stretch band of the imidazole ring was also observed in the spectrum. This indicates an existence of an isomer in which MeOH is bounded to the carboxyl group of HisH$^+$. Furthermore, it is found that a protonated position of His is influenced by a hydrogen bonding position of MeOH. In the case of $n = 2$ cluster, one MeOH molecule is bonded to the amino group, while the other MeOH molecule is separately bonded to the carboxyl group in the most stable isomer. However, there is a possibility that other conformers also exist in our experimental condition. The details of the experimental and theoretical results will be presented in the paper.

WH07 **3:12 – 3:27**

THEORETICAL INVESTIGATION OF THE UV/VIS PHOTODISSOCIATION DYNAMICS OF $ICN^{-}(Ar)_n$ and $BrCN^{-}(Ar)_n$

BERNICE OPOKU-AGYEMAN, ANNE B McCOY, *Department of Chemistry and Biochemistry, The Ohio State University, Columbus, OH, USA.*

An interesting experimental observation in the photodissociation dynamics of $ICN^{-}(Ar)_n$ is that, even in Ar–ICN^{-}, a small fraction of the products recombine to form ICN^{-} following electronic excitation.[a] The two electronic states that are experimentally accessible dissociate into X^{*} + CN^{-} and X^{-} + CN (X=I or Br). The energy differences between these two asymptotes are roughly 0.14 eV and -0.04 eV for ICN^{-} and $BrCN^{-}$, respectively.[a,b] The addition of an argon atom is expected to shift the relative energies of these potential energy surfaces, and provide a mechanism for dissipating some of the excess energy from the electronically excited ICN^{-} and $BrCN^{-}$, altering the product branching.

In this study, the effects of argon solvation are investigated using classical dynamics approaches. In order to simulate the dynamics, potential energy surfaces for the argon clusters are developed using the results obtained from electronic structure calculations of the fragments in the clusters. Specifically, the potential energies are approximated as the interaction in the bare anion and pair-wise interactions between the argon and the dissociation products. The dynamics are then carried out using classical mechanics. Non-adiabatic effects are treated by incorporating surface hopping into the dynamics.[c] To assess the accuracy of the approach, the branching ratios for the bare anions are calculated using classical dynamics and the results are then compared to the previously reported quantum dynamics results.[a,b] Once the results from both the quantum and classical dynamics are shown to be consistent, classical dynamics simulations are then carried out on the clusters.

[a] A. S. Case, E. M. Miller, J. P. Martin, Y. J. Lu, L. Sheps, A. B. McCoy, and W. C. Lineberger, *Angew. Chem., Int. Ed.* **51**, 2651 (2012).
[b] B. Opoku-Agyeman, A. S. Case, J. H. Lehman, W. Carl Lineberger and A. B. McCoy, *J. Chem Phys.* **141**, 084305 (2014).
[c] J. C. Tully, *J. Chem Phys.* **93**, 1061 (1990).

WH08 **3:29 – 3:44**

DISPERSION-DOMINATED π-STACKED COMPLEXES CONSTRUCTED ON A DYNAMIC SCAFFOLD

DEACON NEMCHICK, MICHAEL COHEN, PATRICK VACCARO, *Department of Chemistry, Yale University, New Haven, CT, USA.*

The non-covalent interactions responsible for π-stacking play crucial roles in many fields of modern chemistry, influencing topics ranging from assembly and recognition in biomolecular systems to the design and function of nano-materials. Owing to the propensity for stronger non-aryl forces (*e.g.*, hydrogen bonding) to dominate complex formation, the number of detailed laser-spectroscopic studies on π-bound species is surprisingly limited, with most reported examples focusing on adducts involving combinations of benzene (Bz) and/or substituted-benzene derivatives. A concerted experimental and theoretical effort has been undertaken to explore novel π-stacking motifs based on the non-benzoidal framework of tropolone (TrOH), where the potentially frustrated (tunneling-mediated) transfer of a proton between donor and acceptor sites can afford an *in situ* probe of non-covalent binding. Laser-induced fluorescence spectra acquired for the binary TrOH·Bz complex synthesized under cryogenic free-jet expansion conditions show extensive vibronic features that are red-shifted from the intense $\tilde{A}^1B_2$–$\tilde{X}^1A_1$ absorption resonance of the bare TrOH substrate and display intensity patterns indicative of changing *intermolecular* potential-surface topography upon $\pi^* \leftarrow \pi$ electron promotion. These results, as well as spectral signatures from *intramolecular* TrOH reaction dynamics, will be discussed, with complimentary quantum-chemical calculations serving to provide new insights into the nature of weak, dispersion-dominated interactions.

Intermission

WH09 **4:03 – 4:18**

THE COMPETITION BETWEEN INSERTION AND SURFACE BINDING OF BENZENE TO THE WATER HEPTAMER

PATRICK S. WALSH, *Department of Chemistry, Purdue University, West Lafayette, IN, USA*; DANIEL P. TABOR, EDWIN SIBERT, *Department of Chemistry, The Univeristy of Wisconsin, Madison, WI, USA*; TIMOTHY S. ZWIER, *Department of Chemistry, Purdue University, West Lafayette, IN, USA.*

Previous work on the benzene-(water)$_n$ clusters with n=7 have focused attention on the main conformer, whose S_0-S_1 $6^1{}_0$ R2PI transition appears +138 cm^{-1}above the benzene monomer. Using resonant ion-dip infrared spectroscopy with a higher resolution IR source, we have recently returned to this cluster to record improved OH stretch infrared spectra and more thoroughly consider the possible Bz-$(H_2O)_7$ structures that might give rise to it. Analysis of that spectrum led to its assignment as an inserted cube structure with pseudo-S_4 symmetry. This talk will consider the spectrum and structure of a minor conformer of Bz-$(H_2O)_7$ with R2PI transition 65 $\mathrm{cm}^{-1}$to the blue of the monomer. This spectrum, recorded for the first time, shows a distinctive OH stretch infrared spectrum that is best matched with an expanded prism structure in which the seventh water molecule inserts into one edge of the hexamer prism. In this case, benzene acts as acceptor for an OH$\cdots\pi$ H-bond, sitting on the surface of a preformed water heptamer structure. The infrared spectra of the two Bz-$(H_2O)_7$ structures are compared, and the results of a local mode Hamiltonian model are applied to make an assignment for the observed structure. The monomer Hamiltonians resulting from this model shed light on the unique two- and three-coordinate water molecules found in this structure.

WH10 **4:20 – 4:35**

VIBRATIONAL SPECTROSCOPY OF BENZENE-(WATER)$_N$ CLUSTERS WITH $N = 6, 7$

DANIEL P. TABOR, EDWIN SIBERT, *Department of Chemistry, The Univeristy of Wisconsin, Madison, WI, USA*; RYOJI KUSAKA, *Chemistry, Hiroshima University, Higashi-Hiroshima, Japan*; PATRICK S. WALSH, TIMOTHY S. ZWIER, *Department of Chemistry, Purdue University, West Lafayette, IN, USA.*

The investigation of benzene-water clusters (Bz-$(H_2O)_n$) provides insight into the relative importance π-hydrogen bond interactions in cluster formation. Taking advantage of the higher resolution of current IR sources, isomer-specific resonant ion-dip infrared (RIDIR) spectra were recorded in the OH stretch region (3000-3750 cm^{-1}). A local mode Hamiltonian for describing the OH stretch vibrations of water clusters is applied to Bz-$(H_2O)_6$ and Bz-$(H_2O)_7$ and compared with the RIDIR spectra. These clusters are the smallest water clusters in which three-dimensional H-bonded networks containing three-coordinate water molecules begin to be formed, and are therefore particularly susceptible to re-ordering or re-shaping in response to the presence of a benzene molecule. The spectrum of Bz-$(H_2O)_6$ is assigned to an inverted book structure while the major conformer of Bz-$(H_2O)_7$ is assigned to an S_4-derived inserted cubic structure in which the benzene occupies one corner of the cube. The local mode model is used to extract monomer Hamiltonians for individual water molecules, including stretch-bend Fermi resonance and intra-monomer couplings. The monomer Hamiltonians divide into sub-groups based on their local H-bonding architecture (DA, DDA, DAA) and the nature of their interaction with benzene.

WH11 **4:37 – 4:52**

THEORETICAL STUDY OF THE IR SPECTROSCOPY OF BENZENE-(WATER)$_N$ CLUSTERS

DANIEL P. TABOR, EDWIN SIBERT, *Department of Chemistry, The Univeristy of Wisconsin, Madison, WI, USA*; RYOJI KUSAKA, *Chemistry, Hiroshima University, Higashi-Hiroshima, Japan*; PATRICK S. WALSH, TIMOTHY S. ZWIER, *Department of Chemistry, Purdue University, West Lafayette, IN, USA*.

The local mode Hamiltonian that assigns RIDIR spectra for Bz-$(H_2O)_6$ and Bz-$(H_2O)_7$ is explored in detail for Bz-$(H_2O)_n$ with $n = 3 - 7$. In addition to contributions from OH stretches, the Hamiltonian includes the anharmonic coupling of each water monomer's bend overtone and its OH stretch fundamentals, which is necessary for accurately modeling 3150-3300 cm^{-1} region of the spectra. The parameters of the Hamiltonian can be calculated using either MP2 or density functional theory. The relative strengths and weaknesses of these two electronic structure approaches are examined to gain further physical understanding. Initial assignments of Bz-$(H_2O)_6$ and Bz-$(H_2O)_7$ were based on a linear scaling of M06-2X harmonic frequencies. In most cases, counterpoise-corrected MP2 calculations obtain similar frequencies (across all cluster sizes) if stretch anharmonicity is taken into account. Individual "monomer Hamiltonians" are constructed via the application of fourth order Van Vleck perturbation theory to MP2 potential energy surfaces. These calculations elucidate the sensitivity of intra-monomer couplings to chemical environment. The presence of benzene has particularly important consequences for the spectra of the Bz-$(H_2O)_{3-5}$ clusters, in which the symmetry of the water cycles is broken by π-H-bonding to benzene. The nature of these perturbations is discussed.

WH12 **4:54 – 5:09**

SPECTROSCOPIC INVESTIGATION OF TEMPERATURE EFFECTS ON THE HYDRATION STRUCTURE OF THE PHENOL CLUSTER CATION

REONA YAGI, YASUTOSHI KASAHARA, HARUKI ISHIKAWA, *Department of Chemistry, School of Science, Kitasato University, Sagamihara, Japan*.

Owing to recent technical developments of various spectroscopies, microscopic hydration structures of various clusters in the gas phase have been determined so far. The next step for further understanding of the microscopic hydration is to reveal the temperature effect, such as a fluctuation of the hydration structure. Thus, we are carrying out photodissociation spectroscopy on the hydrated phenol cation clusters, $[PhOH(H_2O)_n]^+$. Since electronic spectra of $[PhOH(H_2O)_n]^+$ have been reported already[a], this system is suitable for our purpose.

In the present study, we use our temperature-variable 22-pole ion trap apparatus[b]. The ions in the trap become thermal equilibrium condition by multiple collisions with temperature-controlled He buffer gas. By this way, the temperature of the ions can be controlled.

In the electronic spectrum of the $n = 5$ cluster measured at 30 K, a sharp band is observed. It shows that the temperature of ions are well-controlled. Contrary to the $n = 5$ cluster, the $n = 6$ cluster exhibits a wider band shape. The temperature dependence of the band shape indicates the existence of several, at least two, isomers in the present experimental condition.

[a] S. Sato, N. Mikami *J. Phys. Chem.* **100**, 4765 (1996).
[b] H. Ishikawa, T. Nakano, T. Eguchi, T. Shibukawa, K. Fuke *Chem. Phys. Lett.* **514**, 234 (2011).

WH13 **5:11 – 5:26**

ULTRAVIORET AND INFRARED PHOTODISSOCIATION SPECTROSCOPY OF HYDRATED ANILINIUM ION

ITARU KURUSU, REONA YAGI, YASUTOSHI KASAHARA, HARUKI ISHIKAWA, *Department of Chemistry, School of Science, Kitasato University, Sagamihara, Japan.*

To understand the temperature effect on the microscopic hydration, we have been carrying out the laser spectroscopy of temperature-controlled hydrated phenol cation clusters using our temperature-variable ion trap apparatus[a,b]. In the present study, we have chosen an anilinium ion (AnH^+) as a solute. Since the phenol cation has $(\pi)^{-1}$ configuration, the phenyl ring does not play as a proton-acceptor. On the contrary, the π-orbitals in the AnH^+ are fulfilled and both the NH_3^+ and phenyl groups can behave as hydrogen-bonding sites. Thus, hydration structures around the AnH^+ are expected to be different from those of the phenol cation. Since there is no spectroscopic report on the hydrated AnH^+ clusters, we have carried out the UV and IR photodissociation spectroscopy of $AnH^+(H_2O)$ clusters.

In the present study, the $AnH^+(H_2O)$ is produced by an electrospray ionization method. As the first step, spectroscopic measurements are carried out without temperature control. In the UV photodissociation spectrum, the 0-0 band appears at 36351 cm^{-1} which is red-shifted by 1863 cm^{-1} from that of the AnH^+ monomer[c]. The band pattern is similar to that of the AnH^+ monomer. This indicates that the structure of the AnH^+ is not so affected by the single hydration. In the IR photodissociation spectrum, OH stretching band of the H_2O moiety and free NH stretching band of AnH^+ moiety are observed. Comparison with the results of the DFT calculation at M05-2X/6-31++G(d,p) level, we determined the structure of the $AnH^+(H_2O)$ cluster.

[a] R. Yagi, Y. Kasahara, H. Ishikawa, the 70th International Symposium on Molecular Spectroscopy (2015).
[b] H. Ishikawa, T. Nakano, T. Eguchi, T. Shibukawa, K. Fuke *Chem. Phys. Lett.* **514**, 234 (2011).
[c] G. Féraud, *et al. Phys. Chem. Chem. Phys.* **16**, 5250 (2014).

WH14 **5:28 – 5:43**

WATER-NETWORK MEDIATED, ELECTRON INDUCED PROTON TRANSFER IN ANIONIC $[C_5H_5N{\cdot}(H_2O)_n]^-$ CLUSTERS: SIZE DEPENDENT FORMATION OF THE PYRIDINIUM RADICAL FOR $n \geq 3$

ANDREW F DeBLASE, *Department of Chemistry, Purdue University, West Lafayette, IN, USA*; GARY H WEDDLE, *Department of Chemistry and Biochemistry, Fairfield University, Fairfield, CT, USA*; KAYE A ARCHER, KENNETH D. JORDAN, *Department of Chemistry, University of Pittsburgh, Pittsburgh, PA, USA*; MARK JOHNSON, *Department of Chemistry, Yale University, New Haven, CT, USA.*

As an isolated species, the radical anion of pyridine (Py^-) exists as an unstable transient negative ion, while in aqueous environments it is known to undergo rapid protonation to form the neutral pyridinium radical [$PyH^{(0)}$] along with hydroxide. Furthermore, the negative adiabatic electron affinity (AEA) of Py^- can become diminished by the solvation energy associated with cluster formation. In this work, we focus on the hydrates $[Py{\cdot}(H_2O)\_n]^-$ with n = 3-5 and elucidate the structures of these water clusters using a combination of vibrational predissociation and photoelectron spectroscopies. We show that H-trasfer to form $PyH^{(0)}$ occurs in these clusters by the infrared signature of the nascent hydroxide ion and by the sharp bending vibrations of aromatic ring CH bending.

WI. Astronomy

Wednesday, June 24, 2015 – 1:30 PM

Room: 274 Medical Sciences Building

Chair: Holger S. P. Müller, Universität zu Köln, Köln, NRW, Germany

WI01 **1:30 – 1:45**

THE NEW ALMA PROTOTYPE 12 M TELSCOPE OF THE ARIZONA RADIO OBSERVATORY: TRANSPORT, RECOMMISSIONING, AND FIRST LIGHT

LUCY ZIURYS, *Department of Chemistry and Biochemistry, University of Arizona, Tucson, AZ, USA*; N J EMERSON, T W FOLKERS, R W FREUND, D FORBES, G P REILAND, M McCOLL, S C KEEL, S H WARNER, J KINGSLEY, *Steward Observatory, University of Arizona, Tucson, AZ, USA*; DeWAYNE T HALFEN, *Department of Chemistry and Astronomy, University of Arizona, Tucson, AZ, USA.*

In March 2013, the Arizona Radio Observatory (ARO) acquired the European 12 m prototype antenna of the Atacama Large Millimeter Array (ALMA) from the European Southern Observatory (ESO). The antenna was located at the Very Large Array (VLA) site near Socorro, New Mexico. During the summer of 2013, the antenna was prepared for the move to the ARO Kitt Peak site in Arizona, and in November 2013, the actual transport began. The 97 ton antenna was transported to Arizona in two major parts: the 40 ft. reflector and the base/receiver cabin, which were reassembled in the dome at Kitt Peak in December 2013. Recommissioning began in January 2014, and "first light" observations occurred in September 2014 at 115 GHz. Scientific observations began in December 2014.

WI02 **1:47 – 2:02**

FIRST SCIENTIFIC OBSERVATIONS WITH THE NEW ALMA PROTOTYPE ANTENNA OF THE ARIZONA RADIO OBSERVATORY: HCN AND CCH IN THE HELIX NEBULA

LUCY ZIURYS, *Department of Chemistry and Biochemistry, University of Arizona, Tucson, AZ, USA*; DEBORAH SCHMIDT, *Department of Astronomy, University of Arizona, Tucson, AZ, USA.*

Observations have been conducted with the new 12 m antenna of the Arizona Radio Observatory (ARO) at 3 mm towards the Helix Nebula. This object is the oldest known planetary nebula. The $J = 1 \rightarrow 0$ transition of HCN at 88 GHz and two hyperfine components of the $N = 1 \rightarrow 0$ line of CCH at 87 GHz were observed towards nine positions sampling different regions across the nebula. Both molecules were detected at all positions at the 5 – 30 mK intensity level. The line profiles exhibited multiple velocity components, as also seen in HCO^+ and H_2CO towards the same positions. The widespread distribution of HCN and CCH at this late stage of stellar evolution is further evidence that polyatomic molecules are being dispersed into the ISM. It also suggests that the progenitor star in the Helix is carbon-rich.

WI03 **2:04 – 2:19**

CCH AND HNC IN PLANETARY NEBULAE

DEBORAH SCHMIDT, *Department of Astronomy, University of Arizona, Tucson, AZ, USA*; LUCY ZIURYS, *Department of Chemistry and Biochemistry, University of Arizona, Tucson, AZ, USA.*

A survey of CCH and HNC has been conducted towards a sample of ten planetary nebulae of varying ages using the Submillimeter Telescope (SMT) of the Arizona Radio Observatory (ARO) at 1 mm. The $N = 3 \rightarrow 2$ transition of CCH at 262 GHz and the $J = 3 \rightarrow 2$ line of HNC at 272 GHz were observed using the ALMA Band 6 receiver at the SMT. The molecules were detected in most of the sources where HCN and HCO^+ had been identified in a previous survey. Molecular abundances for CCH and HNC have been determined in these nebulae, as well as [HCN]/[HNC] ratios. These observations further support the notion that the chemistry in planetary nebulae remains active despite the ultraviolet radiation field from the central white dwarf star.

WI04 **2:21 – 2:36**

MAPPING THE SPATIAL DISTRIBUTION OF METAL-BEARING OXIDES IN VY CANIS MAJORIS

ANDREW BURKHARDT, S. TOM BOOTH, *Department of Astronomy, The University of Virginia, Charlottesville, VA, USA*; ANTHONY REMIJAN, *ALMA, National Radio Astronomy Observatory, Charlottesville, VA, USA*; BRANDON CARROLL, *Division of Chemistry and Chemical Engineering, California Institute of Technology, Pasadena, CA, USA*; LUCY ZIURYS, *Department of Astronomy, University of Arizona, Tucson, AZ, USA.*

The formation of silicate-based dust grains is not well constrained. Despite this, grain surface chemistry is essential to modern astrochemical formation models. In carbon-poor stellar envelopes, such as the red hypergiant VY Canis Majoris (VY CMa), metal-bearing oxides, the building blocks of silicate grains, dominate the grain formation, and thus are a key location to study dust chemistry. TiO_2, which was only first detected in the radio recently (Kaminski et al., 2013a), has been proposed to be a critical molecule for silicate grain formation, and not oxides containing more abundant metals (eg. Si, Fe, and Mg) (Gail and Sedlmayr, 1998). In addition, other molecules, such as SO_2, have been found to trace shells produced by numerous outflows pushing through the expanding envelope, resulting in a complex velocity structure (Ziurys et al., 2007). With the advanced capabilities of ALMA, it is now possible to individually resolve the velocity structure of each of these outflows and constrain the underlying chemistry in the region. Here, we present high resolution maps of rotational transitions of several metal-bearing oxides in VY CMa from the ALMA Band 7 and Band 9 Science Verification observations. With these maps, the physical parameters of the region and the formation chemistry of metal-bearing oxides will be studied.

WI05 **2:38 – 2:53**

C^+ AND THE CONNECTION BETWEEN DIFFERENT TRACERS OF THE DIFFUSE INTERSTELLAR MEDIUM

STEVEN FEDERMAN, JOHNATHAN S RICE, *Physics and Astronomy, University of Toledo, Toledo, OH, USA*; JORGE L PINEDA, WILLIAM D LANGER, PAUL F GOLDSMITH, *Jet Propulsion Laboratory, California Institute of Technology, Pasadena, CA, USA*; NICOLAS FLAGEY, *Institute for Astronomy, University of Hawaii, Hilo, Hawaii, USA.*

Using radio, microwave, sub-mm, and optical data, we analyze several lines of sight toward stars generally closer than 1 kpc on a component by component basis. We derive the component structure seen in emission from C^+, H I, and CO and its isotopoloques, along with those for CH^+, CH, CN, Ca II, and Ca I in absorption. We study how these tracers are related to the CO-Dark H_2 gas being probed by C^+ emission and discuss the kinematic connections among the species. Physical conditions of the various components seen in absorption, especially density, are inferred from a simple chemical analysis based on the column densities of CH^+, CH, and CN.

WI06 **2:55 – 3:10**

INFERRING THE TEMPERATURE AND DENSITY OF DIFFUSE INTERSTELLAR CLOUDS FROM C_3 OBSERVATIONS

NICOLE KOEPPEN, *Department of Chemistry, University of Illinois at Urbana-Champaign, Urbana, IL, USA*; BENJAMIN J. McCALL, *Departments of Chemistry and Astronomy, University of Illinois at Urbana-Champaign, Urbana, IL, USA.*

Observations of carbon chain molecules are useful in determining the number densities and temperatures of diffuse interstellar clouds. In 2003, C_3 was observed towards ten different sightlines and the rotational distributions were determined using the oscillator strengths available at that time.[a] The population of each rotational level was adjusted individually in order to obtain the best fit for all of the P, Q, and R branch lines. This past year, the effect of perturbing states on the C_3 spectrum was elucidated, and improved oscillator strengths determined.[b] With these new values, we have redetermined the rotational distribution of C_3 in these ten sightlines, and used a rotational excitation model analogous to that of Roueff et al.[c] and collisional cross sections from Smith et al.[d] to infer the kinetic temperatures and number densities.

[a] Adamkovics et al. Ap.J., 595, 235 (2003)
[b] Schmidt et al. MNRAS, 441, 1134 (2014)
[c] Roueff et al. A&A, 384, 629 (2002)
[d] Smith et al. J. Phys. Chem. A, 118, 6351 (2014)

WI07 **3:12 – 3:27**

NEW BACKGROUND INFRARED SOURCES FOR STUDYING THE GALACTIC CENTER'S INTERSTELLAR GAS

THOMAS R. GEBALLE, *Gemini Observatory, Hilo, HI, USA*; TAKESHI OKA, *Department of Astronomy and Astrophysics, Chemistry, The University of Chicago, Chicago, IL, USA*; E. LAMBRIDES, *Astrophysics, American Museum of Natural History, New York, NY, USA*; S. C. C. YEH, , *Subaru Telescope, Hilo, HI, USA*; B. SCHLEGELMILCH, *Astronomy, UCLA, Los Angeles, CA, USA*; MIWA GOTO, , *Max Planck Institute for Extraterrestrial Physics, Munich, Germany*; C W WESTRICK, *College of Dupage, Glen Ellyn, IL, USA.*

We are nearing completion of a low-resolution 2.0-2.5 μm (4000-5000 cm^{-1}) survey of ~500 very red point-like objects in the Central Molecular Zone (CMZ) of the Milky Way Galaxy. The goal is to find bright objects with intrinsically featureless or nearly featureless spectra that are suitable as background light sources for high-resolution infrared absorption spectroscopy of H_3^+ and CO in the Galactic center's interstellar gas, on sightlines spread across the CMZ. Until recently very few such objects had been known outside of two clusters of hot and luminous stars close to the very center. We have used Spitzer Space Telescope 3.6–8.0-μm photometry and 2-Micron All Sky Survey 1.0–2.5-μm photometry to identify candidates with a significant probability of being stars embedded in circumstellar dust, and over the last several years have been acquiring low resolution spectra of them to determine their natures. The low resolution spectra, which encompass the wavelengths of the first overtone band heads of CO, which are prominent in cool stellar photospheres, show that by far the majority of candidates are very cool and/or highly reddened red giants, which are unsuitable as background sources because of their complex spectra . However, approximately ten percent of the candidates have featureless or nearly featureless spectra and are useful for investigations of the interstellar gas. Most of these have continua rising steeply to longer wavelengths and are luminous, dust embedded stars.

WI08 **3:29 – 3:44**

CO SPECTRAL LINE ENERGY DISTRIBUTIONS IN ORION SOURCES: TEMPLATES FOR EXTRAGALACTIC OBSERVATIONS

NICK INDRIOLO, EDWIN BERGIN, *Department of Astronomy, University of Michigan, Ann Arbor, MI, USA.*

The *Herschel Space Observatory* has enabled the observation of CO emission lines originating in the $J = 5$ through $J = 48$ rotational levels. Surveys of active galaxies (e.g., starbursts, Seyferts, ULIRGs) detect emission from levels as high as $J = 30$, but the precise excitation mechanisms responsible for producing the observed CO SLEDs (Spectral Line Energy Distribution) remain ambiguous. To better constrain the possible excitation mechanisms in extragalactic sources, we investigate the CO SLEDs arising from sources with known characteristics in the nearby Orion region. Targets include Orion-KL (high-mass star forming region containing a hot core, embedded protostars, outflows, and shocks), Orion South (high-mass star forming region containing embedded protostars, outflows, and a photodissociation region), Orion H_2 Peak 1 (molecular shock), and the Orion Bar (a photodissociation region). Emission lines from complex sources are decomposed using velocity information from high spectral resolution observations made with *Herschel*-HIFI (Heterodyne Instrument for the Far-Infrared). Each source and/or component is taken as a template for a particular excitation mechanism, and then applied to interpret excitation in more distant regions within the Galaxy, as well as external galaxies.

Intermission

WI09 **4:03 – 4:18**

THE DISTRIBUTION, EXCITATION, AND ABUNDANCE OF C^+, CH^+, AND CH IN ORION KL

HARSHAL GUPTA, PATRICK MORRIS, *Infrared Processing and Analysis Center, California Institute of Technology, Pasadena, CA, USA*; ZSOFIA NAGY, *Department of Physics and Astronomy, University of Toledo, Toledo, OH, USA*; JOHN PEARSON, *Jet Propulsion Laboratory, California Institute of Technology, Pasadena, CA, USA*; VOLKER OSSENKOPF, *I. Physikalisches Institut, Universität zu Köln, Köln, Germany.*

The CH^+ ion was one of the first molecules identified in the interstellar gas over 75 years ago, and is postulated to be a key species in the initial steps of interstellar carbon chemistry. The high observed abundances of CH^+ in the interstellar gas remain a puzzle, because the main production pathway of CH^+, *viz.*, $C^+ + H_2 \rightarrow CH^+ + H$, is so endothermic (4640 K), that it is unlikely to proceed at the typical temperatures of molecular clouds. One way in which the high endothermicity may be overcome, is if a significant fraction of the H_2 is vibrationally excited, as is the case in molecular gas exposed to intense far-ultraviolet radiation fields. Elucidating the formation of CH^+ in molecular clouds requires characterization of its spatial distribution, as well as that of the key participants in the chemical pathways yielding CH^+. Here we present high-resolution spectral maps of the two lowest rotational transitions of CH^+, the fine structure transition of C^+, and the hyperfine-split fine structure transitions of CH in a $\sim 3' \times 3'$ region around the Orion Kleinmann-Low (KL) nebula, obtained with the *Herschel Space Observatory's* Heterodyne Instrument for the Far-Infrared (HIFI).[a] We compare these maps to those of CH^+ and C^+ in the Orion Bar photodissociation region (PDR), and discuss the excitation and abundance of CH^+ toward Orion KL in the context of chemical and radiative transfer models, which have recently been successfully applied to the Orion Bar PDR.[b]

[a]These observations were done as part of the Herschel observations of EXtraordinary sources: the Orion and Sagittarius star-forming regions (HEXOS) Key Programme, led by E. A. Bergin at the University of Michigan, Ann Arbor, MI.

[b]Nagy, Z. et al. 2013, A&A 550, A96

WI10 **4:20 – 4:35**

THE DISTRIBUTION OF SH^+ AROUND ORION KL

HARSHAL GUPTA, *Infrared Processing and Analysis Center, California Institute of Technology, Pasadena, CA, USA*; KARL M. MENTEN, *Millimeter- und Submillimeter-Astronomie, Max-Planck-Institut für Radioastronomie, Bonn, NRW, Germany*; ZSOFIA NAGY, *Department of Physics and Astronomy, University of Toledo, Toledo, OH, USA*; PATRICK MORRIS, *Infrared Processing and Analysis Center, California Institute of Technology, Pasadena, CA, USA*; VOLKER OSSENKOPF, *I. Physikalisches Institut, Universität zu Köln, Köln, Germany*; NATHAN CROCKETT, *Geological and Planetary Sciences , California Institute of Techonolgy, Pasadena, CA, USA*; JOHN PEARSON, *Jet Propulsion Laboratory, California Institute of Technology, Pasadena, CA, USA.*

The SH^+ ion is thought to probe energetic processes such as shocks, turbulence, and intense UV fields in interstellar clouds, because its principal formation route is endothermic by a temperature equivalent of 10117 K—more than twice as endothermic as that of the well-known ion CH^+. Here we present spectral maps of the lowest fine-structure transitions of SH^+ obtained with the Atacama Pathfinder EXperiment (APEX) telescope over a small ($\sim 1' \times 1'$) region in the vicinity of the Orion Kleinmann-Low (KL) nebula, the closest chemically rich star-forming region known to contain hot young stars and shocked gas. Observations of SH^+ provide complementary information to those of CH^+ (see, e.g., abstract P1303), and may help assess the role of UV-irradiation vs shocks in the production of SH^+ and CH^+, as well as the molecular fraction and electron density in the regions traced by SH^+ and CH^+. We also present chemical and radiative transfer models to help elucidate the production of SH^+ and CH^+, and assess the utility of these ions as probes of processes that regulate the thermal balance of the interstellar gas and influence star formation in molecular clouds.

WI11 **4:37 – 4:52**

CARMA 1 CM LINE SURVEY OF ORION-KL

DOUGLAS FRIEDEL, LESLIE LOONEY, *Department of Astronomy, University of Illinois at Urbana-Champaign, Urbana, IL, USA*; JOANNA F. CORBY, *Department of Astronomy, The University of Virginia, Charlottesville, VA, USA*; ANTHONY REMIJAN, *ALMA, National Radio Astronomy Observatory, Charlottesville, VA, USA.*

We have conducted the first 1 cm (27-35 GHz) line survey of the Orion-KL region by an array. With a primary beam of ~4.5 arcminutes, the survey looks at a region ~166,000 AU (0.56 pc) across. The data have a resolution of ~6 arcseconds on the sky and 97.6 kHz(1.07-0.84 km/s) in frequency. This region of frequency space is much less crowded than at 3mm or 1mm frequencies and contains the fundamental transitions of several complex molecular species, allowing us to probe the largest extent of the molecular emission. We present the initial results, and comparison to 3mm results, from several species including, dimethyl ether [$(CH_3)_2O$], ethyl cyanide [C_2H_5CN], acetone [$(CH_3)_2CO$], SO, and SO_2.

WI12 **4:54 – 5:09**

CHEMICAL COMPLEXITY IN THE SHOCKED OUTFLOW L1157-B REVEALED BY CARMA

NIKLAUS M DOLLHOPF, *Department of Astronomy, The University of Virginia, Charlottesville, VA, USA*; BRETT A. McGUIRE, *NAASC, National Radio Astronomy Observatory, Charlottesville, VA, USA*; BRANDON CARROLL, *Division of Chemistry and Chemical Engineering, California Institute of Technology, Pasadena, CA, USA*; ANTHONY REMIJAN, *ALMA, National Radio Astronomy Observatory, Charlottesville, VA, USA.*

We present results from a targeted chemical search toward the prototypical shocked outflow L1157. L1157-B0, -B1, and -B2 are shocked regions within the outflow from the Class 0 low-mass protostar L1157-mm. We have mapped a variety of molecular tracers in the region with typical spatial resolutions of $\sim 3''$ using CARMA, and find differences in the chemical makeups between shocked regions within the same precursor outflow material. We present observations of CH_3OH, HCO^+, HCN, and the first maps of HNCO in the source. We will examine the utility of HNCO as a sensitive tracer of the shocks in this source, and finally, we will discuss what insights we can gain into the chemical evolution, and evolutionary time scales, that have given rise to the differentiation we see between the shocks.

WI13 **5:11 – 5:26**

THE CURIOUS CASE OF NH_2OH: HUNTING A DIRECT AMINO ACID PRECURSOR SPECIES IN THE INTERSTELLAR MEDIUM

BRETT A. McGUIRE, *NAASC, National Radio Astronomy Observatory, Charlottesville, VA, USA*; BRANDON CARROLL, *Division of Chemistry and Chemical Engineering, California Institute of Technology, Pasadena, CA, USA*; NIKLAUS M DOLLHOPF, *Department of Astronomy, The University of Virginia, Charlottesville, VA, USA*; NATHAN CROCKETT, *Geological and Planetary Sciences , California Institute of Techonolgy, Pasadena, CA, USA*; GEOFFREY BLAKE, *Division of Chemistry and Chemical Engineering, California Institute of Technology, Pasadena, CA, USA*; ANTHONY REMIJAN, *ALMA, National Radio Astronomy Observatory, Charlottesville, VA, USA.*

Despite the detection of amino acids, the building blocks of the proteins that support life, in cometary and meteoritic samples, we do not yet understand the conditions under which these life-essential species have formed. Hydroxylamine (NH_2OH) is potentially a direct precursor to the formation of the amino acids glycine and alanine in the ISM, through reaction with acetic and propionic acids. Recent laboratory and modeling work has shown that there are a variety of pathways to the formation of NH_2OH in interstellar ices both efficiently and in high abundance. Here, we present the result of a deep, multi-telescope search for NH_2OH in the shocked, complex molecular source L1157. We find no evidence suggesting the presence of this important precursor, and discuss the implications of this non-detection on the reactivity of NH_2OH both within the ices, and in the gas-phase ISM. We will also discuss how these observations should inform the direction of future studies, both in the laboratory and with state-of-the-art telescopes such as ALMA.

WI14 **5:28 – 5:43**

NEW RESULTS FROM THE CARMA LARGE-AREA STAR FORMATION SURVEY (CLASSY)

ROBERT J HARRIS, LESLIE LOONEY, DOMINIQUE M. SEGURA-COX, *Department of Astronomy, University of Illinois at Urbana-Champaign, Urbana, IL, USA*; MANUEL FERNANDEZ-LOPEZ, *Instituto Argentino de Radioastronomía, Centro Científico Tecnológico La Plata, Villa Elisa, Argentina*; LEE MUNDY, SHAYE STORM, MAXIME RIZZO, *Department of Astronomy, University of Maryland, College Park, MD, USA*; KATHERINE LEE, *Radio and Geoastronomy Division, Harvard-Smithsonian Center for Astrophysics, Cambridge, MA, USA*; HÉCTOR ARCE, *Astronomy Department, Yale University, New Haven, CT, USA*.

Interferometric imaging spectroscopy of molecular clouds permits the physical and thermodynamic structure, kinematics, and chemistry of molecular clouds to be probed over a wide range of spatial scales, from entire clouds to the individual cores where stars are born. As such, it allows the study of what fundamental physical processes are at play in star formation. The CARMA Large Area Star-formation Survey (CLASSy) Key Project surveyed dense gas tracers (the HCN, HCO+, and N_2H^+ J=1-0 emission) and dust continuum emission over 700 square arc-minutes from 3 fields in Perseus (NGC 1333, Barnard 1, and L1451) and 2 fields in Serpens (Serpens Main and Serpens South), with sensitivity to structures on spatial scales ranging from 1000 AU to several parsecs. We have used these data to characterize the importance of turbulence and magnetic fields in star formation on physical scales ranging from the largest clouds to the immediate environment of individual young stellar objects. We present results from both CLASSy and a significant extension of this project, CLASSy Prime, to more deeply survey these regions in both the same and different tracers, including several organic molecules. In particular, we focus on discrepancies between the structure of filaments seen in line emission – particularly N_2H^+ – and the same filaments seen in dust emission, and we suggest that these might be due to excitation conditions and/or chemical effects. We also discuss how emission from the different molecules that have been observed with CLASSy Prime highlight kinematics of different substructures within these regions.

WI15 **5:45 – 6:00**

ADMIT: ALMA DATA MINING TOOLKIT

DOUGLAS FRIEDEL, LESLIE LOONEY, *Department of Astronomy, University of Illinois at Urbana-Champaign, Urbana, IL, USA*; LISA XU, *NCSA, University of Illinois at Urbana-Champaign, Urbana, IL, USA*; MARC W. POUND, PETER J. TEUBEN, KEVIN P. RAUCH, LEE MUNDY, *Department of Astronomy, University of Maryland, College Park, MD, USA*; JEFFREY S. KERN, *NRAO, NRAO, Socorro, NM, USA*.

ADMIT (ALMA Data Mining Toolkit) is a toolkit for the creation and analysis of new science products from ALMA data. ADMIT is an ALMA Development Project written purely in Python. While specifically targeted for ALMA science and production use after the ALMA pipeline, it is designed to be generally applicable to radio-astronomical data. ADMIT quickly provides users with a detailed overview of their science products: line identifications, line 'cutout' cubes, moment maps, emission type analysis (e.g., feature detection), etc. Users can download the small ADMIT pipeline product (<20MB), analyze the results, then fine-tune and re-run the ADMIT pipeline (or any part thereof) on their own machines and interactively inspect the results. ADMIT will have both a GUI and command line interface available for this purpose. By analyzing multiple data cubes simultaneously, data mining between many astronomical sources and line transitions will be possible. Users will also be able to enhance the capabilities of ADMIT by creating customized ADMIT tasks satisfying any special processing needs. Future implementations of ADMIT may include EVLA and other instruments.

WJ. Non-covalent interactions

Wednesday, June 24, 2015 – 1:30 PM

Room: 217 Noyes Laboratory

Chair: Wolfgang Jäger, University of Alberta, Edmonton, AB, Canada

WJ01 **1:30 – 1:45**

FORMATION OF COMPLEXES c-$C_3H_6\cdots$MCl (M = Ag or Cu) AND THEIR CHARACTERIZATION BY BROADBAND ROTATIONAL SPECTROSCOPY

DANIEL P. ZALESKI, JOHN CONNOR MULLANEY, NICK WALKER, *School of Chemistry, Newcastle University, Newcastle-upon-Tyne, United Kingdom*; ANTHONY LEGON, *School of Chemistry, University of Bristol, Bristol, United Kingdom.*

New molecules formed by the non-covalent interaction of cyclopropane (c-C_3H_6) with MCl, where M is either Ag or Cu, have been detected and characterized by means of broadband rotational spectroscopy. They were synthesized by laser ablation of a silver or copper rod in the presence of a gaseous sample containing 1% each of c-C_3H_6 and CCl_4, with the remainder argon. Spectra of several isotopologues of each complex have been analysed. The title molecules are found to have C_{2v} symmetry, and the geometry can be described by the MCl subspecies coordinating "edge on" to the cyclopropane ring. Experimental structures will be compared with those from ab initio calculations and those of related species.

WJ02 **1:47 – 2:02**

ROTATIONAL SPECTROSCOPY OF MONOFLUOROETHANOL AGGREGATES WITH ITSELF AND WITH WATER

JAVIX THOMAS, WENYUAN HUANG, XUNCHEN LIU[a], WOLFGANG JÄGER, YUNJIE XU, *Department of Chemistry, University of Alberta, Edmonton, AB, Canada.*

Fluoroalcohols are used as common cosolvents for studies of the secondary and tertiary substructures of polypeptides and proteins in aqueous solution. It has been proposed that small fluoroalcohol aggregates are crucial for the protein structural altering process.[1] A rotational spectroscopic study of the monofluoroethanol (MFE) dimer was reported by our group before.[2] In this presentation, we report our recent results on the MFE trimer and MFE-water clusters. We analyze the competitive formation of intra- and intermolecular hydrogen bonds, processes that may be crucial for the changes in protein structure that occur in fluoroalcohol-water solution. We show that the MFE trimer takes on a much different binding topology from the recently reported phenol trimer.[3] The results will also be compared to the closely related 2,2,2-trifluoroethanol systems.

[1] H. Reiersen, A. R. Rees, Protein Eng. 2000, 13, 739 – 743. [2] X. Liu, N. Borho, Y. Xu, Chem. Eur. J. 2009, 15, 270 – 277. [3] a) N. A. Seifert, A. L. Steber, J. L. Neill, C. Pérez, D. P. Zaleski, B. H. Pate, A. Lesarri, Phys. Chem. Chem. Phys., 2013, 15, 11468; b) T. Ebata, T. Watanabe, N. Mikami, J. Phys. Chem., 1995, 99, 5761.

[a] Permanent address: Mechanical Department, Shanghai Jiao Tong University, Shanghai, P. R. China.

WJ03 **2:04 – 2:19**

O-TOLUIC ACID MONOMER AND MONOHYDRATE: ROTATIONAL SPECTRA, STRUCTURES, AND ATMOSPHERIC IMPLICATIONS

ELIJAH G SCHNITZLER, BRANDI L M ZENCHYZEN, WOLFGANG JÄGER, *Department of Chemistry, University of Alberta, Edmonton, AB, Canada.*

Clusters of carboxylic acids with water, sulfuric acid, and other atmospheric species potentially increase the rate of new particle formation in the troposphere.[a,b] Here, we present high-resolution pure rotational spectra of *o*-toluic acid and its complex with water in the range of 5-14 GHz, measured with a cavity-based molecular beam Fourier-transform microwave spectrometer. In both the monomer and the complex, the carboxylic acid functional group adopts a *syn*- conformation, with the acidic proton oriented away from the aromatic ring. In the complex, water participates in two hydrogen bonds, forming a six-membered intermolecular ring. Despite its large calculated *c*-dipole moment, no *c*-type transitions were observed for the complex, because of a large amplitude "wagging" motion of the unbound hydrogen of water, similar to the case of the benzoic acid-water complex.[c] No methyl internal rotation splittings were observed, consistent with a high barrier (7 kJ mol^{-1}) calculated for the monomer at the B3LYP/6-311++G(d,p) level of theory. Using statistical thermodynamics, experimental rotational constants were combined with a theoretical frequency analysis and binding energy to give an estimate of the percentage of hydrated acid in the atmosphere under various conditions.

[a] F. Riccobono, *et al.*, *Science*, **344**, 717 (2014).
[b] R. Zhang, *et al.*, *Science*, **304**, 1487 (2004).
[c] E. G. Schnitzler and W. Jäger, *Phys. Chem. Chem. Phys.*, **16**, 2305 (2014).

WJ04 **2:21 – 2:36**

A ROVIBRATIONAL ANALYSIS OF THE WATER BENDING VIBRATION IN OC-H_2O AND A MORPHED POTENTIAL OF THE COMPLEX

LUIS A. RIVERA-RIVERA, SEAN D. SPRINGER, BLAKE A. McELMURRY, *Department of Chemistry, Texas A & M University, College Station, TX, USA*; IGOR I LEONOV, *Microwave Spectroscopy, Institute of Applied Physics, Nizhny Novgorod, Russia*; ROBERT R. LUCCHESE, JOHN W. BEVAN, *Department of Chemistry, Texas A & M University, College Station, TX, USA*; L. H. COUDERT, *LISA, CNRS, Universités Paris Est Créteil et Paris Diderot, Créteil, France.*

Rovibrational transitions associated with tunneling states in the water bending vibration in OC-H_2O complex have been recorded using a supersonic jet quantum cascade laser spectrometer at 6.2 μm. Analysis of the resulting spectra is facilitated by incorporating fits of previously recorded microwave and submillimeter data accounting for Coriolis coupling to obtain the levels of the ground vibrational state. The results were then used to confirm assignment of the vibration and explore the nature of tunneling dynamics in associated vibrationally excited states of the complex. A seven-dimension *ab initio* interaction potential is constructed for the complex. The available spectroscopic data is used to generated a morphed potential. Previous prediction of the D_0 of the complex will be incorporated in the analysis.

WJ05 **2:38 – 2:53**

THE MICROWAVE SPECTRUM AND UNEXPECTED STRUCTURE OF THE BIMOLECULAR COMPLEX FORMED BETWEEN ACETYLENE AND (*Z*)-1-CHLORO-2-FLUOROETHYLENE

NAZIR D. KHAN, HELEN O. LEUNG, MARK D. MARSHALL, *Chemistry Department, Amherst College, Amherst, MA, USA.*

In all previously studied complexes between protic acids and chlorofluoroethylenes in our laboratory, the acidic hydrogen atom forms the primary intermolecular interaction with a fluorine atom on the ethylene subunit. This has been rationalized by the greater electronegativity of the fluorine atom leading to a stronger, hydrogen-bond like interaction, than would be formed with the chlorine atom. With (*Z*)-1-chloro-2-fluoroethylene, however, *ab initio* calculations for its complex with acetylene indicate that participation of the chlorine atom in the intermolecular interaction leads to lower energy configurations. This is confirmed by observation of the rotational spectrum of the complex by chirped-pulse and Balle-Flygare Fourier transform microwave spectroscopy. The complex is determined to be planar with one interaction between an acetylenic hydrogen and the chlorine atom and a second between the triple bond and the hydrogen atom geminal to chlorine.

WJ06 **2:55 – 3:10**

CHLORINE NUCLEAR QUADRUPOLE HYPERFINE STRUCTURE IN THE VINYL CHLORIDE–HYDROGEN CHLORIDE COMPLEX

HELEN O. LEUNG, MARK D. MARSHALL, JOSEPH P. MESSINGER, *Chemistry Department, Amherst College, Amherst, MA, USA.*

The microwave spectrum of the vinyl chloride–hydrogen chloride complex, presented at last year's symposium, is greatly complicated by the presence of two chlorine nuclei as well as an observed, but not fully explained tunneling motion. Indeed, although it was possible at that time to demonstrate conclusively that the complex is nonplanar, the chlorine nuclear quadrupole hyperfine splitting in the rotational spectrum resisted analysis. With higher resolution, Balle-Flygare Fourier transform microwave spectra, the hyperfine structure has been more fully resolved, but appears to be perturbed for some rotational transitions. It appears that knowledge of the quadrupole coupling constants will provide essential information regarding the structure of the complex, specifically the location of the hydrogen atom in HCl. Our progress towards obtaining values for these constants will be presented.

WJ07 **3:12 – 3:27**

ELECTRONIC COMMUNICATION IN COVALENTLY *vs.* NON-COVALENTLY BONDED POLYFLUORENE SYSTEMS: THE ROLE OF THE COVALENT LINKER.

BRANDON UHLER, NEIL J REILLY, MARAT R TALIPOV, MAXIM IVANOV, QADIR TIMERGHAZIN, RAJENDRA RATHORE, SCOTT REID, *Department of Chemistry, Marquette University, Milwaukee, WI, USA.*

benzyl radical $\tilde{X}\,^2B_1$ | **tropyl radical** $\tilde{X}\,^2E_2''$ | **norboradienyl radical** $\tilde{X}\,^2B_1$

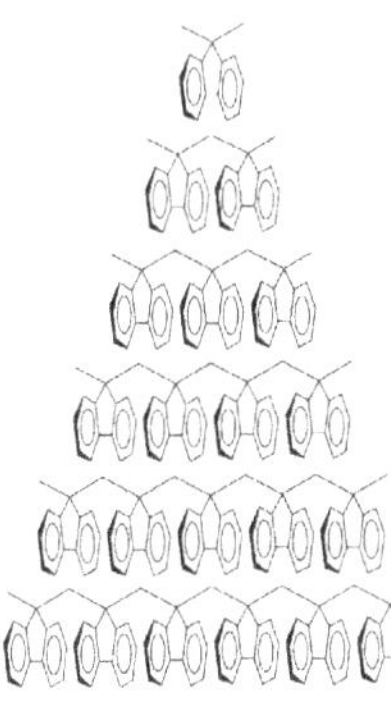

The covalently linked polyfluorene molecules F1-F6 (see left) are prototypical molecular wires by virtue of their favorable electron/hole transport properties brought about by π-stacking. To understand the role of the covalent linker in facilitating electron transport in these systems, we have investigated several van der Waals (vdW) analogues by resonant mass spectroscopy. Electronic spectra and ion yield curves are reported for jet-cooled vdW clusters containing up to six fluorene units. The near-coincidence of the electronic band origins for the dimer and larger clusters suggests that a structure containing a central dimer chromophore is the predominant conformational motif. As for F1-F6, the threshold ionization potentials extracted from the ion yield measurements decrease linearly with inverse cluster size. Importantly, however, the rate of decrease is significantly smaller in the vdW clusters, indicating more efficient hole stabilization in the covalently bound systems. Results for similar vdW clusters that are locked into specific conformations by steric effects will also be reported.

Intermission

WJ08 **3:46 – 4:01**

A GENERAL TRANSFORMATION TO CANONICAL FORM FOR POTENTIALS IN PAIRWISE INTERMOLECULAR INTERACTIONS

JAY R. WALTON, *Department of Mathematics, Texas A & M University, College Station, TX, USA*; LUIS A. RIVERA-RIVERA, ROBERT R. LUCCHESE, JOHN W. BEVAN, *Department of Chemistry, Texas A & M University, College Station, TX, USA.*

A generalized formulation of explicit transformations is introduced to investigate the concept of a canonical potential in both fundamental chemical and intermolecular bonding. Different classes of representative ground electronic state pairwise interatomic interactions are referenced to a single canonical potential illustrating application of explicit transformations. Specifically, accurately determined potentials of the diatomic molecules H_2, H_2^+, HF, LiH, argon dimer, and one-dimensional dissociative coordinates in Ar-HBr, OC-HF, and OC-Cl_2 are investigated throughout their bound potentials. The advantages of the current formulation for accurately evaluating equilibrium dissociation energies and a fundamentally different unified perspective on nature of intermolecular interactions will be emphasized. In particular, this canonical approach has relevance to previous assertions that *there is no very fundamental distinction between van der Waals bonding and covalent bonding* or for that matter hydrogen and halogen bonds.

WJ09 **4:03 – 4:18**

THREE-DIMENSIONAL WATER NETWORKS SOLVATING AN EXCESS POSITIVE CHARGE: NEW INSIGHTS INTO THE MOLECULAR PHYSICS OF ION HYDRATION

CONRAD T. WOLKE, JOSEPH FOURNIER, *Department of Chemistry, Yale University, New Haven, CT, USA*; GARY H WEDDLE, *Department of Chemistry and Biochemistry, Fairfield University, Fairfield, CT, USA*; EVANGELOS MILIORDOS, SOTIRIS XANTHEAS, *Physical Sciences Division, Pacific Northwest National Laboratory, Richland, WA, USA*; MARK JOHNSON, *Department of Chemistry, Yale University, New Haven, CT, USA.*

In a recent effort our group investigated the vibrational mechanics of water using the cage of 20 water molecules surrounding an alkali ion as a paradigm system. The $M^+(H_2O)_{20}$ clusters are well known “magic number” species (for the larger alkali metals) and are thought to form a pentagonal dodecahedral web encapsulating the ion. We are attracted to these systems because they are sufficiently large to display broad OH fundamental envelopes in a manner similar to that found in bulk water, but do so with a relatively small number of structurally distinct, three coordinated sites in a finite assembly that, although challenging, can be analyzed with electronic structure calculations in the context of a “supermolecule”. We show how this arrangement can provide an ideal platform on which to unambiguously identify the spectral signatures of particular binding sites, information that is invoked to explain the bulk (and interface) spectrum of water but cannot be directly measured in bulk water.

Although this behavior is most relevant to simulations of interfacial water, a future direction of this study will be gaining site-specific information for water in an extended two dimensional structure, and the elucidation of the paths of spectral diffusion associated with this arrangement. This unprecedented work will clarify a number of open questions regarding the site-specificity of ground and vibrationally excited state dynamics.

WJ10 **4:20 – 4:35**

MATRIX ISOLATION INFRARED SPECTROSCOPY OF A SERIES OF 1:1 PHENOL-WATER COMPLEXES

PUJARINI BANERJEE, TAPAS CHAKRABORTY, *Physical Chemistry, Indian Association for the Cultivation of Science, Kolkata, India.*

We report here the FTIR spectra of 1:1 complexes of eight fluorophenol derivatives with water measured under matrix isolation condition. In all the complexes, oxygen of water is the hydrogen bond acceptor and phenolic O-H the hydrogen bond donor. The attributes of the O-H...O linkage in the complexes are tuned remotely by fluorine substitutions at different aromatic sites of phenol. The goal of the study is to find the intermolecular interactions that correlate best with the sequence of spectral shifts of the donor O-H stretching (ν_{OH}) frequencies. Measurements reveal that the probe phenolic ν_{OH} shifts vary by nearly 90% from unsubstituted phenol to pentafluorophenol. Interestingly, this large variation correlates poorly with the predicted binding energies of the complexes. Secondly, although electrostatic interaction is considered to dominate the overall stabilization of such classical hydrogen bonds, we see that the shifts do not display any correlation with this interaction at the hydrogen bonding sites. On the other hand, the purely quantum mechanical charge-transfer interaction energies, as obtained from Natural Bond Orbital analysis,are found to display very good correspondence with the spectral shifts. Thus, we propose that such local charge-transfer type interactions are better descriptors of weakening of the hydrogen bond donor than electrostatic energy parameters.

WJ11 **4:37 – 4:52**

MATRIX ISOLATION IR SPECTROSCOPY AND QUANTUM CHEMISTRY STUDY OF 1:1 Π-HYDROGEN BONDED COMPLEXES OF BENZENE WITH A SERIES OF FLUOROPHENOLS

PUJARINI BANERJEE, TAPAS CHAKRABORTY, *Physical Chemistry, Indian Association for the Cultivation of Science, Kolkata, India.*

O-H stretching infrared fundamentals (ν_{OH}) of phenol and a series of fluorophenol monomers and their 1:1 complexes with benzene have been measured under a matrix isolation condition (8K). For the phenol-benzene complex the measured shift of ν_{OH} is 78 $\mathrm{cm}^{-1}$and for 3, 4, 5-trifluorophenol it is 98 cm^{-1}. Although the cold matrix isolation environment is very different from an aqueous medium, the measured spectral shifts display an interesting linear correlation with the aqueous phase acid dissociation constants (pK_a) of the phenols. The spectral shifts predicted by quantum chemistry calculations at several levels of theory are consistent with the observed values. Correlations of the shifts are also found with respect to energetic, geometric and several other electronic structure parameters of the complexes. Partitioning of binding energies of the complexes into components following the Morokuma-Kitaura scheme shows that dispersion is the predominant component of attractive interaction, and electrostatics, polarization and charge-transfer terms also have contributions to overall binding stability. NBO analysis reveals that hyperconjugative charge-transfers from the filled π-orbitals of the hydrogen bond acceptor (benzene) to the anti-bonding σ*(O–H) orbital of the donors (phenols) display correlations which are fully consistent with the observed variations of spectral shifts. The analysis also shows that the O-H bond dipole moments of all the phenolic species are nearly the same, implying that local electrostatics has only a little effect at the site of hydrogen bonding.

WJ12 **4:54 – 5:09**

MATRIX ISOLATION IR SPECTROSCOPY OF 1:1 COMPLEXES OF ACETIC ACID AND TRIHALOACETIC ACIDS WITH WATER AND BENZENE

PUJARINI BANERJEE, TAPAS CHAKRABORTY, *Physical Chemistry, Indian Association for the Cultivation of Science, Kolkata, India.*

A comparative study of infrared spectral effects for 1:1 complex formation of acetic acid (AA), trifluoroacetic acid (TFAA) and trichloroacetic acid (TFAA) with water and benzene has been carried out under a matrix isolation environment. Despite the large difference in aqueous phase acidities of the three acids, the measured ν_{OH}stretching frequencies of the monomers of the three molecules are found to be almost same, and in agreement with gas phase electronic structure calculations. Intrinsic acidities are expressed only in the presence of the proton acceptors, water or benzene. Although electronic structure calculations predict distinct ν_{OH} red-shifts for all three acids, the measured spectral features for TCAA and TFAA in this range do not allow unambiguous assignments for the 1:1 complex. On the other hand, the spectral changes in the $\nu_{C=O}$ region are more systematic, and the observed changes are consistent with predictions of theory. Components of overall binding energy of each complex have been obtained from energy decomposition analysis, which allows determination of the relative contributions of various physical forces towards overall stability of the complexes, and the details will be discussed in the talk.

WJ13 **5:11 – 5:21**

SPECTROSCOPIC INVESTIGATION OF THE EFFECTS OF ENVIRONMENT ON NEWLY-DEVELOPED NEAR INFRARED EMITTING DYES

LOUIS E. McNAMARA, NALAKA LIYANAGE, *Chemistry and Biochemistry, University of Mississippi, Oxford, MS, USA*; JARED DELCAMP, *Chemistry, University of Mississippi, Oxford, MS, USA*; NATHAN I HAMMER, *Chemistry and Biochemistry, University of Mississippi, Oxford, MS, USA.*

The effects of environment on the photophysical properties of a series of newly-developed near infrared emitting dyes was studied spectroscopically. Properties of interest include fluorescence emission, fluorescence lifetime, and quantum yield. Tracking how the photophysics of these compounds are affected in the solid phase, in thin films, in solution, and at the single molecule level with changing environment will provide a deeper insight into how dye structure affects their function.

WJ14 **5:23 – 5:38**

SPECTROSCOPIC SIGNATURES AND STRUCTURAL MOTIFS IN ISOLATED AND HYDRATED XANTHINE AND ITS METHYLATED DERIVATIVES

VIPIN BAHADUR SINGH, *Department of Physics, Udai Pratap Autonomous College, Varanasi, India.*

The conformational landscapes of neutral xanthine and its methylated derivatives and their hydrated complexes have been investigated by MP2 and DFT methods. We investigated the low-lying excited states of bare xanthine , theophylline, theobromine and caffeine by means of coupled cluster singles and approximate doubles (CC2) and TDDFT methods and a satisfactory interpretation of the electronic absorption spectra1 is obtained. One striking feature is the coexistence of the blue and red shift of the vertical excitation energy of the optically bright state S1 (1pipi*) of xanthine, caffeine2 and theophylline3 upon forming complex with a water at C2=O and C6=O carbonyl sites, respectively. The lowest singlet pipi* excited-state of the caff1-(H2O)1 complex involving isolated carbonyl are strongly blue shifted which is in agreement with the result of R2PI spectra of singly hydrated caffeine.4 While for the most stable and the second most stable caff1-(H2O)1 complexes involving conjugated carbonyl, the lowest singlet pipi* excited-state is red shifted. The effect of hydration on S1 (1pipi*) excited state due to bulk water environment was mimicked by a combination of PCM and COSMO models, which also shows a blue shift in accordance with the result of electronic absorption spectra in aqueous solution.1 This hypsochromic shift, is expected to be the result of the changes in the pi-electron delocalization extent of molecule because of hydrogen bond formation. We have also clarified that the intermolecular hydrogen-bond strengthening and weakening correspond to red shifts and blue shifts, respectively, in the electronic spectra. It is expected that the radiation less deactivation is dramatically influenced through the regulation of electronic states by strong hydrogen bonding interaction. The optimized structure of newly characterized theophylline dimer Form IV,computed the first time by MP2 and DFT methods. The binding energy of this dimer linked by double N-H…O=C hydrogen bonds was found to be 88 kj/mole at the MP2/6-311++G(d,p) level of theory.3 Computed IR spectra is found in remarkable agreement with the experiment and the out of phase (C=O)2 stretching mode shows tripling of intensity upon dimerisation. 1. J. Chen and B.Kohler, Phys. Chem. Chem. Phys., 2012, 14, 10677-10682 2. Vipin Bahadur Singh, RSC.Adv.,2014,4,58116-58126 3. Vipin Bahadur Singh, RSC.Adv.,2015,5,11433-11444 4. D. Kim, H. M. Kim, K. Y.Yang, S. K. Kim and N. J. Kim, J.Chem.Phys. 2008, 128, 134310-134316

RA. Metal containing

Thursday, June 25, 2015 – 8:30 AM

Room: 116 Roger Adams Lab

Chair: Damian L Kokkin, Arizona State University, Tempe, Arizona, USA

RA01 **8:30 – 8:45**

HYPERFINE RESOLVED PURE ROTATIONAL SPECTROSCOPY OF ScN, YN, AND BaNH ($X^1\Sigma^+$): INSIGHT INTO METAL-NITROGEN BONDING

LINDSAY N. ZACK, *Department of Chemistry, Wayne State University, Detroit, MI, USA*; MATTHEW BUCCHINO, *Department of Chemistry and Astronomy, University of Arizona, Tucson, AZ, USA*; JUSTIN YOUNG, MARSHALL BINNS, PHILLIP M. SHERIDAN, *Department of Chemistry and Biochemistry, Canisius College, Buffalo, NY, USA*; LUCY ZIURYS, *Department of Chemistry and Biochemistry, University of Arizona, Tucson, AZ, USA.*

Fourier transform microwave spectroscopy coupled with a discharge-assisted laser ablation source (DALAS) has been used to record the J = 1 → 0 pure rotational transitions of $Sc^{14}N$, $Sc^{15}N$, $Y^{14}N$, $Y^{15}N$, and $Ba^{14}NH$ ($X^1\Sigma^+$). Each species was synthesized by the reaction of the ablated metal with either NH_3 or $^{15}NH_3$ in the presence of a DC discharge. For each species hyperfine structure was resolved. In the case of ScN and YN hyperfine parameters (quadrupole and nuclear spin-rotation) for the metal and nitrogen were determined and for BaNH the nitrogen quadrupole coupling constant was measured. These hyperfine constants are interpreted to gain insight into the metal-nitrogen bonding in each species. In addition, DFT calculations were performed to assist with the assignment of each spectrum and the characterization of the metal-nitrogen bond.

RA02 **8:47 – 9:02**

THE SUBMILLIMETER/THz SPECTRUM OF AlH ($X^1\Sigma^+$), CrH ($X^6\Sigma^+$), and SH^+ ($X^3\Sigma^-$)

DeWAYNE T HALFEN, *Department of Chemistry and Astronomy, University of Arizona, Tucson, AZ, USA*; LUCY ZIURYS, *Department of Chemistry and Biochemistry, University of Arizona, Tucson, AZ, USA.*

The N = 2 ← 1 transition of the CrH ($X^6\Sigma^+$) radical near 730-734 GHz and the J = 2 ← 1 line of AlH ($X^1\Sigma^+$) near 755 GHz have been measured using submillimeter/Terahertz direct absorption spectroscopy. CrH was created in an AC discharge of $Cr(CO)_6$ vapor and H_2 in the presence of argon. AlH was produced from $Al(CH_3)_3$ vapor and H_2 in argon with an AC discharge. In addition, three fine structure components of the N = 1 ← 0 transition of the SH^+ ($X^3\Sigma^-$) cation from 345-683 GHz were recorded. SH^+ was generated from H_2S and argon in an AC discharge. The data have been analyzed, and spectroscopic constants for these species have been refined. These parameters are in excellent agreement with past millimeter, infrared, and optical data. SH^+ is a known interstellar molecule, and these measurements confirm recent observations of this species. The new data for CrH and AlH could facilitate the detection of these species in interstellar/circumstellar gas.

RA03 **9:04 – 9:19**

FORMATION OF M-C≡C-Cl (M = Ag or Cu) AND CHARACTERIZATION BY ROTATIONAL SPECTROSCOPY

DANIEL P. ZALESKI, *School of Chemistry, Newcastle University, Newcastle-upon-Tyne, United Kingdom*; DAVID PETER TEW, *School of Chemistry, University of Bristol, Bristol, United Kingdom*; NICK WALKER, *School of Chemistry, Newcastle University, Newcastle-upon-Tyne, United Kingdom*; ANTHONY LEGON, *School of Chemistry, University of Bristol, Bristol, United Kingdom.*

The new linear molecule Ag-C≡C-Cl has been detected and characterized by means of rotational spectroscopy. It was synthesized by laser ablation of a silver rod in the presence of a gaseous sample containing a low concentration of CCl_4 in argon, cooled to a rotational temperature approaching 2 K through supersonic expansion and analyzed by chirped pulse Fourier transform microwave spectroscopy. Substitution coordinates are available for the silver and chlorine positions and will be compared to ab initio calculations at the CCSD(T)/aug-cc-pV5Z level of theory. The Ag-^{13}C≡^{13}C-Cl isotopologue was also observed using a similar gas mixture containing $^{13}CCl_4$. The Cu analogue Cu-C≡C-Cl was similarly identified and characterized.

RA04 **9:21 – 9:36**

$(CH_3)_3N\cdots AgI$ AND $H_3N\cdots AgI$ STUDIED BY BROADBAND ROTATIONAL SPECTROSCOPY AND *AB INITIO* CALCULATIONS

DROR M. BITTNER, DANIEL P. ZALESKI, SUSANNA L. STEPHENS, NICK WALKER, *School of Chemistry, Newcastle University, Newcastle-upon-Tyne, United Kingdom*; ANTHONY LEGON, *School of Chemistry, University of Bristol, Bristol, United Kingdom.*

The pure rotational spectra of 8 isotopologues of $H_3N\cdots AgI$ and 6 isotopologues of $(CH_3)_3N\cdots AgI$ were measured in a chirped pulse Fourier-transform microwave spectrometer. The complexes were synthesized in a molecular beam from a gas sample containing H_3N or $(CH_3)_3N$ and CF_3I precursors diluted in argon. Laser ablation was used to introduce silver atoms to the gas phase. The rotational constant B_0, centrifugal distortion constants D_J and D_{JK}, and the nuclear quadrupole coupling constant χ_{aa}(I) have been determined for $(CH_3)_3{}^{14/15}N\ldots{}^{107/109}AgI$, $(CD_3)_3N\ldots{}^{107/109}AgI$, $H_3{}^{14/15}N\ldots{}^{107/109}AgI$ and $D_3N\ldots{}^{107/109}AgI$ by fitting the measured transitions to a symmetric top Hamiltonian. The spectroscopic constants (B_0+ C_0), Δ_J and χ_{aa}(I) have been determined for $D_2HN\cdots{}^{107/109}AgI$ through fits that employed a Hamiltonian appropriate for a very near prolate asymmetric rotor. Partial effective (r_0) and substitution (r_s) structures have been determined.

RA05 **9:38 – 9:53**

MICROWAVE SPECTRA AND GEOMETRIES OF $C_2H_2\cdots AgI$ and $C_2H_4\cdots AgI$

SUSANNA L. STEPHENS, *Department of Chemistry, University of Manitoba, Winnipeg, MB, Canada*; DAVID PETER TEW, *School of Chemistry, University of Bristol, Bristol, United Kingdom*; NICK WALKER, *School of Chemistry, Newcastle University, Newcastle-upon-Tyne, United Kingdom*; ANTHONY LEGON, *School of Chemistry, University of Bristol, Bristol, United Kingdom.*

A chirped-pulse Fourier transform microwave spectrometer has been used to measure the microwave spectra of both $C_2H_2\cdots AgI$ and $C_2H_4\cdots AgI$. These complexes are generated via laser ablation at 532 nm of a silver surface in the presence of CF_3I and either C_2H_2 or C_2H_4 and argon and are stabilized by a supersonic expansion. Rotational (A_0, B_0, C_0) and centrifugal distortion constants (Δ_J and Δ_{JK}) of each molecule have been determined as well the nuclear electric quadrupole coupling constants the iodine atom (χ_{aa} (I) and $\chi_{bb}-\chi_{cc}$ (I)). The spectrum of each molecule is consistent with a C_{2v} structure in which the metal atom interacts with the π-orbital of the ethene or ethyne molecule. Isotopic substitutions of atoms within the C_2H_2 or C_2H_4 subunits are in progress and in conjunction with high level *ab initio* calculations will allow for accurate determination of the geometry of each molecule. These to complexes are put in the context of the recently studied $H_2S\cdots AgI$,[a] $OC\cdots AgI$,[b], $H_3N\cdots AgI$ and $(CH_3)_3N\cdots AgI$.[c]

[a]S.Z. Riaz, S.L. Stephens, W. Mizukami, D.P. Tew, N.R. Walker, A.C. Legon, Chem. Phys. Let., **531**, 1-12 (2012)
[b]S.L. Stephens, W. Mizukami, D.P. Tew, N.R. Walker, A.C. Legon, J. Chem. Phys., **136(6)**, 064306 (2012)
[c]D.M. Bittner, D.P. Zaleski, S.L. Stephens, N.R. Walker, A.C. Legon, Study in progress.

Intermission

RA06 **10:12 – 10:27**

EVALUATION OF THE EXOTHERMICITY OF THE CHEMI-IONIZATION REACTION Sm + O $\rightarrow$ SmO^+ + e^-

RICHARD M COX, JUNGSOO KIM, PETER ARMENTROUT, *Department of Chemistry, University of Utah, Salt Lake City, UT, USA*; JOSHUA BARTLETT, ROBERT A. VANGUNDY, MICHAEL HEAVEN, *Department of Chemistry, Emory University, Atlanta, GA, USA*; JOSHUA J. MELKO, *Department of Chemistry, University of North Florida, Jacksonville, FL, USA*; SHAUN ARD, NICHOLAS S. SHUMAN, ALBERT VIGGIANO, *Space Vehicles Directorate, Air Force Research Lab, Kirtland AFB, NM, USA.*

The chemi-ionization reaction Sm + O $\rightarrow$ SmO^+ + e^- has been used for chemical release experiments in the thermosphere. This reaction was chosen, in part, because the best available data indicated that it is exothermic by 0.35 $\pm$ 0.12 eV. Low ion yields in the initial atmospheric release experiments raised questions concerning the accuracy of the ionization energy (IE) for SmO and the bond dissociation energy (BDE) of SmO^+. New measurements of the BDE, obtained using a selected ion flow tube and guided ion beam techniques, yielded a more precise value of 5.73 $\pm$ 0.07 eV. The ionization energy of SmO was reexamined using pulsed-field ionization zero kinetic energy (ZEKE) photoelectron spectroscopy. The value obtained, 5.7427 $\pm$ 0.0006 eV, was significantly higher than the literature value. Combined with literature bond energies of SmO, this IE indicates an exothermicity for Sm + O $\rightarrow$ SmO^+ + e^- of 0.14 $\pm$ 0.17 eV, independent from and in agreement with the value deduced from the guided ion beam measurements. The implications of these results for interpretation of chemical release experiments are considered.

RA07 **10:29 – 10:44**

The PERMANENT ELECTRIC DIPOLE MOMENT AND HYPERFINE INTERACTION IN GOLD SULFIDE, AuS

RUOHAN ZHANG, DAMIAN L KOKKIN, *Department of Chemistry and Biochemistry, Arizona State University, Tempe, AZ, USA*; THOMAS D. VARBERG, *Chemistry Department , Macalester College, Saint Paul, Minnesota, USA*; TIMOTHY STEIMLE, *Department of Chemistry and Biochemistry, Arizona State University, Tempe, AZ, USA.*

The bonding and electrostatic properties of gold containing molecules are highly influenced by the large relativistic and electron correlation effects.[a] Here we report on the electricpermanent dipole moment measurement and hyperfine interaction analysis of the $^2\Delta_{3/2}$-$^2\Pi_{3/2}$ and $^2\Delta_{5/2}$-$^2\Pi_{3/2}$ bands of AuS. A cold molecular beam sample of gold sulfide was generated using a supersonic laser ablation source. The electronic bands were recorded at high resolution (35 MHz, FWHM) using laser excitation spectroscopy both field-free and in the presence of a static electric field. The observed hyperfine spectral features were assigned and a set of spectroscopic parameters for the $^2\Delta$ and $^2\Pi$ states were obtained. The Stark induced shifts of selected low-rotational features were analyzed to determine the permanent electric dipole moments in both the ground and excited states.

[a] P. Pyykko; *Angew Chem. Int[43]* , **4412**,(2004).

RA08 **10:46 – 11:01**

HIGH-ACCURACY CALCULATION OF Cu ELECTRIC-FIELD GRADIENTS: A REVISION OF THE Cu NUCLEAR QUADRUPOLE MOMENT VALUE

LAN CHENG, *Department of Chemistry, The University of Texas, Austin, TX, USA*; DEVIN A. MATTHEWS, *Department of Chemistry and Biochemistry, The University of Texas, Austin, TX, USA*; JÜRGEN GAUSS, *Institut für Physikalische Chemie, Universität Mainz, Mainz, Germany*; JOHN F. STANTON, *Department of Chemistry, The University of Texas, Austin, TX, USA.*

A revision of the value for the Cu nuclear quadrupole moment (NQM) is reported based on high-accuracy ab initio calculations on the Cu electric field gradients in the CuF and CuCl molecules. Electron-correlation effects have systematically been taken into account using a hierarchy of coupled-cluster methods including up to quadruple excitations. It is shown that the $CCSD(T)_\Lambda$ method provides a more reliable treatment of triples corrections for Cu electric-field gradients than the ubiquitously applied CCSD(T) method, which is tentatively attributed to the importance of the wavefunction relaxation in the calculations of a core property. Augmenting large-basis-set $CCSD(T)_\Lambda$ results with the remaining corrections obtained using additive schemes, including full triples contributions, quadruples contributions, zero-point vibrational corrections, spin-orbit corrections, as well as the correction from the Gaunt term, a new value of 209.7(50) mbarn for the Cu NQM has been obtained. The new value substantially reduces the uncertainty of this parameter in comparison to the standard value of 220(15) mbarn obtained from a previous muonic experiment.

RA09 **11:03 – 11:18**

CATION-π AND CH-π INTERACTIONS IN THE COORDINATION AND SOLVATION OF Cu^+ (ACETYLENE)$_n$ (n=1-6) COMPLEXES INVESTIGATED VIA INFRARED PHOTODISSOCIATION SPECTROSCOPY

ANTONIO DAVID BRATHWAITE, *College of Science and Mathematics, University of the Virgin Islands, St. Thomas, USVI*; RICHARD S. WALTERS, TIMOTHY B WARD, MICHAEL A DUNCAN, *Department of Chemistry, University of Georgia, Athens, GA, USA.*

Mass-selected copper-acetylene cation complexes of the form $Cu(C_2H_2)_n+$ are produced by laser ablation and studied via infrared laser photodissociation spectroscopy in the C-H stretching region (3000-3500 cm^{-1}). Spectra for larger species are measured via ligand elimination, whereas argon tagging is employed to enhance dissociation yields in smaller complexes. The number of infrared active bands, their frequency positions and their relative intensities provide insight into the structure and bonding of these ions. Density functional theory calculations are carried out in support of this work. The combined data show that cation-π bonds are formed for the n=1-3 species, resulting in red-shifted C-H stretches on the acetylene ligands. Three acetylene ligands complete the coordination of the copper cation. Additional ligands (n=4-6) solvate the n=3 core by forming CH-*pi* bonds. Distinctive vibrational patterns are exhibited for coordinated vs. solvent ligands. Theory reproduces these results.

RA10 **11:20 – 11:35**

ANION PHOTOELECTRON SPECTROSCOPIC STUDIES OF $NbCr(CO)_n^-$ (n = 2,3) HETEROBIMETALLIC CARBONYL COMPLEXES

MELISSA A. BAUDHUIN, PRAVEENKUMAR BOOPALACHANDRAN, DOREEN LEOPOLD, *Chemistry Department, University of Minnesota, Minneapolis, MN, USA.*

Anion photoelectron spectra and density functional calculations are reported for $NbCr(CO)_2^-$ and $NbCr(CO)_3^-$ complexes prepared by addition of $Cr(CO)_6$ vapor to a flow tube equipped with a niobium cathode discharge source. Electron affinities ($\pm$ 0.007 eV) are measured to be 1.668 eV for $NbCr(CO)_2$ and 1.162 eV for $NbCr(CO)_3$, values which exceed the 0.793 eV electron affinity previously measured for ligand-free NbCr. The vibrationally-resolved 488 nm photoelectron spectra are compared with Franck-Condon spectra predicted for various possible isomers and spin states of the anionic and neutral metal carbonyl complexes. Results are also compared with photoelectron spectra of the corresponding chromium carbonyl complexes and of NbCr and $NbCr^-$, which have formal bond orders of 5.5 ($^2\Delta$) and 6 ($^1\Sigma^+$), respectively. These comparisons help to elucidate the effects of sequential carbonylation on this multiple metal-metal bond, and of the formation of this bond on the chromium-carbonyl interactions.

RA11 **11:37 – 11:52**

MASS-ANALYZED THRESHOLD IONIZATION SPECTROSCOPY OF CYCLIC $La(C_5H_6)$ FORMED BY La ATOM ACTIVATION OF PENTYNE AND PENTADIENE

WENJIN CAO, YUCHEN ZHANG, DONG-SHENG YANG, *Department of Chemistry, University of Kentucky, Lexington, KY, USA.*

La atom reactions with 1-pentyne ($CH{\equiv}CCH_2CH_2CH_3$) and 1,4-pentadiene ($CH_2 = CHCH_2CH = CH_2$) were carried out in a laser-ablation molecular beam source. $La(C_5H_6)$ was observed in the two reactions through time-of-flight mass spectrometry and characterized by mass-analyzed threshold ionization spectroscopy. The most stable isomer of $La(C_5H_6)$ was identified as a six-membered metallacycle, $La(CH_2CH = CHCH = CH)$, with La binding to the two terminal carbon atoms of the unsaturated hydrocarbon. The metallacycle is formed by hydrogen elimination and migration induced by the La atom. The neutral complex with C_1 symmetry has a doublet ground state, and the corresponding ion has a singlet state generated by the removal of a La 6s-based electron. The adiabatic ionization energy of the metallacycle was determined to be 37941 (5) cm^{-1}. Three vibrational modes of the ion were measured to be 318, 407, and 538 cm^{-1}, which correspond to the La-ligand stretching, carbon skeleton bending with CH_2 rocking, and carbon skeleton bending with CH_2 twisting, respectively. In addition, two hot bands were observed at 276 and 367 cm-1 below the origin band and identified to be the vibrational frequencies of the La-ligand stretching and carbon skeleton bending with CH_2 rocking modes of the neutral complex.

RB. Mini-symposium: Accelerator-Based Spectroscopy

Thursday, June 25, 2015 – 8:30 AM

Room: 100 Noyes Laboratory

Chair: Gert von Helden, Fritz Haber Institute - MPG, Berlin, Germany

RB01 *INVITED TALK* **8:30 – 9:00**

PROBING INTRA- AND INTER- MOLECULAR INTERACTIONS VIA IRMPD EXPERIMENTS AND COMPUTATIONAL CHEMISTRY

SCOTT HOPKINS, TERRY McMAHON, *Department of Chemistry, University of Waterloo, Waterloo, ON, Canada.*

Experiments carried out at the CLIO Free Electron Facility have been used to probe a range of novel bonding motifs and dissociation dynamics in a variety of chemical systems. Among these are species which exhibit anion-pi interactions in complexes of halide ions with aromatic ring systems with electron withdrawing substituents; charge solvated and zwitterionic clusters of protonated methylamines with phenylalanines; hydrogen bonded dimers of nucleic acid analogues and Pd complexes potentially involving agnostic hydrogen bond interactions. Accompanying DFT computational work is used to assist in identifying the most probable structure(s) present in the IRMPD experiments.

RB02 **9:05 – 9:20**

EXPLORING CONFORMATION SELECTIVE FAR INFRARED ACTION SPECTROSCOPY OF ISOLATED MOLECULES AND SOLVATED CLUSTERS

DANIËL BAKKER, ANOUK RIJS, *FELIX Laboratory, Radboud University Nijmegen, Nijmegen, The Netherlands*; JÉRÔME MAHÉ, MARIE-PIERRE GAIGEOT, *Laboratoire Analyse et Modélisation pour la Biologie et l'Environnement, Université d'Evry val d'Essonne, Evry, France.*

Far-Infrared (IR) spectroscopy has been labeled as a promising method for identifying structural motifs in large molecules. However, several hurdles have kept this promising spectral region from breaking through to widespread use for gas phase experiments. Normal modes in the far-IR mostly have weak intensities, and high brightness sources of far-IR radiation are rare. Moreover, standard density functional theory - applied to identify the specific molecular structure responsible for the measured IR spectra - does not reproduce features in the far-IR well. This mismatch can be attributed to the high degree of anharmonicity of many of the normal modes present in the far-IR. We have overcome these hurdles by combining an advanced laser source with novel experiments and high-level dynamical calculations.

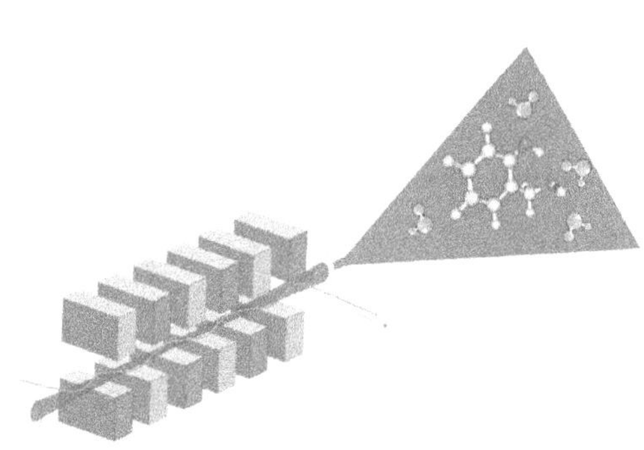

We present far-IR spectra of a family of phenolic molecules and solvated clusters, obtained using the free electron laser FELIX. By employing IR-UV ion-dip spectroscopy in the gas phase, we are able to obtain conformer specific far-IR spectra of isolated molecules or solvated clusters. The studied systems display both intra- and intermolecular hydrogen bonding, enabling us to study the merits of far-IR action spectroscopy for direct probing of these weak interactions. Moreover, the combination of far-IR experiments with quantum chemical calculations allows us to test the limits of the harmonic approximation in DFT calculations, and to test the possibilities of employing a more sophisticated technique, namely Born-Oppenheimer molecular dynamics.

RB03 **9:22 – 9:37**

FIRST INFRARED PREDISSOCIATION SPECTRA OF He-TAGGED PROTONATED PRIMARY ALCOHOLS AT 4 K

ALEXANDER STOFFELS[a], BRITTA REDLICH, J. OOMENS, *Institute for Molecules and Materials (IMM), Radboud University Nijmegen, Nijmegen, Netherlands*; OSKAR ASVANY, SANDRA BRÜNKEN, PAVOL JUSKO, SVEN THORWIRTH, STEPHAN SCHLEMMER, *I. Physikalisches Institut, Universität zu Köln, Köln, Germany*.

Cryogenic multipole ion traps have become popular devices in the development of sensitive action-spectroscopic techniques. The low ion temperature leads to enhanced spectral resolution, and less congested spectra. In the early 2000s, a 22-pole ion trap was coupled to the Free-Electron Laser for Infrared eXperiments (FELIX), yielding infrared Laser Induced Reaction (LIR) spectra of the molecular ions $C_2H_2^+$ and CH_5^{+b}. This pioneering work showed the great opportunities combining cold mass-selected molecular ions with widely tunable broadband IR radiation.

In the past year a cryogenic (T>3.9 K) 22-pole ion trap designed and built in Cologne (FELion) has been successfully coupled to FELIX, which in its current configuration provides continuously tunable infrared radiation from 3 μm to 150 μm, hence allowing to probe characteristic vibrational spectra in the so-called "fingerprint region" with a sufficient spectral energy density also allowing for multiple photon processes (IR-MPD). Here we present the first infrared predissociation spectra of He-tagged protonated methanol and ethanol ($MeOH_2^+$/$EtOH_2^+$) stored at 4 K. These vibrational spectra were recorded with both a commercial OPO and FELIX, covering a total spectral range from 3700 $cm^{-1}$to 550 cm^{-1}at a spectral resolution of a few cm^{-1}. The H-O-H stretching and bending modes clearly distinguish the protonated alcohols from their neutral analoga. For $EtOH_2^+$, also IR-MPD spectra of the bare ion could be recorded. The symmetric and antisymmetric H-O-H stretching bands at around 3 μm show no significant shift within the given spectral resolution in comparison to those recorded with He predissociation, indicating a rather small perturbation caused by the attached He. The vibrational bands were assigned using quantum-chemical calculations on different levels of theory. The computed frequencies correspond favorably to the experimental spectra. Subsequent high resolution measurements could lead to a better structural characterization of these protonated alcohols.

[a] also at: I.Physikalisches Institut, Universität zu Köln, Köln, Germany.

[b] Asvany et al.: Phys. Rev.Lett. 94, 073001 (2005), Asvany et al.: Science 309, 1219-1222 (2005)

RB04 **9:39 – 9:49**

METAL ION INDUCED PAIRING OF CYTOSINE BASES: FORMATION OF I-MOTIF STRUCTURES IDENTIFIED BY IR ION SPECTROSCOPY

JUEHAN GAO, GIEL BERDEN, J. OOMENS, *Institute for Molecules and Materials (IMM), Radboud University Nijmegen, Nijmegen, Netherlands*.

While the Watson-Crick structure of DNA is among the most well-known molecular structures of our time, alternative base-pairing motifs are also known to occur, often depending on base sequence, pH, or presence of cations. Pairing of two cytosine (C) bases induced by the sharing of a single proton (C-H^+-C) gives rise to the so-called i-motif, occurring particularly in the telomeric region of DNA, and particularly at low pH. At physiological pH, silver cations were recently suggested to form cytosine dimers in a C-Ag^+-C structure analogous to the hemiprotonated cytosine dimer, which was later confirmed by IR spectroscopy.[1] Here we investigate whether Ag^+ is unique in this behavior.

Using infrared action spectroscopy employing the free-electron laser FELIX and a tandem mass spectrometer in combination with quantum-chemical computations, we investigate a series of C-M^+-C complexes, where M is Cu, Li and Na. The complexes are formed by electrospray ionization (ESI) from a solution of cytosine and the metal chloride salt in acetonitrile/water. The complexes of interest are mass-isolated in the cell of a FT ion cyclotron resonance mass spectrometer, where they are irradiated with the tunable IR radiation from FELIX in the 600 - 1800 cm^{-1}range. Spectra in the H-stretching range are obtained with a LaserVision OPO.

Both experimental spectra as well as theoretical calculations indicate that while Cu behaves as Ag, the alkali metal ions induce a clearly different dimer structure, in which the two cytosine units are parallelly displaced. In addition to coordination to the ring nitrogen atom, the alkali metal ions coordinate to the carbonyl oxygen atoms of both cytosine bases, indicating that the alkali metal ion coordination favorably competes with hydrogen bonding between the two cytosine sub-units of the i-motif like structure.

1. Berdakin, Steinmetz, Maitre, Pino, J. Phys. Chem. A 2014, 118, 3804

RB05 **9:51 – 10:06**

MOLECULAR PROPERTIES OF THE "ANTI-AROMATIC" SPECIES CYCLOPENTADIENONE, C_5H_5=0

THOMAS ORMOND, BARNEY ELLISON, JOHN W DAILY, *Department of Chemistry and Biochemistry, University of Colorado, Boulder, CO, USA*; JOHN F. STANTON, *Department of Chemistry, The University of Texas, Austin, TX, USA*; MUSAHID AHMED, *UXSL, Chemical Sciences Division, Lawrence Berkeley National Laboratory, Berkeley, CA, USA*; TIMOTHY S. ZWIER, *Department of Chemistry, Purdue University, West Lafayette, IN, USA*; PATRICK HEMBERGER, *General Energy, Paul Scherrer Institute, Villigen, Switzerland.*

4 electrons — "anti-aromatic"

A common intermediate in the high temperature combustion of benzene is cyclopentadienone, C_5H_4=O. Cyclopentadienone is considered to be an "anti-aromatic" molecule. It is certainly a metastable species; samples persist at LN_2 temperatures but dimerize upon warming to -80°C. It is of great interest to physically characterize this "anti-aromatic" species. The microwave spectrum, the infrared spectrum, the ionization energy, and the electron affinity of cyclopentadienone have been measured. Flash pyrolysis of o–phenylene sulfite ($C_6H_4O_2SO$) provides molecular beams of C_5H_4=O entrained in a rare gas carrier. The beams are interrogated with time-of-flight photoionization mass spectrometry, confirming the clean, intense production of C_5H_4=O. a) Chirped-pulse Fourier transform microwave spectroscopy and CCSD(T) electronic structure calculations have combined to determine[a] the r_e molecular structure of C_5H_4=O. b) Guided by CCSD(T) electronic structure calculations, the matrix infrared absorbance spectrum of C_5H_4=O isolated in a 4 K neon matrix has been used[b] to assign 20 of the 24 fundamental vibrational frequencies. c) Imaging photoelectron photoion coincidence (iPEPICO) spectra[c] of cyclopentadienone establishes the ionization energy, IE(C_5H_4=O), to be 9.41 $\pm$ 0.01 eV. d) Prof. A. Sanov's group[d] has reported the electron affinity, EA(C_5H_4=O), to be 1.06 $\pm$ 0.01 eV.

[a] Kidwell *et al.* J. Phys. Chem. Letts. 2201 (2014)
[b] Ormond *et al.* J. Phys. Chem. A **118**, 708 (2014)
[c] Ormond *et al.* Mol. Phys. **in press** (2015)
[d] Khuseynov *et al.* J. Phys. Chem. A **118**, 6965 (2014)

Intermission

RB06 **10:25 – 10:40**

HIGH-RESOLUTION SYNCHROTRON INFRARED SPECTROSCOPY OF THIOPHOSGENE: THE ν_1, ν_5, $2\nu_4$, and ν_2 + $2\nu_6$ bands

BOB McKELLAR, *Steacie Laboratory, National Research Council of Canada, Ottawa, ON, Canada*; BRANT E BILLINGHURST, *EFD, Canadian Light Source Inc., Saskatoon, Saskatchewan, Canada.*

Thiophosgene (Cl_2CS) is a favorite model system for studies of photophysics, vibrational dynamics, and intersystem interactions. But at high resolution its infrared spectrum is very congested due to hot bands and multiple isotopic species. Previously, we reported the first high resolution IR study of this molecule, analyzing the ν_2 (504 cm^{-1}) and ν_4 (471 cm^{-1}) fundamental bands.[a] Here we continue, with analysis of the ν_1 (1139 cm^{-1}) and ν_5 (820 cm^{-1}) fundamentals for the two most abundant isotopologues, $^{35}Cl_2CS$ and $^{35}Cl^{37}ClCS$, based on spectra with a resolution of about 0.001 cm^{-1}obtained at the Canadian Light Source far-infrared beamline using synchrotron radiation and a Bruker IFS125 Fourier transform spectrometer. The ν_2 + ν_4 (942 cm^{-1}) and ν_2 + $2\nu_6$ (1104 cm^{-1}) bands are also studied here. But so far the ν_2 + ν_6 combination band (795 cm^{-1}) resists analysis, as do the weak ν_3 (292.9 cm^{-1}) and ν_6 ($\approx$300? cm^{-1}) fundamentals.

[a] A.R.W. McKellar, B.E.Billinghurst, J. Mol. Spectrosc. **260**, 66 (2010).

RB07 **10:42 – 10:57**

THE SOLEIL VIEW ON SULFUR OXIDES: THE S_2O BENDING MODE ν_2 AT 380 cm^{-1} AND ITS ANALYSIS USING AN AUTOMATED SPECTRAL ASSIGNMENT PROCEDURE (ASAP)

MARIE-ALINE MARTIN-DRUMEL, *Spectroscopy Lab, Harvard-Smithsonian Center for Astrophysics, Cambridge, MA, USA*; CHRISTIAN ENDRES, OLIVER ZINGSHEIM, T. SALOMON, *I. Physikalisches Institut, Universität zu Köln, Köln, Germany*; JENNIFER VAN WIJNGAARDEN, *Department of Chemistry, University of Manitoba, Winnipeg, MB, Canada*; OLIVIER PIRALI, SÉBASTIEN GRUET, *AILES beamline, Synchrotron SOLEIL, Saint Aubin, France*; FRANK LEWEN, STEPHAN SCHLEMMER, *I. Physikalisches Institut, Universität zu Köln, Köln, Germany*; MICHAEL C McCARTHY, *Atomic and Molecular Physics, Harvard-Smithsonian Center for Astrophysics, Cambridge, MA, USA*; SVEN THORWIRTH, *I. Physikalisches Institut, Universität zu Köln, Köln, Germany*.

The fundamental vibrational bending mode ν_2 of disulfur monoxide, S_2O, and the associated hot band $2\nu_2 - \nu_2$ have been observed at high spectral resolution for the first time at the SOLEIL synchrotron facility using Fourier-transform far-infrared spectroscopy. This transient species has been produced using a radio-frequency discharge by flowing SO_2 over elemental sulfur. The spectroscopic analysis has been performed using an Automated Spectral Assignment Procedure (ASAP) which has enabled the accurate determination of more than 3500 energy levels of the $v_2 = 1$ and $v_2 = 2$ vibrational states. In addition to the high-resolution synchrotron study, pure rotational spectra of S_2O in the $v_2 = 1$ and 2 vibrational states were observed in the frequency range 250 – 500 GHz in a long-path absorption cell.

RB08 **10:59 – 11:14**

THE SOLEIL VIEW ON SULFUR RICH OXIDES: THE ν_3 MODE OF S_2O REVISITED

SVEN THORWIRTH, *I. Physikalisches Institut, Universität zu Köln, Köln, Germany*; MARIE-ALINE MARTIN-DRUMEL, *Spectroscopy Lab, Harvard-Smithsonian Center for Astrophysics, Cambridge, MA, USA*; CHRISTIAN ENDRES, OLIVER ZINGSHEIM, T. SALOMON, *I. Physikalisches Institut, Universität zu Köln, Köln, Germany*; JENNIFER VAN WIJNGAARDEN, *Department of Chemistry, University of Manitoba, Winnipeg, MB, Canada*; OLIVIER PIRALI, SÉBASTIEN GRUET, *AILES beamline, Synchrotron SOLEIL, Saint Aubin, France*; FRANK LEWEN, STEPHAN SCHLEMMER, *I. Physikalisches Institut, Universität zu Köln, Köln, Germany*; MICHAEL C McCARTHY, *Atomic and Molecular Physics, Harvard-Smithsonian Center for Astrophysics, Cambridge, MA, USA*.

In the course of our recent study of the ν_2 bending mode of S_2O (Martin-Drumel el al.; see Talk P1190), the S-S stretching mode ν_3 located at 679cm^{-1} and first studied by Lindenmayer *et al.* in 1986 (*J. Mol. Spectrosc.* **119**, 56) has been re-investigated at the French national synchrotron facility SOLEIL using Fourier-transform far-infrared spectroscopy. In addition to the vibrational fundamental, evidence for at least one more hot band, most likely $\nu_3 + \nu_2 - \nu_2$, was found. Complementary submillimeter wave measurements of the pure rotational spectrum in the $v_3 = 1$ state were also performed.

RB09 **11:16 – 11:31**

FT-IR MEASUREMENTS OF NH_3 LINE INTENSITIES IN THE 60 – 550 CM^{-1} USING SOLEIL/AILES BEAMLINE

KEEYOON SUNG, *Jet Propulsion Laboratory, California Institute of Technology, Pasadena, CA, USA*; SHANSHAN YU, *Molecular Spectroscopy, Jet Propulsion Laboratory, Pasadena, CA, USA*; JOHN PEARSON, *Jet Propulsion Laboratory, California Institute of Technology, Pasadena, CA, USA*; LAURENT MANCERON, *Beamline AILES, Synchrotron SOLEIL, Saint-Aubin, France*; F. KWABIA TCHANA, *LISA, CNRS, Universités Paris Est Créteil et Paris Diderot, Créteil, France*; OLIVIER PIRALI, *AILES beamline, Synchrotron SOLEIL, Saint Aubin, France.*

Ammonia (NH_3) has been found ubiquitous, *e.g.*, in the interstellar medium, low-mass stars, Jovian planets of our solar system, and possibly in the low temperature exoplanets. Their spectroscopic line parameters are essential in the accurate interpretation of the planetary and astrophysical spectra observed with Herschel, SOFIA, ALMA, and JWST.

In our previous paper[a], the NH_3 line positions in the far-IR region were studied for the ground state and ν_2 in an unprecedented accuracy, which revealed significant deficiencies in the NH_3 intensities, for instance, some weak $\Delta K = 3$ lines were predicted to be 100 times stronger. Measurement of line intensity for these lines in a consistent manner is demanded because the $\Delta K = 3$ forbidden lines are only way other than collisions and l-doubled states to excite NH_3 to $K > 0$ levels. Recalling that NH_3 transition lines in the high J and K up to 18 were detected toward the galactic center in the star forming region of Sgr B_2, their accurate intensity measurements are critical in explaining the observed high K excitation, which will provide insights into radiative-transfer *vs.* collision excitation mechanics of interstellar NH_3.

For this, we obtained a series of spectra of $^{14}NH_3$ in the 50 – 550 cm^{-1} using a Fourier-transform spectrometer, Bruker 125HR, and AILES beam line at Synchrotron SOLEIL, France. Line positions, intensities, and pressure-broadened half-widths have been measured using non-linear least squares spectrum fitting algorithm. In this presentation we report and discuss preliminary results of line position and intensity measurements for the inversion transitions in the ground state, ν_2, $2\nu_2$, ν_4 and for the vibration-rotation transitions of ν_2, $2\nu_2$, ν_4, $2\nu_2 - \nu_2$, $\nu_4 - \nu_2$ and $\nu_4 - 2\nu_2$ in this region. Comparison of the new measurements with the current databases and *ab initio* calculations will be discussed.

[a]S. Yu, et al. J. Chem. Phys. (2010) 174317/1-174317/14.

RB10 ***Post-Deadline Abstract*** **11:33 – 11:48**

THE H_2O-CH_3F COMPLEX: A COMBINED MICROWAVE AND INFRARED SPECTROSCOPIC STUDY SUPPORTED BY STRUCTURE CALCULATIONS

SHARON PRIYA GNANASEKAR, *Department of Inorganic and Physical Chemistry, Indian Institute of Science, Bangalore, India*; MANUEL GOUBET, *Laboratoire PhLAM, Université de Lille 1, Villeneuve de Ascq, France*; ELANGANNAN ARUNAN, *Department of Inorganic and Physical Chemistry, Indian Institute of Science, Bangalore, India*; ROBERT GEORGES, *IPR UMR6251, CNRS - Université Rennes 1, Rennes, France*; PASCALE SOULARD, PIERRE ASSELIN, *MONARIS UMR8233, CNRS - UNiversité Paris 6 UPMC, Paris, France*; T. R. HUET, *Laboratoire PhLAM, UMR8523 CNRS - Université Lille 1, Villeneuve d'Ascq, France*; OLIVIER PIRALI, *AILES beamline, Synchrotron SOLEIL, Saint Aubin, France.*

The H_2O-CH_3F complex could have two geometries, one with a hydrogen bond and one with the newly proposed carbon bond[a]. While in general carbon bonds are weaker than hydrogen bonds, this complex appears to have comparable energies for the two structures. Infrared (IR) and microwave (MW) spectroscopic measurements using, respectively, the Jet-AILES apparatus[b] and the FTMW spectrometer at the PhLAM laboratory[c], have been carried out to determine the structure of this complex. The IR spectrum shows the formation of the CH_3F- H_2O hydrogen bonded complex and small red-shifts in OH frequency most probably due to $(CH_3F)_m$-$(H_2O)_n$ clusters. Noticeably, addition of CH_3F in the mixture promotes the formation of small water clusters. Preliminary MW spectroscopic measurements indicate the formation of the hydrogen bonded complex. So far, we have no experimental evidence for the carbon bonded structure. However, calculations of the Ar-CH_3F complex show three energetically equivalent structures: a T-shape, a "fluorine" bond and a carbon bond. The MW spectrum of the $(Ar)_n$-CH_3F complexes is currently under analysis.

[a]Mani, D; Arunan, E. Phys. Chem. Chem. Phys. 2013, 15, 14377.

[b]Cirtog, M; Asselin, P; Soulard, P; Tremblay, B; Madebene, B; Alikhani, M. E; Georges, R; Moudens, A; Goubet, M; Huet, T.R; Pirali, O; Roy, P. J. Phys. Chem. A. 2011, 115, 2523

[c]Kassi, S; Petitprez, D; Wlodarczak, G. J. Mol. Struct. 2000, 517-518, 375

RC. Mini-symposium: Spectroscopy in the Classroom

Thursday, June 25, 2015 – 8:30 AM

Room: B102 Chemical and Life Sciences

Chair: Kristopher J Ooms, The King's University, Edmonton, Alberta, Canada

RC01 *INVITED TALK* **8:30 – 9:00**

DIRECT DIGITAL SYNTHESIS CHIRPED PULSE MICROWAVE SPECTROMETERS FOR THE CLASSROOM AND RESEARCH

GEOFFREY BLAKE, BRANDON CARROLL, IAN A FINNERAN, *Division of Chemistry and Chemical Engineering, California Institute of Technology, Pasadena, CA, USA.*

By combining the rapid development in direct digital synthesis circuitry and Field Programmable Gate Arrays (FPGAs) coupled to fast A/D samplers, it is possible to construct high performance chirped pulse microwave spectrometers suitable for gas-phase rotational spectroscopy experiments in undergraduate physical chemistry labs as well as graduate level research. The technology is highly tailorable, and sufficiently robust that extensive experimentation is feasible in the teaching environment. The time domain nature of the experiment has strong ties to concepts in Nuclear Magnetic Resonance (NMR) widely discussed in undergraduate curricula, and the software environment for the instrument control and spectral assignment can be integrated with ab initio quantum chemistry predictions of molecular structure and dynamics.

RC02 **9:05 – 9:20**

A SIMPLE, COST EFFECTIVE RAMAN-FLUORESCENCE SPECTROMETER FOR USE IN LABORATORY AND FIELD EXPERIMENTS

FRANK E MARSHALL, MICHAEL A PRIDE, MICHELLLE ROJO, KATELYN R. BRINKER, ZACHARY WALKER, *Department of Chemistry, Missouri University of Science and Technology, Rolla, MO, USA*; MICHAEL STORRIE-LOMBARDI, *Department of Physics, Harvey Mudd College and Kinohi Institute, Inc., Claremont, CA, USA*; MELANIE R. MORMILE, *Department of Biological Sciences, Missouri University of Science and Technology, Rolla, MO, USA*; G. S. GRUBBS II, *Department of Chemistry, Missouri University of Science and Technology, Rolla, MO, USA.*

Research, design, construction, and operation of a portable mixed Raman and Fluorescence type spectrometer implemented by the Missouri University of Science and Technology's Mars Rover Design Team will be presented. This spectrometer has been built for the team's annual competition. The spectrometer, completely built by undergraduates, is designed to use a 50 mW, 532 nm constant waveform laser to probe a sample of soil to find bacteria or bio-markers. However, initial tests of the spectrometer were carried out in a laboratory environment making the spectrometer also suitable for simple undergraduate physical chemistry or chemical physics laboratory experiments. The final cost of the device is roughly $2100, weighs 1.4 kg, and is 22.9 cm x 22.6 cm in size. Integrating the spectrometer with a computer database, results from the competition, complications of fitting mixed Raman-Fluorescence spectra, and future ideas/improvements will also be discussed.

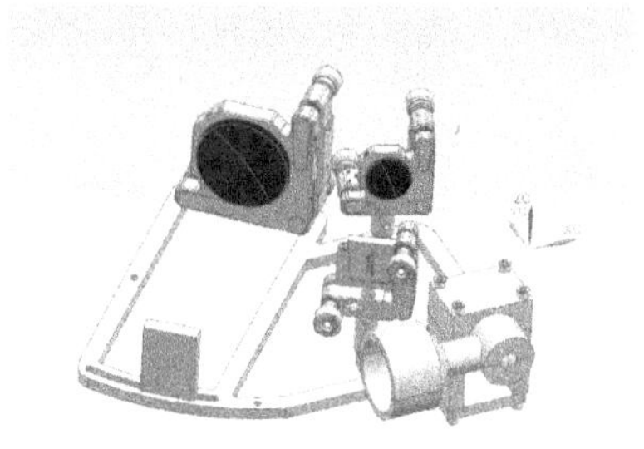

RC03 **9:22 – 9:37**

LIF AND RAMAN SPECTROSCOPY IN UNDERGRADUATE LABS USING GREEN DIODE-PUMPED SOLID-STATE LASERS

JEFFREY A. GRAY, *Department of Chemistry, Ohio Northern University, Ada, OH, USA.*

Electronic spectroscopy of molecular iodine vapor has long been studied in undergraduate physical chemistry teaching laboratories, but the effectiveness of emission work has typically been limited by availability of instrumentation. This talk shows how to make inexpensive green diode-pumped solid-state (DPSS) lasers easily tunable for efficient, selective excitation of I_2. Miniature fiber-optic spectrometers then enable rotationally resolved fluorescence spectroscopy up to v" = 42 near 900 nm with acquisition times of less than one minute. DPSS lasers are also versatile excitation sources for vibrational Raman spectroscopy, which is another common exercise that has been limited by lack of proper instrumentation in the teaching laboratory. This talk shows how to construct a simple accessory for commercial fluorimeters to record vibrational Raman spectra and depolarization ratios for CCl_4 and C_2Cl_4 as part of a lab exercise featuring molecular symmetry.

RC04 **9:39 – 9:54**

SPECFITTER: A LEARNING ENVIRONMENT FOR THE ROTATIONAL SPECTROSCOPIST

YOON JEONG CHOI, WEIXIN WU, *Natural and Social Science, Purchase College SUNY, Purchase, NY, USA*; A. J. MINEI, *Department of Chemistry and Biochemistry, Division of Natural Sciences, College of Mount Saint Vincent, Riverdale, NY, USA*; S. A. COOKE, *Natural and Social Science, Purchase College SUNY, Purchase, NY, USA*.

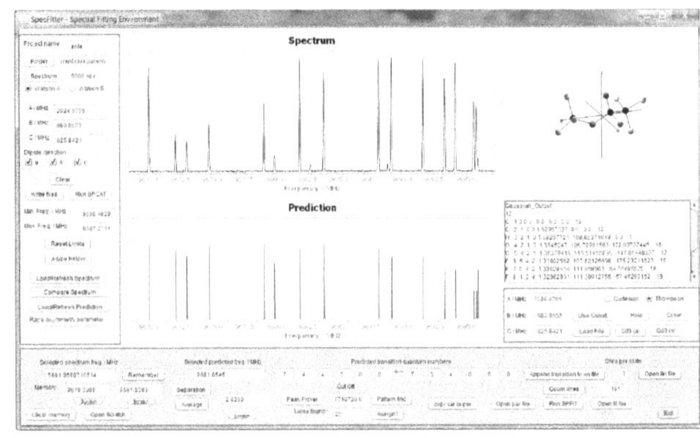

A windows based, mouse-event driven software program that acts a graphical user interface to Pickett's fitting programs has been created and improved. The program, SpecFitter, is aimed at walking users through the process of assigning pure rotational spectra. Rotational spectra, in XY format, may be viewed and inspected and the user is provided with tools for observing and recording repeating, similar patterns of transitions. The structure of these patterns is interpreted into "guesses" at rotational constants which the user may then use to predict a spectrum. Observed transition frequencies may then be assigned quantum number transitions and appended to the .lin file through mouse clicks. Although the thrust of this project is to develop a users ability to assign spectra without knowing the molecule producing the spectra the program can also read in and display calculated structures of target molecules produced by the Gaussian03/09 software, or alternatively the user can draw their own structures. Structures can be edited allowing users to observe the relationship between molecular structure and (i) the direction of dipole moment components and (ii) the relationships between structure and rotational constants. Users may also easily predict spectra from the molecules structure and further relate rotational constants to observed spectra. Students in CHEM 3510 at Purchase College have been vital in developing the software.

RC05 **9:56 – 10:11**

APPLICATIONS OF GROUP THEORY: INFRARED AND RAMAN SPECTRA OF THE ISOMERS OF *cis*- AND *trans*-1,2-DICHLOROETHYLENE

NORMAN C. CRAIG, *Department of Chemistry and Biochemistry, Oberlin College, Oberlin, OH, USA*.

A study of the vibrational spectra of *cis*- and *trans*-1,2-dichloroethylene provides an excellent way for undergraduates to gain experience with the application of group theory in the physical chemistry laboratory. Although the group vibrations are similar for these two molecules, the selection rules for infrared (IR) and Raman spectra differ significantly. Most of the transitions for the fundamentals of the cis isomer of C_{2v} symmetry are both IR and Raman active. Mutual exclusion for the vibrational transitions applies to the centrosymmetric trans isomer of C_{2h} symmetry. Thus, half the transitions for the trans isomer are IR active and half are Raman active. The two isomers are volatile enough that gas-phase IR spectra can be recorded at room temperature. Band shapes in gas-phase IR spectra provide additional evidence for assignments of fundamentals. The two isomers are small enough that good quality quantum chemical calculations of harmonic frequencies can be done by students with commercial software.

Intermission

RC06 **10:30 – 10:40**

INFRARED ANALYSIS OF COMBUSTION PRODUCTS AND INTERMEDIATES OF HYDROCARBON COMBUSTION FOR SEVERAL SPECIES

ALLEN WHITE, *Department of Mechanical Engineering, Rose-Hulman Institute of Technology, Terre Haute, IN, USA*; REBECCA DEVASHER, *Department of Chemistry, Rose-Hulman Institute of Technology, Terre Haute, IN, USA.*

Hydrocarbons, especially large ones such as isooctane, have infrared active species that give insight into combustion stoichiometry and temperature. Here a Fourier-transform infrared spectrometer is utilized to study the IR active species for a number of stociometric conditions for several fuels including isooctane, kerosene, and ethanol. Special attention is given to intermediate species in different flame regions.

RC07 **10:42 – 10:52**

CHIRPED-PULSE MICROWAVE SPECTROSCOPY IN THE UNDERGRADUATE CHEMISTRY CURRICULUM

SYDNEY A GASTER, TAYLOR M HALL, SEAN ARNOLD, GORDON G BROWN, *Department of Science and Mathematics, Coker College, Hartsville, SC, USA.*

The use of chirped-pulse Fourier transform microwave (CP-FTMW) spectroscopy as a tool for training undergraduates will be discussed. Coker College's inexpensive, versatile CP-FTMW spectrometer has been applied both in the undergraduate teaching laboratory and the undergraduate research laboratory. In both cases, the education of the students is a central priority of the project. The study of 3-iodopyridine, a project recently completed by Coker undergraduate students, will be discussed. Details of the Coker CP-FTMW spectrometer will also be presented.

RC08 **10:54 – 11:04**

USB SPECTROMETERS AND THE TEMPERATURE OF THE SUN: MEASURING BLACK BODY RADIATION IN THE PALM OF YOUR HAND

DANIEL P. ZALESKI, BENJAMIN R HORROCKS, NICK WALKER, *School of Chemistry, Newcastle University, Newcastle-upon-Tyne, United Kingdom.*

A new experiment appropriate for both general chemistry and physical chemistry students will be described. The experiment utilizes "pocket size" USB spectrometers (operating in the UV/vis region) coupled with fiber optic cables to record a solar spectrum. A further extension of the experiment involves recording spectra of a light bulb at several voltages (and thus resistances). Using provided software, students can fit black body distributions to their obtained spectra. The software will display the acquired spectrum, a simulation based on their guess temperature, a simulation based on their fit, and OMC^2 for both. Students can then compare their results to the known temperature of the sun and the known temperature vs resistance curve of tungsten.

RC09 **11:06 – 11:16**

VIBRATION-ROTATION ANALYSIS OF THE $^{13}CO_2$ ASYMMETRIC STRETCH FUNDAMENTAL BAND IN AMBIENT AIR FOR THE PHYSICAL CHEMISTRY TEACHING LABORATORY

DAVID A DOLSON, CATHERINE B ANDERS[a], *Department of Chemistry, Wright State University, Dayton, OH, USA.*

The CO_2 asymmetric stretch fundamental band near 4.3 μm is one of the strongest infrared absorption transitions of all small molecules. This band is an undesired interference in most infrared spectra, but it also serves as a potential choice for a vibration-rotation analysis experiment in the physical chemistry teaching laboratory. Due to the strength of this band and the 1.1% natural abundance of carbon-13, the asymmetric stretch fundamental band of $^{13}CO_2$ is readily observable in a typical ambient air background spectrum and is shifted sufficiently from the stronger $^{12}CO_2$ fundamental such that the $^{13}CO_2$ P-branch lines are almost completely free of interferences and are easily assigned. All of the $^{13}CO_2$ R-branch lines appear within the $^{12}CO_2$ P-branch, which creates assignment challenges. Students in our program have analyzed the $^{13}CO_2$ fundamental asymmetric stretch band over a two-year period. Analyses of the P-branch line positions enabled the prediction of additional R-branch line positions, which guided line identification and measurements in the $^{13}CO_2$ R-branch. C=O bond lengths determined from analyses of the $^{13}CO_2$ spectra improved when R-branch lines were added to the initial P-branch data sets. Spectral appearance, analyses and results will be presented for spectra obtained at 0.5 cm^{-1} resolution and at 0.125 cm^{-1} resolution. The challenge of predicting and finding the $^{13}CO_2$ R-branch lines among other interfering lines adds an element of realism to this experiment that is not found in many student experiments of this type.

[a]present address: Department of Biomolecular Sciences, Boise State University, Boise, ID, USA

RC10 **11:18 – 11:33**

UTILIZING SPECTROSCOPIC RESEARCH TOOLS AND SOFTWARE IN THE CLASSROOM

G. S. GRUBBS II, *Department of Chemistry, Missouri University of Science and Technology, Rolla, MO, USA.*

Given today's technological age, it has become crucial to be able to reach the student in a more "tech-savvy" way than traditional classroom methods afford. Given this, there are already a vast range of software packages available to the molecular spectroscopist that can easily be introduced to the classroom with success. This talk will highlight taking a few of these tools (Gaussian09, SPFIT/SPCAT, the AABS Package, LabView^{TM}, etc.) and implementing them in the classroom to teach subjects such as Quantum Mechanics and Thermodynamics as well as to aid in the linkage between these subjects. Examples of project implementation on both undergraduate and graduate level students will be presented with a discussion on the successes and failures of such attempts.

RD. Astronomy

Thursday, June 25, 2015 – 8:30 AM

Room: 274 Medical Sciences Building

Chair: Brett A. McGuire, California Institute of Technology, Pasadena, CA, USA

RD01 **8:30 – 8:40**

NEW INSTRUMENTAL TOOLS FOR ADVANCED ASTROCHEMICAL APPLICATIONS

AMANDA STEBER, *The Centre for Ultrafast Imaging (CUI), Universität Hamburg, Hamburg, Germany*; SABRINA ZINN, MELANIE SCHNELL, *CoCoMol, Max-Planck-Institut für Struktur und Dynamik der Materie, Hamburg, Germany*; ANOUK RIJS, *Institute for Molecules and Materials (IMM), Radboud University Nijmegen, Nijmegen, Netherlands.*

Astrochemistry has been a growing field over the past several years. As the data from the Atacama Large Millimeter Array (ALMA) becomes publicly available, new and fast techniques for the analysis of the data will need to be developed, as well as fast, sensitive laboratory techniques. This lab is in the process of building up instrumentation that will be dedicated to the measurement of astrochemically relevant species, both in the microwave and the millimeter wave regimes. Discharge experiments, laser ablation experiments, as well as time of flight measurements will be possible with this instrumentation. Coupled with instrumentation capabilities will be new software aimed at a speeding up the analysis. The laboratory data will be used to search for new molecular signatures in the interstellar medium (ISM), and help to elucidate molecular reaction pathways occurring in the ISM.

RD02 **8:42 – 8:57**

DOPPLER AND SUB-DOPPLER MILLIMETER AND SUB-MILLIMETER WAVE SPECTROSCOPY OF KEY ASTRONOMICAL MOLECULES: HNC AND CS

OLIVER ZINGSHEIM, THOMAS SCHMITT, FRANK LEWEN, STEPHAN SCHLEMMER, SVEN THORWIRTH, *I. Physikalisches Institut, Universität zu Köln, Köln, Germany.*

In the course of ongoing efforts to determine accurate pure rotational transition frequencies for the astronomical community, the millimeter- and submillimeter-wave spectra of HNC and selected isotopic species have been investigated using a radio-frequency discharge of (isotopically enriched) methyl cyanide. Besides the ground vibrational state, vibrational satellites from the first excited bending mode were targeted. In part, rotational transitions were observed employing the Lamb-Dip technique to obtain sub-Doppler resolution. The Lamb-dip technique has also been applied to other short-lived molecules such as carbon monosulfide, CS.

RD03 **8:59 – 9:14**

MILLIMETRE-WAVE SPECTRUM OF ISOTOPOLOGUES OF ETHANOL FOR RADIO ASTRONOMY

ADAM WALTERS, *IRAP, Université de Toulouse 3 - CNRS - OMP, Toulouse, France*; MIRKO SCHÄFER, MATTHIAS H. ORDU, FRANK LEWEN, STEPHAN SCHLEMMER, HOLGER S. P. MÜLLER, *I. Physikalisches Institut, Universität zu Köln, Köln, Germany.*

Complex molecules have been identified in star-forming regions and their formation is linked to the specific physical and chemical conditions there. They are suspected to form a role in the origins of life. Amongst these, ethanol is a fairly abundant molecule in warmer regions.

For this reason, we have recently carried out laboratory measurements and analyses of the rotational spectra of the three mono-substituted deuterium isotopologues of ethanol (one of which, CH_2DCH_2OH, exists as two distinct conformers according to the position of the deuterium atom with respect to the molecular skeleton). Measurements were taken between 35-500 GHz, allowing accurate predictions in the range of radio telescopes. We have concentrated on the lowest energy *anti* conformers. The dataset was constrained for fitting with a standard Watson-S reduction Hamiltonian by rejecting transitions from high-lying states, which appear to be perturbed by the gauche states, and by averaging some small methyl torsional splits. This treatment is compatible with the needs for a first search in the interstellar medium, in particular in spectra taken by ALMA. For this purpose an appropriate set of predictions will be included on the Cologne Database for Molecular Spectroscopy.

Previous results on the two mono-substituted ^{13}C isotopologues[a] which led to a tentative detection in Sgr B2(N)[b] will be briefly summarized and compared with the latest measurements.

The usefulness of studying different isotopologues in the interstellar medium will also be rapidly addressed.

[a]Bouchez et al, JQSRT 113 (11), pp. 1148-1154, 2012.
[b]Belloche et al. A&A 559, id.A47, 187pp., 2013.

RD04 **9:16 – 9:31**

TERAHERTZ SPECTROSCOPY OF DEUTERATED METHYLENE BI-RADICAL, CD_2

HIROYUKI OZEKI, *Department of Environmental Science, Toho University, Funabashi, Japan*; STEPHANE BAILLEUX, *Laboratoire PhLAM, Université de Lille - Sciences et Technologies, Villeneuve d'Ascq, France.*

Methylene, the parent of the carbene compounds, plays a crucial role in many chemical reactions. This bi-radical is a known interstellar molecule that has been detected towards hot cores in dense interstellar clouds. CH_2 is also thought to be present in cometary atmospheres. In the gas phase chemical models of both dense and diffuse molecular clouds, CH_2 is a key intermediate in interstellar carbon chemistry which is produced primarily by dissociative recombination of the methyl ion, CH_3^+. Recently tentative detection of the mono-deuterated methyl ion, CH_2D^+ has been reported toward an infrared source in the vicinity of Orion.[a] Deuterated methylene CHD and CD_2 can be produced from this ion or its counterpart CHD_2^+ by dissociative recombination with an electron:

$$CH_2D^+ + e^- \rightarrow CHD + H \text{ or } CH_2 + D, \quad (1)$$

$$CHD_2^+ + e^- \rightarrow CHD + D \text{ or } CD_2 + H. \quad (2)$$

Thus, both CHD and CD_2 can be observed in warm interstellar clouds, where the deuterium fractionation process is important. Precise laboratory reference data are desirable for radioastronomical observation of these molecules.

Here we report on our high-resolution spectroscopic investigation on the deuterated methylene radical, CD_2 ($X\ ^3B_1$) up to 1.45 THz. At present time, eleven out of the twelve fine-structure components of four pure rotational transitions have been newly recorded, and these measurements double the number of previously observed transitions.[b] CD_2 was generated in a discharge in CD_2CO which was obtained from the flash pyrolysis of acetic anhydride-d6 ($(CD_3CO)_2O$). Effort is currently made to measure the astronomically important $1_{11} - 0_{00}$ transition whose fine-structure components are predicted to occur at 1.224,1.228 and 1.234 THz.

[a]D. C. Lis, P. F. Goldsmith, E. A. Bergin et al. 2009, in Submillimeter Astrophysics and Technology, ASP Conf. Ser., 417, 23.
[b]H. Ozeki and S. Saito J. Chem. Phys. 1996, 104, 2167.

RD05 **9:33 – 9:48**

THZ SPECTROSCOPY OF D_2H^+

SHANSHAN YU, JOHN PEARSON, TAKAYOSHI AMANO, *Jet Propulsion Laboratory, California Institute of Technology, Pasadena, CA, USA.*

Pure rotational transitions of D_2H^+ observed by high-resolution spectroscopy have been limited so far to the J = 1_{10}-1_{01} transition at 691.7 GHz,[a] $J = 2_{20} - 2_{11}$ at 1.370 THz, and $J = 1_{11} - 0_{00}$ at 1.477 THz.[bc] As this ion is a light asymmetric-top molecule, spectroscopic characterization and prediction of other rotational transition frequencies are not straightforward.

In this presentation, we extended the measurements up to 2 THz by using the JPL frequency multiplier chains, and observed three new THz lines and re-measured the three known transitions. D_2H^+ was generated in an extended negative glow discharge cell cooled to liquid nitrogen temperature. Six rotational transition frequencies together with the combination differences derived from three fundamental bands were subject to least square analysis to determine the molecular constants. New THz measurements are definitely useful for better characterization of spectroscopic properties. The improved molecular constants provide better predictions of other unobserved rotational transitions.

[a]T. Hirao and T. Amano, *Ap. J.*,**597**, L85 (2003)
[b]K. M. Evenson et al cited by O. L. Polyansky and A. R. W. McKellar, *J. Chem. Phys.*, **92**, 4039 (1990)
[c]O. Asvany et al, *Phys. Rev. Lett.*, **100**, 233004 (2008)

RD06 **9:50 – 10:05**

THZ SPECTROSCOPY OF $^{12}CH^+$, $^{13}CH^+$, AND $^{12}CD^+$

SHANSHAN YU, BRIAN DROUIN, JOHN PEARSON, TAKAYOSHI AMANO, *Jet Propulsion Laboratory, California Institute of Technology, Pasadena, CA, USA.*

In 1937, Dunham[a] detected a couple of unidentified lines in near-UV, and later Douglas and Herzberg[b] identified them based on their laboratory observations to be low-J electronic transitions of CH^+. The electronic spectra, in particular the $A^1\Pi - X^1\Sigma^+$ band, have been investigated extensively. On the other hand, the pure rotational transitions have not been studied so extensively. Only the lowest rotational transition, $J = 1 - 0$, was observed in the laboratory for the normal species, $^{13}CH^+$, and CD^+.[cd]

Based on the laboratory frequency, CH^+ was detected in star forming regions with the Hershel space observatory. Cernicharo et al identified pure rotational transitions from $J = 2 - 1$ to $J = 6 - 5$ in the far-infrared region in the ISO spectrum of the planetary nebula NGC 7027[e]. The ISO spectra, however, were of low-resolution, so high-resolution spectroscopic observation is highly desirable.

In this presentation, we have extended the measurements to higher-J lines up to 2 THz. For production of CH^+, an extended negative glow discharge in a gas mixture of CH_4 ($\sim$ 0.5 mTorr) diluted in He ($\sim$ 60 mTorr) was used. The optimum discharge current was about 15 mA and the axial magnetic filed to 160 Gauss was applied up. The discharge cell was cooled down to liquid nitrogen temperature. Several frequency multiplier chains, developed at JPL and purchased from Virginia Diodes, were used as THz radiation sources.

New THz measurements are not only useful for providing better characterization of spectroscopic properties but also will serve as starting point for astronomical observations.

[a]T. Dunham, *Publ. Astron. Soc. Pac.*, **49**, 26 (1937)
[b]A. E. Douglas and G. Herzberg, *Ap. J.* **94**, 381 (1941)
[c]T. Amano, *Ap.J.Lett.*, **716**, L1 (2010)
[d]T. Amano, *J. Chem. Phys.*, **133**, 244305 (2010)
[e]J. Cernicharo et al., *Ap. J. Lett.*, **483**, L65 (1997)

RD07 **10:07 – 10:22**

ROTATIONAL SPECTROSCOPY OF VIBRATIONALLY EXCITED N_2H^+ and N_2D^+ UP TO 2 THZ

SHANSHAN YU, JOHN PEARSON, BRIAN DROUIN, TIMOTHY J CRAWFORD, ADAM M DALY, BEN ELLIOTT, TAKAYOSHI AMANO, *Jet Propulsion Laboratory, California Institute of Technology, Pasadena, CA, USA.*

Terahertz absorption spectroscopy was employed to extend the measurements on the pure rotational transitions of N_2H^+, N_2D^+ and their ^{15}N-containing isotopologues in the ground state and first excited vibrational states for the three fundamental vibrational modes. In total 88 new pure rotational transitions were observed in the range of 0.7–2.0 THz. The observed transition frequencies were fit to experimental accuracy, and the improved molecular parameters were obtained. The new measurements and predictions will support the analysis of high-resolution astronomical observations made with facilities such as SOFIA and ALMA where laboratory rest frequencies with uncertainties of 1 MHz or smaller are required for proper analysis of velocity resolved astrophysical components.

Intermission

RD08 **10:41 – 10:56**

NEW ACCURATE WAVENUMBERS OF $H^{35}Cl^+$ AND $H^{37}Cl^+$ ROVIBRATIONAL TRANSITIONS IN THE $v = 0 - 1$ BAND OF THE $^2\Pi$ STATE.

JOSE LUIS DOMENECH, MAITE CUETO, VICTOR JOSE HERRERO, ISABEL TANARRO, *Molecular Physics, Instituto de Estructura de la Materia (IEM-CSIC), Madrid, Spain*; JOSE CERNICHARO, *Molecular Astrophysics, ICMM, Madrid, Spain.*

HCl^+ is a key intermediate in the interstellar chemistry of chlorine. It has been recently identified in space from *Herschel's* spectra[a] and it has also been detected in the laboratory through its optical emission[b], infrared[c] and mm-wave spectra[d]. Now that *Hershchel* is decomissioned, further astrophysical studies on this radical ion will likely rely on ground-based observations in the mid-infrared. We have used a difference frequency laser spectrometer coupled to a hollow cathode discharge to measure the absorption spectrum of $H^{35}Cl^+$ and $H^{37}Cl^+$ in the $v = 0 - 1$ band of the $^2\Pi$ state with Dopppler limited resolution. The accuracy of the individual measurements ($\sim$ 10 MHz (3σ)) relies on a solid state wavemeter referenced to an iodine-stabilized Ar^+ laser.

[a]M. De Luca et al., *Astrophys. J. Lett.* **751**, L37 (2012)
[b]W. D. Sheasley and C. W. Mathews, *J. Mol. Spectrosc.* **47**, 420 (1973)
[c]P. B. Davies, P. A. Hamilton, B. A. Johnson, *Mol. Phys.* **57**, 217 (1986)
[d]H. Gupta, B. J. Drouin, and J. C. Pearson, *Astrophys. J. Lett.* **751**, L37 (2012)

RD09 **10:58 – 11:13**

OSCILLATOR STRENGTHS AND PREDISSOCIATION RATES FOR $W - X$ BANDS AND THE 4P5P COMPLEX IN $^{13}C^{18}O$

MICHELE EIDELSBERG, JEAN LOUIS LEMAIRE, *Meudon, Observatoire de Paris, Paris, France*; STEVEN FEDERMAN, *Physics and Astronomy, University of Toledo, Toledo, OH, USA*; GLENN STARK, *Department of Physics, Wellesley College, Wellesley, MA, USA*; ALAN HEAYS, *Leiden Observatory, University of Leiden, Leiden, Netherlands*; LISSETH GAVILAN, *Institut d'Astrophysique Spatiale, Campus de l'Université Paris XI, Orsay, France*; JAMES R LYONS, *School of Earth and Space Exploration, Arizona State University, Tempe, AZ, USA*; PETER L SMITH, *93 Pleasant St., 93 Pleasant St., Watertown, MA, USA*; NELSON DE OLIVEIRA, DENIS JOYEUX, *DESIRS Beamline, Synchrotron SOLEIL, Saint Aubin, France.*

In our ongoing experiments on the DESIRS beam-line at the SOLEIL Synchrotron, we are acquiring the necessary data on oscillator strengths and predissociation rates for modeling CO photochemistry in astronomical environments. A VUV Fourier Transform Spectrometer with a resolving power of about 350,000 allows us to discern individual lines in electronic transitions. Here we focus on results obtained from absorption spectra of $^{13}C^{18}O$, for the $W\ ^1\Pi\ -\ X\ ^1\Sigma^+$ bands with $v' = 0, 2$, and 3 and $v'' = 0$ and three resolved bands involving transitions to the upper levels 4pπ(2), 5pπ(0), and 5pσ(0) of the 4p(2) and 5p(0) complexes. We compare our results with earlier determinations for this isotopologue of CO, as well as with our SOLEIL measurements on $^{12}C^{16}O$, $^{13}C^{16}O$, and $^{12}C^{18}O$.

RD10 **11:15 – 11:30**

LINE STRENGTHS OF ROVIBRATIONAL AND ROTATIONAL TRANSITIONS IN THE $X^2\Pi$ GROUND STATE OF OH

JAMES S.A. BROOKE, *Department of Chemistry, University of York, York, United Kingdom*; PETER F. BERNATH, *Department of Chemistry and Biochemistry, Old Dominion University, Norfolk, VA, USA*; COLIN WESTERN, *School of Chemistry, University of Bristol, Bristol, United Kingdom*; CHRIS SNEDEN, *Department of Astronomy, The University of Texas at Austin, Austin, TX, USA*; GANG LI, IOULI E GORDON, *Atomic and Molecular Physics, Harvard-Smithsonian Center for Astrophysics, Cambridge, MA, USA.*

For cool stellar and substellar objects including brown dwarfs and exoplanets, atomic lines weaken and detailed elemental and isotopic abundances are often derived from molecular absorption features. We have embarked on a project to provide molecular line lists based on combining experimental observations for line positions and ab initio calculations for line strengths. In this talk we present a new line list including positions and absolute intensities (in the form of Einstein A values and oscillator strengths) that has been produced for the OH ground $X^2\Pi$ state rovibrational (Meinel system) and pure rotational transitions. All possible allowed transitions are included with v up to 13, and J up to between 9.5 and 59.5, depending on the band. A new fit to determine molecular constants has been performed, based on the fit of Bernath and Colin [J. Mol. Spectrosc. 257, 20 (2009)], but includes some new rotational data and a simultaneous fitting of all molecular constants. The line intensities are based on a new dipole moment function, which is a combination of two high level ab initio calculations. The calculations show good agreement with an experimental v=1 lifetime, experimental dipole moment values and Δv=2 line intensity ratios from an observed laboratory spectrum. A new partition function was computed suitable for use up to about 6000 K. The new line list was also evaluated by determining the oxygen abundance for a number of cool stars from high resolution observations in the near infrared region.

RD11 ***Post-Deadline Abstract*** **11:32 – 11:42**

CLASS I METHANOL MASER CONDITIONS NEAR SNRS

BRIDGET C. McEWEN, YLVA M. PIHLSTRÖM, *Physics and Astronomy, The University of New Mexico, Albuquerque, NM, USA*; LORÁNT O. SJOUWERMAN, *NRAO, NRAO, Socorro, NM, USA.*

We present results from calculations of the physical conditions necessary for the occurrence of 36.169 ($4_{-1} - 3_0\,E$), 44.070 ($7_0 - 6_1\,A^+$), 84.521 ($5_{-1} - 4_0\,E$), and 95.169 ($8_0 - 7_1\,A^+$) GHz methanol (CH_3OH) maser emission lines near supernova remnants (SNRs), using the MOLPOP-CEP program. The calculations show that given a sufficient methanol abundance, methanol maser emission arises over a wide range of densities and temperatures, with optimal conditions at $n \sim 10^4 - 10^6$ cm^{-3} and $T > 60$ K. The 36 GHz and 44 GHz transitions display more significant maser optical depths compared to the 84 GHz and 95 GHz transitions over the majority of physical conditions. It is also shown that line ratios are an important and applicable probe of the gas conditions. The line ratio changes are largely a result of the E-type transitions becoming quenched faster at increasing densities. The modeling results will be discussed using recent observations of CH_3OH masers near the SNRs G1.4−0.1, W28, and Sgr A East and used as a diagnostic tool to estimate densities and temperatures of the regions in which the CH_3OH masers are observed.

RD12 *Post-Deadline Abstract* **11:44 – 11:59**

THE MISSING LINK: ROTATIONAL SPECTRUM AND GEOMETRICAL STRUCTURE OF DISILICON CARBIDE, Si_2C

MICHAEL C McCARTHY, *Atomic and Molecular Physics, Harvard-Smithsonian Center for Astrophysics, Cambridge, MA, USA*; JOSHUA H BARABAN, *Department of Chemistry, University of Colorado, Boulder, CO, USA*; BRYAN CHANGALA, *JILA, National Institute of Standards and Technology and Univ. of Colorado Department of Physics, University of Colorado, Boulder, CO, USA*; JOHN F. STANTON, *Department of Chemistry, The University of Texas, Austin, TX, USA*; MARIE-ALINE MARTIN-DRUMEL, *Spectroscopy Lab, Harvard-Smithsonian Center for Astrophysics, Cambridge, MA, USA*; SVEN THORWIRTH, *I. Physikalisches Institut, Universität zu Köln, Köln, Germany*; NEIL J REILLY, *Department of Chemistry, Marquette University, Milwaukee, WI, USA*; CARL A GOTTLIEB, *Radio and Geoastronomy Division, Harvard-Smithsonian Center for Astrophysics, Cambridge, MA, USA.*

Disilicon carbide Si_2C is one of the most fascinating small molecules for both fundamental and applied reasons. Like C_3, it has a shallow bending angle, and may therefore also serve as a classic example of a quasilinear species. Si_2C is also thought to be quite stable. Mass spectrometric studies conclude that it is one of the most common gas-phase fragments in the evaporation of silicon carbide at high temperature. For these same reasons, it may be abundant in certain evolved carbon stars such as IRC+12016. Its electronic spectrum was recently studied by several of us, but its ground state geometry and rotational spectrum remain unknown until now. Using sensitive microwave techniques and high-level coupled cluster calculations, Si_2C has been detected in the radio band, and is found to be highly abundant. Its more common rare isotopic species have also be observed either in natural abundance or using isotopically-enriched samples, from which a highly precise semi-experimental structure has been derived. This talk will summarize recent work, and discuss the prospects for astronomical detection. Now that all four of the Si_mC_n clusters with $m + n = 3$ has been detected experimentally, a rigorous comparison of their structure and chemical bonding can be made.

RE. Instrument/Technique Demonstration

Thursday, June 25, 2015 – 8:30 AM

Room: 217 Noyes Laboratory

Chair: Arthur Suits, Wayne State University, Detroit, MI, USA

RE01 8:30 – 8:45

OPTIMIZATION OF EXTREME ULTRAVIOLET LIGHT SOURCE FROM HIGH HARMONIC GENERATION FOR CONDENSED-PHASE CORE-LEVEL SPECTROSCOPY

MING-FU LIN, MAX A VERKAMP, ELIZABETH S RYLAND, KRISTIN BENKE, KAILI ZHANG, MICHAELA CARLSON, JOSH VURA-WEIS, *Department of Chemistry, University of Illinois at Urbana-Champaign, Urbana, IL, USA.*

Extreme ultraviolet (XUV) light source from high-order harmonic generation has been shown to be a powerful tool for core-level spectroscopy. In addition, this light source provides very high temporal resolution (10^{-18} s to 10^{-15} s) for time-resolved transient absorption spectroscopy. Most applications of the light source have been limited to the studies of atomic and molecular systems, with technique development focused on optimizing for shorter pulses (i.e. tens of attoseconds) or higher XUV energy (i.e. keV range). For the application to general molecular systems in solid and liquid forms, however, the XUV photon flux and stability are highly demanded due to the strong absorption by substrates and solvents. In this case, the main limitation is due to the stability of the high order generation process and the limited bandwidth of the XUV source that gives only discrete even/odd order peaks. Consequently, this results in harmonic artifact noise that overlaps with the resonant signal. In our current study, we utilize a semi-infinite cell for high harmonic generation from two quantum trajectories (i.e. short and long) at over-driven NIR power. This condition, produces broad XUV spectrum without using complicated optics (e.g. hollow-core fibers and double optical gating). This light source allows us to measure the static absorption spectrum of the iron M-edge from a $Fe(acac)_3$ molecular solid film, which shows a resonant feature of 0.01 OD (2.3% absorption). Moreover, we also investigate how sample roughness affects the static absorption spectrum. We are able to make smooth solar cell precursor materials (i.e. PbI_2 and $PbBr_2$) by spin casting and observe iodine (50 eV) and bromine (70 eV) absorption edges in the order of 0.05 OD with minimal harmonic artifact noise.

RE02 8:47 – 9:02

DEVELOPMENT OF TWO-PHOTON PUMP POLARIZATION SPECTROSCOPY PROBE TECHNIQUE (TPP-PSP) FOR MEASUREMENTS OF ATOMIC HYDROGEN .

AMAN SATIJA, ROBERT P. LUCHT, *Mechanical Engineering, Purdue University, West Lafayette, IN, USA.*

Atomic hydrogen (H) is a key radical in combustion and plasmas. Accurate knowledge of its concentration can be used to better understand transient phenomenon such as ignition and extinction in combustion environments. Laser induced polarization spectroscopy is a spatially resolved absorption technique which we have adapted for quantitative measurements of H atom. This adaptation is called two-photon pump, polarization spectroscopy probe technique (TPP-PSP) and it has been implemented using two different laser excitation schemes. The first scheme involves the two-photon excitation of 1S-2S transitions using a linearly polarized 243-nm beam. An anisotropy is created amongst Zeeman states in 2S-3P levels using a circularly polarized 656-nm pump beam. This anisotropy rotates the polarization of a weak, linearly polarized probe beam at 656 nm. As a result, the weak probe beam "leaks" past an analyzer in the detection channel and is measured using a PMT. This signal can be related to H atom density in the probe volume. The laser beams were created by optical parametric generation followed by multiple pulse dye amplification stages. This resulted in narrow linewidth beams which could be scanned in frequency domain and varied in energy. This allowed us to systematically investigate saturation and Stark effect in 2S-3P transitions with the goal of developing a quantitative H atom measurement technique. The second scheme involves the two-photon excitation of 1S-2S transitions using a linearly polarized 243-nm beam. An anisotropy is created amongst Zeeman states in 2S-4P transitions using a circularly polarized 486-nm pump beam. This anisotropy rotates the polarization of a weak, linearly polarized probe beam at 486 nm. As a result the weak probe beam "leaks" past an analyzer in the detection channel and is measured using a PMT. This signal can be related to H atom density in the probe volume. A dye laser was pumped by third harmonic of a Nd:YAG laser to create a laser beam at 486 nm. The 486-nm beam was frequency doubled to a 243-nm beam. Use of the second scheme simplifies the TPP-PSP technique making it more convenient for diagnostics in practical systems.

RE03 **9:04 – 9:19**

DEVELOPMENT OF COMBINED DUAL-PUMP VIBRATIONAL AND PURE-ROTATIONAL COHERENT ANTI-STOKES RAMAN SCATTERING TECHNIQUE.

AMAN SATIJA, ROBERT P. LUCHT, *Mechanical Engineering, Purdue University, West Lafayette, IN, USA.*

Coherent anti-Stokes Raman scattering is a parametric, four-wave mixing process. CARS, as a diagnostic technique, has been used extensively for obtaining accurate temperature and species concentration information in non-reacting and reacting flows. Dual-pump vibrational CARS (DPVCARS) can provide quantitative temperature and concentration information on multiple species in the probe volume. Mole-fraction information on molecules such as N_2, O_2, H_2 and CO_2 have been obtained in flames with peak temperature in excess of 2000 K. Although DPVCARS provides high accuracy at higher temperatures it has low sensitivity at lower temperatures (below 800 K). Typically, pure-rotational CARS (PRCARS) provides excellent sensitivity and precision at lower temperatures. We have combined DPVCARS and two-beam PRCARS into a single system which employs three laser beams at different wavelengths. The accuracy and precision of the new combined CARS system has been characterized in laminar flames. The system's single-shot precision is better than 5.5 % between 295-2200 K, indicating its suitability for diagnostics in turbulent flames. The new system has been applied towards understanding flame structure of CH_4/H_2/air laminar flames, stabilized in a counter-flow burner. Here, we present results detailing the development and application of the new combined CARS technique.

RE04 **9:21 – 9:36**

VELOCITY MAP IMAGING STUDY OF THE PHOTOINITIATED CHARGE-TRANSFER DISSOCIATION OF $Cu^+(C_6H_6)$ AND $Ag^+(C_6H_6)$

JON MANER, DANIEL MAUNEY, MICHAEL A DUNCAN, *Department of Chemistry, University of Georgia, Athens, GA, USA.*

$M^+(C_6H_6)$ (M = Cu, Ag) complexes are generated in the gas phase by laser vaporization and detected in a reflectron time-of-flight mass spectrometer. Excitation of $M^+(C_6H_6)$ at 355 nm results exclusively in dissociative charge transfer, leading to neutral M and $C_6H_6{}^+$ products for both Cu and Ag complexes. Kinetic energy release in translationally hot $C_6H_6{}^+$ fragments is detected using a new apparatus designed for photofragment imaging of mass-selected ion beams. Velocity map imaging and slice imaging techniques are employed. Analysis of the data provide new information on the binding energies of $Cu^+(C_6H_6)$ and $Ag^+(C_6H_6)$.

Intermission

RE05 **9:55 – 10:10**

MID-IR CAVITY RINGDOWN SPECTROSCOPY FOR ATMOSPHERIC ETHANE ABUNDANCE MEASUREMENTS

LINHAN SHEN, THINH QUOC BUI, *Division of Chemistry and Chemical Engineering, California Institute of Technology, Pasadena, CA, USA*; LANCE CHRISTENSEN, *Science Division, Jet Propulsion Laboratory/Caltech, Pasadena, CA, USA*; MITCHIO OKUMURA, *Division of Chemistry and Chemical Engineering, California Institute of Technology, Pasadena, CA, USA.*

We demonstrate a mid-IR (3.3 μm) cw cavity ringdown spectrometer capable of measuring atmospheric ethane abundances. This technique can measure atmospheric ethane concentration as low as 100 ppb. The atmospheric ethane to methane ratio could also be observed by measuring methane concentration using a high precision near-IR (1.65 μm) cavity ringdown spectrometer. We will also discuss the daily variation of ethane abundance and ethane to methane ratio in Pasadena observed using this tecnhique.

RE06 **10:12 – 10:27**

STRONG THERMAL NONEQUILIBRIUM IN HYPERSONIC CO AND CH_4 PROBED BY CRDS

MAUD LOUVIOT, *Laboratoire ICB, CNRS/Université de Bourgogne, DIJON, France*; NICOLAS SUAS-DAVID, *IPR UMR6251, CNRS - Université Rennes 1, Rennes, France*; VINCENT BOUDON, *Laboratoire ICB, CNRS/Université de Bourgogne, DIJON, France*; ROBERT GEORGES, *IPR UMR6251, CNRS - Université Rennes 1, Rennes, France*; MICHAEL REY, *Groupe de Spectrométrie Moléculaire et Atmosphérique, UMR CNRS 7331, Université de Reims, Reims Cedex 2, France*; SAMIR KASSI, *UMR5588 LIPhy, Université Grenoble 1/CNRS, Saint Martin D'heres, France.*

A new experimental set-up coupling a High Enthalpy Source (HES) reaching 2000 K to a cw Cavity Ring-Down Spectrometer has been developed to investigate rotationnally cold hot bands of polyatomic molecules in the $[1.5, 1.7]$ μm region. The rotational and vibrational molecular degrees of freedom are strongly decoupled in the hypersonic expansion produced by the HES and probed by Cavity Ring-Down Spectroscopy. Carbon monoxide has been used as a first test molecule to validate the experimental approach. Its expansion in argon led to rotational and vibrational temperatures of 6.7 ± 0.8 K and 2006 ± 476 K, respectively. The Tetradecad polyad of methane (1.67 μm) was investigated under similar conditions leading to rotational and vibrational temperatures of 13 ± 5 K and 750 ± 100 K, respectively. The rotationally cold structure of the spectra reveals many hot bands involving highly excited vibrational states of methane.

RE07 **10:29 – 10:44**

ROTATIONALLY-RESOLVED INFRARED SPECTROSCOPY OF THE ν_{16} BAND OF 1,3,5-TRIOXANE

BRADLEY M. GIBSON, NICOLE KOEPPEN, *Department of Chemistry, University of Illinois at Urbana-Champaign, Urbana, IL, USA*; BENJAMIN J. McCALL, *Departments of Chemistry and Astronomy, University of Illinois at Urbana-Champaign, Urbana, IL, USA.*

1,3,5-trioxane is the simplest cyclic form of polyoxymethylene (POM), a class of formaldehyde polymers that has been proposed as the origin of distributed formaldehyde formation in comet comae and a potential source of formaldehyde in prebiotic chemistry. Although claimed POM detections have since been proven to be inconclusive, laboratory simulations of cometary conditions have yielded trioxane and other POMs[a].

While the microwave spectrum of 1,3,5-trioxane has been studied extensively[b], to date only one rotationally-resolved ro-vibrational spectrum has been published[c]. Here, we present our studies of the ν_{16} band of gas-phase trioxane centered at 1177 cm^{-1}. Trioxane was entrained in a supersonic expansion of argon and characterized by continuous-wave cavity ringdown spectroscopy using an etalon-stabilized external-cavity quantum cascade laser[d]. Rotationally resolved spectra were obtained with less than 15 MHz resolution.

[a]Cottin, H., Bénilan, Y., Gazeau, M-C., and Raulin, F. Origin of Cometary Extended Sources from Degradation of Refractory Organics on Grains: Polyoxymethylene as Formaldehyde Parent Molecule. *Icarus* **167** (2004), 397-416.

[b]Oka, T., Tsuchiya, K., Iwata, S., and Morino, Y. Microwave Spectrum of s-Trioxane. *Bull. Chem. Soc. Jpn.* **37** (1964), 4-7.

[c]Henninot, J-F., Bolvin, H., Demaison, J., and Lemoine, B. The Infrared Spectrum of Trioxane in a Supersonic Slit Jet. *J. Mol. Spect.* **152** (1992), 62-68.

[d]Gibson, B.M. and McCall, B.J., contribution TJ08, presented at the 69th International Symposium on Molecular Spectroscopy, Urbana, IL, USA, 2014.

RE08 **10:46 – 11:01**

IMPROVING SNR IN TIME-RESOLVED SPECTROSCOPIES WITHOUT SACRIFICING TEMPORAL-RESOLUTION: APPLICATION TO THE UV PHOTOLYSIS OF METHYL CYANOFORMATE

MICHAEL J. WILHELM, JONATHAN M. SMITH, HAI-LUNG DAI, *Department of Chemistry, Temple University, Philadelphia, PA, USA.*

We demonstrate a new analysis for the enhancement of the signal-to-noise ratio (SNR) in time-resolved spectroscopies, termed spectral reconstruction analysis (SRA). As distinct from a simple linear average which produces only a *single* representative spectrum with enhanced SNR, SRA produces a comparable enhancement, but fully preserves the measured time-dependence. Specifically, given a series of (n) time-resolved spectra, SRA yields an approximate *sqrt(n)* SNR enhancement for each of the original n-spectra. SRA operates by eliminating noise in the temporal domain, thereby significantly attenuating noise in the spectral domain, as follows (see Figure): Temporal profiles of each measured frequency are fit to capture the representative temporal evolutions, then time-resolved spectra are reconstructed by replacing the measured profiles with the fit profiles.

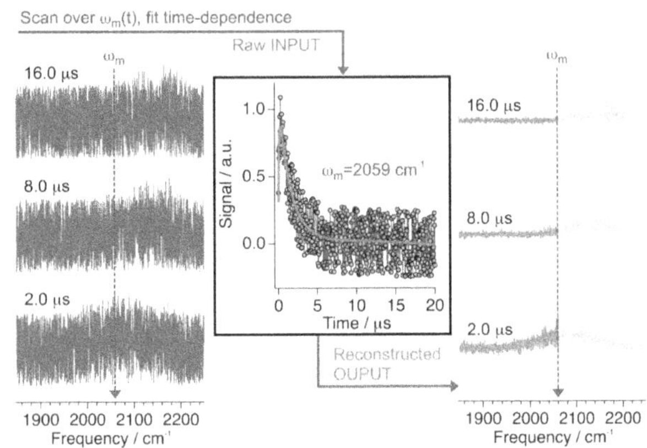

In addition to simulated control data sets, we demonstrate SRA with experimentally measured time-resolved IR emission spectra, collected following the 193 nm photolysis of methyl cyanoformate ($CH_3OC(O)CN$). Of significance, we now show the appearance of resonances assignable to hydrogen cyanide (HCN), which were previously obscured in the noise of the measured spectra. The presence of HCN suggests the occurrence of a previously uncharacterized dissociation channel, likely involving a cyclic 5-center transition state.

RE09 ***Post-Deadline Abstract*** **11:03 – 11:18**

LASER-INDUCED PLASMAS IN AMBIENT AIR FOR INCOHERENT BROADBAND CAVITY-ENHANCED ABSORPTION SPECTROSCOPY

ALBERT A RUTH, SOPHIE DIXNEUF, *Physics Department and Environmental Research Institute, University College Cork, Cork, Ireland*; JOHANNES ORPHAL, *Institute for Meteorology and Climate Research, Karlsruhe Institute of Technology, 76344 Eggenstein-Leopoldshafen, Germany.*

The emission from a laser-induced plasma in ambient air, generated by a high power femtosecond laser, was utilized as pulsed incoherent broadband light source in the center of a quasi-confocal high finesse cavity. The time dependent spectra of the light leaking from the cavity was compared with those of the laser-induced plasma emission without the cavity. It was found that the light emission was sustained by the cavity despite the initially large optical losses of the laser-induced plasma in the cavity. The light sustained by the cavity was used to measure part of the $S_1 \leftarrow S_0$ absorption spectrum of gaseous azulene at its vapour pressure at room temperature in ambient air as well as the strongly forbidden γ–band in molecular oxygen: $b^1\Sigma_g^+(\nu' = 2) \leftarrow X^3\Sigma_g^-(\nu'' = 0)$

RF. Atmospheric science

Thursday, June 25, 2015 – 1:30 PM

Room: 116 Roger Adams Lab

Chair: Joseph Hodges, National Institute of Standards and Technology, Gaithersburg, MD, USA

RF01 **1:30 – 1:40**

PHOTOACOUSTIC SPECTROSCOPY OF THE OXYGEN A-BAND

ELIZABETH M LUNNY, THINH QUOC BUI, *Division of Chemistry and Chemical Engineering, California Institute of Technology, Pasadena, CA, USA*; CAITLIN BRAY, *Department of Chemistry, Wesleyan University, Middletown, CT, USA*; PRIYANKA RUPASINGHE, *Physical Sciences, Cameron University, Lawton, OK, USA*; MITCHIO OKUMURA, *Division of Chemistry and Chemical Engineering, California Institute of Technology, Pasadena, CA, USA.*

The oxygen A-band (760 nm) is used in a number of remote sensing applications due to the precisely known, uniform distribution of molecular oxygen throughout the atmosphere and the spectral isolation of the band. The A-band is used to determine the pathlength of solar radiation for OCO-2, a current NASA mission which seeks to measure the global sources and sinks of carbon dioxide at unprecedented spatial and temporal resolution. The goal of measuring atmospheric carbon dioxide concentrations with a precision of 0.25% requires a precise knowledge of line shape parameters. Currently, the most significant uncertainties in A-band spectroscopy result from line mixing and collision induced absorption, which become more prominent at elevated pressures. Photoacoustic spectroscopy is ideal to observe these phenomena due to the large dynamic range and zero-background advantages of the technique. Photoacoustic spectra of the oxygen A-band over a range of pressures will be presented in addition to line shape parameters extracted from multispectrum fits of the data.

RF02 **1:42 – 1:57**

HIGH PRESSURE OXYGEN A-BAND SPECTRA

BRIAN DROUIN, KEEYOON SUNG, SHANSHAN YU, *Jet Propulsion Laboratory, California Institute of Technology, Pasadena, CA, USA*; ELIZABETH M LUNNY, THINH QUOC BUI, MITCHIO OKUMURA, *Division of Chemistry and Chemical Engineering, California Institute of Technology, Pasadena, CA, USA*; PRIYANKA RUPASINGHE, *Physical Sciences, Cameron University, Lawton, OK, USA*; CAITLIN BRAY, *Department of Chemistry, Wesleyan University, Middletown, CT, USA*; DAVID A. LONG, JOSEPH HODGES, *Chemical Sciences Division, National Institute of Standards and Technology, Gaithersburg, MD, USA*; DAVID ROBICHAUD, *Biomass Molecular Science , National Renewable Energy Laboratory , Golden, CO, USA*; D. CHRIS BENNER, V. MALATHY DEVI, JIAJUN HOO, *Department of Physics, College of William and Mary, Williamsburg, VA, USA.*

Composition measurements from remote sensing platforms require knowledge of air mass to better than the desired precision of the composition. Oxygen spectra allow determination of air mass since the mixing ratio of oxygen is fixed. The OCO-2 mission is currently retrieving carbon dioxide concentration using the oxygen A-band for air mass normalization. The 0.25% accuracy desired for the carbon dioxide concentration has pushed the state-of-the-art for oxygen spectroscopy. To produce atmospheric pressure A-band cross-sections with this accuracy requires a sophisticated line-shape model (Galatry or Speed-Dependent) with line mixing (LM) and collision induced absorption (CIA). Models of each of these phenomena exist, but an integrated self-consistent model must be developed to ensure accuracy.

This presentation will describe the ongoing effort to parameterize these phenomena on a representative data set created from complementary experimental techniques. The techniques include Fourier transform spectroscopy (FTS), photo-acoustic spectroscopy (PAS) and cavity ring-down spectroscopy (CRDS). CRDS data allow long-pathlength measurements with absolute intensities, providing lineshape information as well as LM and CIA, however the subtleties of the lineshape are diminished in the saturated line-centers. Conversely, the short paths and large dynamic range of the PAS data allow the full lineshape to be discerned, but with an arbitrary intensity axis. Finally, the FTS data provides intermediate paths and consistency across a broad pressure range. These spectra are all modeled with the Labfit software using first the spectral line database HITRAN, and then model values are adjusted and fitted for better agreement with the data.

RF03 **1:59 – 2:14**

COLLISION-DEPENDENT LINE AREAS IN THE $a^1\Delta_g \leftarrow X^3\Sigma_g^-$ BAND OF MOLECULAR OXYGEN

VINCENT SIRONNEAU, ADAM J. FLEISHER, JOSEPH HODGES, *Material Measurement Laboratory, National Institute of Standards and Technology, Gaithersburg, MD, USA.*

We report precise line areas for individual rotationally resolved transitions within the $a^1\Delta_g \leftarrow X^3\Sigma_g^-$ electronic band of molecular oxygen recorded as a function of pressure for both neat samples of O_2 as well as samples of O_2 dilute with a variety of collisional partners. Using optical frequency comb referenced frequency-stabilized cavity ring-down spectroscopy (FS-CRDS) near 1.27 μm we measure line areas with a quality-of-fit QF $\leq$ 50,000 using a partially correlated quadratic-speed-dependent Nelkin-Ghatak profile. This spectrometer has achieved this high QF by both suppressing coupled cavity effects and by preserving a high-fidelity frequency axis with absolute frequency accuracy approaching 1 part in 10^9. With this instrument we are also currently exploring collision-induced absorption (CIA) and perturbative line mixing effects in O_2 over the entire 7800-7940 cm^{-1}spectral range.

RF04 **2:16 – 2:31**

ANOMALOUS CENTRIFUGAL DISTORTION IN HDO AND SPECTROSCOPIC DATA BASES

L. H. COUDERT, *CNRS et Universités Paris Est et Paris Diderot, LISA, Creteil, France.*

The HDO molecule is important from the atmospheric point of view as it can be used to study the water cycle in the earth atmosphere.[a] It is also interesting from the spectroscopic point of view as it displays an anomalous centrifugal distortion similar to that of the normal species H_2O. A model developed to treat the anomalous distortion in HDO should account for the fact that it lacks a two-fold axis of symmetry.

A new treatment aimed at the calculation of the rovibrational energy of the HDO molecule and allowing for anomalous centrifugal distortion effects has been developed. It is based on an effective Hamiltonian in which the large amplitude bending ν_2 mode and the overall rotation of the molecule are treated simultaneously.[b] Due to the lack of a two-fold axis of symmetry, this effective Hamiltonian contains terms arising from the non-diagonal component of the inertia tensor and from the Coriolis-coupling between the large amplitude bending ν_2 mode and the overall rotation of the molecule.

This new treatment has been used to perform a line position analysis of a large body of infrared,[c] microwave,[d] and hot water vapor[e] data involving the ground and (010) states up to $J = 22$. For these 4413 data, a unitless standard deviation of 1.1 was achieved. A line intensity analysis was also carried out and allowed us to reproduce the strength of 1316 transitions[c] with a unitless standard deviation of 1.1.

In the talk, the new theoretical approach will be presented. The results of both analyses will be discussed and compared with those of a previous investigation.[f] The new spectroscopic data base built will be compared with HITRAN 2012.[g]

[a] Herbin *et al., Atmos. Chem. Phys.* **9** (2009) 9433; and Schneider and Hase, *Atmos. Chem. Phys.* **11** (2011) 11207.

[b] Coudert, Wagner, Birk, Baranov, Lafferty, and Flaud, *J. Molec. Spectrosc.* **251** (2008) 339.

[c] Johns, *J. Opt. Soc. Am. B* **2** (1985) 1340; Toth, *J. Molec. Spectrosc.* **162** (1993) 20; Paso and Horneman, *J. Opt. Soc. Am. B* **12** (1995) 1813; and Toth, *J. Molec. Spectrosc.* **195** (1999) 73.

[d] Messer, De Lucia, and Helminger, *J. Molec. Spectrosc.* **105** (1984) 139; and Baskakov *et al., Opt. Spectrosc.* **63** (1987) 1016.

[e] Parekunnel *et al., J. Molec. Spectrosc.* **210** (2001) 28; and Janca *et al., J. Molec. Specrosc.* **219** (2003) 132.

[f] Tennyson *et al., J. Quant. Spectrosc. Radiat. Transfer* **111** (2010) 2160.

[g] Rothman *et al., J. Quant. Spectrosc. Radiat. Transfer* **130** (2013) 4.

RF05 **2:33 – 2:48**

SPEED-DEPENDENT BROADENING AND LINE-MIXING IN CH_4 PERTURBED BY AIR NEAR 1.64 μm FOR THE FRENCH/GERMAN CLIMATE MISSION MERLIN

THIBAULT DELAHAYE, THI NGOC HA TRAN, *CNRS et Universités Paris Est et Paris Diderot, Laboratoire Interuniversitaire des Systèmes Atmosphériques (LISA), Créteil, France*; ZACHARY REED, STEPHEN E MAXWELL, JOSEPH HODGES, *Chemical Sciences Division, National Institute of Standards and Technology, Gaithersburg, MD, USA.*

Climate change is one of the greatest challenges presently facing mankind, and methane is one of the most powerful anthropogenic greenhouse gases. For a better understanding of future climate trends, a satellite dedicated to the measurements of atmospheric methane is under joint development by the French and German space research centers (CNES and DLR). The so-called MERLIN mission (Methane Remote Sensing Lidar Mission, 2019) aims at providing global information on atmospheric methane concentration with a relative uncertainty less than 2% and with a spatial resolution of 50 km[a]. Such spectroscopic monitoring of gases in the atmosphere of the Earth, requires a precise description of absorption lines shapes that goes beyond the usual Voigt profile (VP). In the case of methane, the differences between the measured profiles and those given by the VP can be very important [b], making the VP completely incompatible with the reliable detection of sources and sinks from space. In this work, we present the first results on the modeling of methane lines broadened by air in the 1.64 μm region and the associated spectroscopic parameters, taking into account various collisional effects between molecules that are neglected by the VP: collisional interference between the lines (line-mixing), collision-induced velocity changes (Dicke narrowing effect) and speed dependence of the collisional broadening and shifting. These results were obtained by simultaneously fitting the model parameters to high sensitivity and high-resolution cavity ring-down spectroscopy (CRDS) spectra recorded at the National Institute of Standards and Technology (NIST) over a wide pressure range (5 to 100 kPa). These spectroscopic data and the associated model to calculate the spectrum absorption coefficient will be then used to analyze ground-based atmospheric spectra at the TCCON facility in Park Falls, Wisconsin.

[a]C. Kiemle, M. Quatrevalet, G. Ehret et al., Atmos. Meas. Tech. 4 (2011)

[b]H. Tran, J.-M. Hartmann, G. Toon et al., Journal of Quant. Spectrosc. Radia. Trans. 111 (2010)

RF06 **2:50 – 3:05**

MID INFRARED DUAL FREQUENCY COMB SPECTROMETER FOR THE DETECTION OF METHANE IN AMBIENT AIR

HANS SCHUESSLER, FENG ZHU, *Department of Physics and Astronomy, Texas A&M University, College Station, TX, USA*; ALEXANDER KOLOMENSKII, *Department of Physics and Astronomy, Texas A & M University, college station, TX, USA.*

We demonstrate using mid infrared dual frequency comb spectroscopy for the detection of methane in ambient air. The mid infrared frequency comb sources based on femtosecond Er fiber oscillators are produced through difference frequency generation with periodically poled MgO doped lithium niobate crystals and stabilized at slightly different repetition rates at about 250 MHz. We performed dual frequency comb spectroscopy in the spectral range between 2900 $cm^{-1}$and 3150 $cm^{-1}$with 0.07 cm^{-1}resolution using a multipass cell of about 580 meter path length, and achieved the sensitivity about 7.6E-7 $cm^{-1}$with 80 ms data acquisition time. We determined the methane concentration as about 1.5 ppmv in the ambient air of the laboratory, and the minimum detection limit as about 60 ppbv for the current setup.

This work was funded by the Robert A. Welch Foundation, Grant No. A1546 and the Qatar Foundation under Grant No. NPRP 6-465-1-091.

RF07 **3:07 – 3:22**

IMPROVED OZONE AND CARBON MONOXIDE PROFILE RETRIEVALS USING MULTISPECTRAL MEASUREMENTS FROM NASA "A TRAIN", NPP, AND TROPOMI SATELLITES

DEJIAN FU, *Jet Propulsion Laboratory, California Institute of Technology , PASADENA, CAL, USA.*

Tropospheric ozone is at the juncture of air quality and climate. Ozone directly impacts human health, is a phytotoxin that undermines carbon uptake, and directly forces the climate system through absorption of thermal radiation. Carbon monoxide is a chemical precursor of greenhouse gases CO_2 and tropospheric O_3, and is also an ideal tracer of transport processes due to its medium life times (weeks to months). The Aqua-AIRS and Aura-OMI instruments in the NASA "A-Train", CrIS and OMPS instruments on the NOAA Suomi-NPP, IASI and GOME-2 on METOP, TROPOMI aboard the Sentinel 5 precursor (S5p) have the potential to provide the synoptic chemical and dynamical context for ozone necessary to quantify long-range transport at global scales and as an anchor to the near-term constellation of geostationary sounders: NASA TEMPO, ESA Sentinel 4, and the Korean GEMS. We introduce the JPL Multi-Spectral, Multi-Species, Multi-Satellite $(MS)^3$ retrieval algorithm, which ingests panspectral observations across multiple platforms in an non-linear optimal estimation framework. It incorporates the advances in remote sensing science developed during EOS-Aura era including rigorous error analysis diagnostics and observation operators needed for trend analysis, climate model evaluation, and data assimilation. Its performances have been demonstrated through prototype studies for multi-satellite missions (AIRS, CrIS, TROPOMI, TES, OMI, and OMPS). We present the preliminary joint tropospheric ozone retrievals from AIRS/OMI and CrIS/OMPS, and demonstrate the potential of joint carbon monoxide profiles from TROPOMI/CrIS. These results indicate that ozone can be retrieved with 2 degrees of freedom for signal (dofs). The joint ozone retrievals are closer to ozone retrieved from the NASA Tropospheric Emission Spectrometer than any single instruments retrievals. Joint CO profiles have a dofs similar to the MOPITT multispectral retrieval. Consequently, multispectral retrievals show promise in providing continuity with NASA EOS observations and paving the way towards a new advanced atmospheric composition constellation. To further improve the quality of measurements using multiple spectral regions, next generation of ozone and carbon monoxide spectroscopic parameters should mitigate the existing discrepancies among different spectral regions (microwave, thermal infrared, visible and ultraviolet).

RF08 **3:24 – 3:39**

TEMPERATURE DEPENDENCES OF AIR-BROADENING AND SHIFT PARAMETERS IN THE ν_3 BAND OF OZONE

MARY ANN H. SMITH, *Science Directorate, NASA Langley Research Center, Hampton, VA, USA*; V. MALATHY DEVI, D. CHRIS BENNER, *Department of Physics, College of William and Mary, Williamsburg, VA, USA.*

Line parameter errors can contribute significantly to the total errors in retrievals of terrestrial atmospheric ozone concentration profiles using the strong 9.6-μm band, particularly for nadir-viewing experiments[a]. Detailed knowledge of the interfering ozone signal is also needed for retrievals of other atmospheric species[b] in this spectral region. We have determined Lorentz air-broadening and pressure-induced shift coefficients along with their temperature dependences for a number of transitions in the ν_3 fundamental band of $^{16}O_3$. These results were obtained by applying the multispectrum nonlinear least-squares fitting technique[c] to a set of 31 high-resolution infrared absorption spectra of O_3 recorded at temperatures between 160 and 300 K with several different room-temperature and coolable sample cells at the McMath-Pierce Fourier transform spectrometer at the National Solar Observatory on Kitt Peak. We compare our results with other available measurements and with the ozone line parameters in the HITRAN database[d] .

[a] J. Worden *et al., J. Geophys. Res.* **109** (2004) 9308-9319.
[b] R. Beer *et al., Geophys. Res. Lett.* **35** (2008) L09801.
[c] D. Chris Benner *et al., JQSRT* **53** (1995) 705-721.
[d] L. S. Rothman *et al., JQSRT* **130** (2013) 4-50.

Intermission

RF09 **3:58 – 4:13**

MICROWAVE OPTICAL DOUBLE RESONANCE STUDIES OF PERTURBATIONS IN THE SO $A^3\Pi$ STATE

ANDREW RICHARD WHITEHILL, *Earth, Atmospheric, and Planetary Sciences, MIT, Cambridge, MA, USA*; ALEXANDER W. HULL, TREVOR J. ERICKSON, JUN JIANG, CARRIE WOMACK, BARRATT PARK, *Department of Chemistry, MIT, Cambridge, MA, USA*; SHUHEI ONO, *Earth, Atmospheric, and Planetary Sciences, MIT, Cambridge, MA, USA*; ROBERT W FIELD, *Department of Chemistry, MIT, Cambridge, MA, USA.*

There is a possibility that perturbations of the SO $A^3\Pi$ state provide a mechanism for photodissociation of SO by a $1 + 1'$ two-color solar radiation process. The resultant S atom photofragments could depart from the isotopologue natural abundance ratios. The SO $A^3\Pi$–$X^3\Sigma^-$ transition is very weak and the radiative lifetimes of $v < 4$ levels of the $A^3\Pi$ state are > 10 μs. The lowest vibrational levels of the $A^3\Pi$ state are perturbed by high vibrational levels of three metastable states: c $^1\Sigma^-$, A′ $^3\Delta$, and A″ $^3\Sigma^+$. If some upper atmospheric photophysical process were to populate the SO c, A′, and A″ states, emission from these states would escape detection. A$\leftrightarrow$(c, A′, A″) perturbations by the various S isotopologues will culminate at widely separated values of J, with the result that, for some isotopologues, the perturbations are located outside of the region of thermally populated rotational levels. These perturbations are exceptionally difficult to characterize by Laser Induced Fluorescence. In addition to Chirped Pulse millimeter-Wave (CPmmW) spectroscopy in the $A^3\Pi$ state, we are using Coherence Converted Population Transfer (CCPT) in the $X^3\Sigma^-$ state, an especially sensitive form of microwave-detected Microwave Optical Double Resonance, in order to characterize the $A^3\Pi$ state and its perturbers.

RF10 **4:15 – 4:30**

VALIDATION OF A NEW HNO_3 LINE PARAMETERS AT 7.6 μm USING LABORATORY INTENSITY MEASUREMENTS AND MIPAS SATELLITE SPECTRA

MARCO RIDOLFI, *Dipartimento di Fisica e Astronomia , Università di Bologna, Bologna, Italy*; AGNES PERRIN, JEAN-MARIE FLAUD, *LISA, CNRS, Universités Paris Est Créteil et Paris Diderot, Créteil, France*; JEAN VANDER AUWERA, *Service de Chimie Quantique et Photophysique, Université Libre de Bruxelles, Brussels, Belgium*; MASSIMO CARLOTTI, *Dipartimento di Chimica Industriale "Toso Montanari", Università di Bologna, Bologna, Italy.*

A new set of line parameters (positions, intensities and line widths) for nitric acid has been generated in the 7.6 μm region using the results of recent high quality experimental laboratory studies and of theoretical calculations. The validation of this new database was performed thanks to limb emission radiances measured in 2002-2012 by the "Michelson Interferometer for Passive Atmospheric Sounding" (MIPAS) instrument on board the ENVISAT satellite. This study will help to improve HNO_3 satellite retrievals by allowing measurements to be performed using simultaneously 11 μm and 7.6 μm microwindows. Hopefully this will be the case for the forthcoming Infrared Atmospheric Sounding Interferometer New Generation (IASI-NG) instrument developed by CNES. IASI-NG will be the key payload element of the future METOP Second Generation (METOP-SG) series of EUMETSAT meteorological polar-orbit satellites.

RF11 **4:32 – 4:47**

ROTATIONAL SPECTROSCOPY OF NEWLY DETECTED ATMOSPHERIC OZONE DEPLETERS: CF_3CH_2Cl, CF_3CCl_3, AND CF_2ClCCl_3

ZBIGNIEW KISIEL, EWA BIAŁKOWSKA-JAWORSKA, LECH PSZCZÓŁKOWSKI, *ON2, Institute of Physics, Polish Academy of Sciences, Warszawa, Poland*; ICIAR URIARTE, PATRICIA ECIJA, FRANCISCO J. BASTERRETXEA, EMILIO J. COCINERO, *Departamento de Química Física, Universidad del País Vasco (UPV-EHU), Bilbao, Spain.*

In a recent study of unpolluted air samples from Tasmania and of deep firn snow in Greenland four previously overlooked ozone-depleting substances have been identified.[a] These compounds started to emerge in the atmosphere in the 1960s, and two: CF_3CCl_3 (CFC-113a) and CF_3CH_2Cl (HCHF-133a) continue to accumulate in the atmosphere.

Three of the four compounds have non-zero dipole moments and are amenable to study by rotational spectroscopy, establishing the basis for analytic applications. Relatively limited studies have been reported for CF_3CH_2Cl[b] and CF_3CCl_3,[c,d] while CF_2ClCCl_3 has not yet been studied by this technique. We presently report extensive results obtained for all three compounds, resulting from concerted application of supersonic expansion FTMW spectroscopy in chirped pulse and cavity modes, and room-temperature MMW spectroscopy. Among the plentiful results, we have been able to resolve and fit the complex nuclear quadrupole hyperfine splitting.

[a]J.C.Laube, et al., *Nature Geoscience* **7**, 266 (2014).
[b]T.Ogata, et al., *J. Mol. Struct.* **144**, 1 (1986).
[c]R.Holm, et al., *Z. Naturforsch.* **23a**, 1040 (1968).
[d]J.H.Carpenter et al., *J. Mol. Spectrosc.* **154**, 207 (1992); P.J.Seo et al., *J. Mol. Spectrosc.* **169**, 58 (1995).

RF12 **4:49 – 5:04**

CHIRPED PULSE AND CAVITY FT MICROWAVE SPECTROSCOPY OF THE FORMIC ACID – TRIMETHYLAMINE WEAKLY BOUND COMPLEX

BECCA MACKENZIE, CHRIS DEWBERRY, KEN LEOPOLD, *Chemistry Department, University of Minnesota, Minneapolis, MN, USA.*

Amine-carboxylic acid interactions are important in many biological systems and have recently received attention for their role in the formation of atmospheric aerosols. Here, we study the molecular and electronic structure of the formic acid – trimethylamine complex, using it as a model for amine-carboxylic acid interactions. The microwave spectrum of the complex has been observed using chirped pulse and conventional cavity-type Fourier transform microwave spectroscopy. The degree of proton transfer has been assessed using the ^{14}N nuclear quadrupole hyperfine structure. Experimental results will be compared to DFT calculations.

RF13 **5:06 – 5:21**

FORMIC SULFURIC ANHYDRIDE: A NEW CHEMICAL SPECIES WITH POSSIBLE IMPLICATIONS FOR ATMOSPHERIC AEROSOL

BECCA MACKENZIE, CHRIS DEWBERRY, KEN LEOPOLD, *Chemistry Department, University of Minnesota, Minneapolis, MN, USA.*

Aerosols are important players in the Earth's atmosphere, affecting climate, cloud formation, and human health. In this work, we report the discovery of a previously unknown molecule, formic sulfuric anhydride (FSA), that may influence the formation and composition of atmospheric aerosol particles. Five isotopologues of FSA have been observed by microwave spectroscopy and further characterized using DFT calculations. The system has dipole moment components along all three inertial axes, and indeed *a*, *b*, and *c*-type transitions have been observed. A $\pi_2 + \pi_2 + \sigma_2$ cycloaddition reaction between SO_3 and HCOOH is proposed as a possible mechanism for the formation of FSA and calculations indicate that the transformation is effectively barrierless. Facile formation of the anhydride followed by hydrolysis in small water-containing clusters or liquid droplets may provide a mechanism of incorporating volatile organics into atmospheric aerosol. We suggest that FSA and its derivatives be considered in future atmospheric and climate models.

RF14 **5:23 – 5:38**

ROTATIONAL SPECTROSCOPY OF METHYL VINYL KETONE

OLENA ZAKHARENKO, R. A. MOTIYENKO, JUAN-RAMON AVILES MORENO, T. R. HUET, *Laboratoire PhLAM, UMR8523 CNRS - Université Lille 1, Villeneuve d'Ascq, France.*

Methyl vinyl ketone, MVK, along with previously studied by our team methacrolein, is a major oxidation product of isoprene, which is one of the primary contributors to annual global VOC emissions. In this talk we present the analysis of the rotational spectrum of MVK recorded at room temperature in the 50 – 650 GHz region using the Lille spectrometer. The spectroscopic characterization of MVK ground state will be useful in the detailed analysis of high resolution infrared spectra. Our study is supported by high level quantum chemical calculations to model the structure of the two stable *s-trans* and *s-cis* conformers and to obtain the harmonic force field parameters, internal rotation barrier heights, and vibrational frequencies. In the Doppler-limited spectra the splittings due to the internal rotation of methyl group are resolved, therefore for analysis of this molecule we used the Rho-Axis-Method Hamiltonian and RAM36 code to fit the rotational transitions. At the present time the ground state of two conformers is analyzed. Also we intend to study some low lying excited states. The analysis is in progress and the latest results will be presented.

Support from the French Laboratoire d'Excellence CaPPA (Chemical and Physical Properties of the Atmosphere) through contract ANR-10-LABX-0005 of the Programme d'Investissements d'Avenir is acknowledged.

RF15 **5:40 – 5:55**

THE MILLIMETER-WAVE SPECTRUM OF METHACROLEIN. TORSION-ROTATION-VIBRATION EFFECTS IN THE EXCITED STATES

OLENA ZAKHARENKO, R. A. MOTIYENKO, JUAN-RAMON AVILES MORENO, T. R. HUET, *Laboratoire PhLAM, UMR8523 CNRS - Université Lille 1, Villeneuve d'Ascq, France.*

Last year we reported the analysis of the rotational spectrum of *s-trans* conformer of methacrolein CH_2=C(CH_3)CHO in the ground vibrational state[a]. In this talk we report the study of its low lying excited vibrational states. The study is based on room-temperature absorption spectra of methacrolein recorded in the frequency range 150 – 465 GHz using the spectrometer in Lille. The new results include assignment of the first excited torsional state (131 cm^{-1}), and the joint analysis of the $v_t = 0$ and $v_t = 1$ states, that allowed us to improve the model in the frame of Rho-Axis-Method (RAM) Hamiltonian and to remove some strong correlations between parameters. Also we assigned the first excited vibrational state of the skeletal torsion mode (170 cm^{-1}). The inverse sequence of A and E tunneling substates as well as anomalous A-E splittings observed for the rotational lines of $v_{sk} = 1$ state clearly indicate a coupling between methyl torsion and skeletal torsion. However we were able to fit within experimental accuracy the rotational lines of $v_{sk} = 1$ state using the RAM Hamiltonian. Because of the inversion of the A and E tunneling substates the rotational lines of the $v_{sk} = 1$ states were assumed to belong to a virtual first excited torsional state. Finally, we assigned several low-K_a rotational transitions of the excited vibrational states above 200 cm^{-1} but their analysis is complicated by different rotation-vibration interactions. In particular there is an evidence of the Fermi-type resonance between the second excited torsional state and the first excited state of the in-plane skeletal bending mode (265 cm^{-1}).

Support from the French Laboratoire d'Excellence CaPPA (Chemical and Physical Properties of the Atmosphere) through contract ANR-10-LABX-0005 of the Programme d'Investissements d'Avenir is acknowledged.

[a]Zakharenko O. et al., 69th ISMS, 2014, TI01

RG. Vibrational structure/frequencies

Thursday, June 25, 2015 – 1:30 PM

Room: 100 Noyes Laboratory

Chair: John F. Stanton, The University of Texas, Austin, TX, USA

RG01 1:30 – 1:45

ALKYL CH STRETCH VIBRATIONS AS A PROBE OF LOCAL ENVIRONMENT AND STRUCTURE

EDWIN SIBERT, DANIEL P. TABOR, *Department of Chemistry, The Univeristy of Wisconsin, Madison, WI, USA*; NATHANAEL KIDWELL, JACOB C. DEAN, TIMOTHY S. ZWIER, *Department of Chemistry, Purdue University, West Lafayette, IN, USA.*

The CH stretch region is a good candidate as a probe of structure and local environment. The functional groups are ubiquitous and their vibration spectra exhibit a surprising sensitivity to molecular structure. In this talk we briefly review our theoretical model Hamiltonian [J. Chem. Phys. **138** 064308 (2013)] for describing vibrational spectra associated with the CH stretch of CH_2 groups and then describe an extension of it to molecules containing methyl and methoxy groups. Results are compared to the infrared spectroscopy of four molecules studied under supersonic expansion cooling in gas phase conditions. The molecules include 1,1-diphenylethane, 1,1-diphenylpropane, 2-methoxyphenol (guaiacol), and 1,3-dimethoxy-2-hydroxybenzene (syringol). The curvilinear local-mode Hamiltonian predicts most of the major spectral features considered in this study and provides insights into mode mixing. We conclude by returning to CH_2 groups and explain both why the CH stretch spectrum of cyclohexane is substantially modified when it forms a complex with an alkali metal and what these spectra tell us about the structure of the complex.

RG02 1:47 – 2:02

COMPUTING THE VIBRATIONAL ENERGIES OF CH_2O AND CH_3CN WITH PHASE-SPACED LOCALIZED FUNCTIONS AND AN ITERATIVE EIGENSOLVER

JAMES BROWN, TUCKER CARRINGTON, *Department of Chemistry, Queen's University, Kingston, ON, Canada.*

For decades scientists have attempted to use ideas of classical mechanics to choose basis functions for calculating spectra. The hope is that a classically-motivated basis set will be small because it covers only the dynamically important part of phase space. One popular idea is to use phase-space localized (PSL) basis functions. Because the overlap matrix, in the matrix eigenvalue problem obtained by using PSL functions with the variational method, is not an identity, it is costly to use iterative methods to solve the matrix eigenvalue problem. Iterative methods are imperative if one wishes to avoid storing matrices which is important for larger molecules. Recently[a] we showed it was possible to circumvent the orthogonality (overlap) problem and use iterative eigensolvers. Here, we present calculated vibrational energies of CH_2O and CH_3CN using the iterative Arnoldi algorithm and PSL functions, and show that our PSL basis is competitive with other previously used basis sets for these molecules.

[a] J. Brown and T. Carrington Jr., Phys. Rev. Lett. **114**, 058901 (2015).

RG03 **2:04 – 2:19**

A MULTILAYER SUM-OF-PRODUCTS METHOD FOR COMPUTING VIBRATIONAL SPECTRA WITHOUT STORING FULL-DIMENSIONAL VECTORS OR MATRCIES

PHILLIP THOMAS, TUCKER CARRINGTON, *Department of Chemistry, Queen's University, Kingston, ON, Canada.*

By optimizing sum-of-products (SOP) basis functions, it is possible to compute vibrational spectra, using a direct product basis, without storing vectors with as many components as there are product basis functions. These ideas are presented in a recent paper: Leclerc and Carrington, J. Chem. Phys 140 174111 (2014). In that paper, the SOP basis functions are products of factors that depend on a single coordinate. When using factors that depend on one coordinate the number of terms (rank) in the SOP basis functions increases with the size of the molecule and the coupling strength. Using multi-dimensional factors makes it possible to incorporate some of the coupling into the factors and to calculate spectra of molecules with more than a dozen atoms. We use multi-dimensional factors that are eigenfunctions of reduced-dimension Hamiltonians. These can be constructed, in different ways, by organizing the factors into a multiple layer tree. Each node in a layer of a tree represents eigenfunctions of a reduced-dimension Hamiltonian for a group of coordinates. We have done calculations with tensor-train and binary tree structures. Efficiency is significantly enhanced by representing the potential with the same tree structure. The ideas are tested by computing energy levels of a 64-D model coupled oscillator Hamiltonian and of CH3CN (12 dimensions) with a realistic potential.

RG04 **2:21 – 2:36**

QUANTUM MONTE CARLO ALGORITHMS FOR DIAGRAMMATIC VIBRATIONAL STRUCTURE CALCULATIONS

MATTHEW HERMES, SO HIRATA, *Department of Chemistry, University of Illinois at Urbana-Champaign, Urbana, IL, USA.*

Convergent hierarchies of theories for calculating many-body vibrational ground and excited-state wave functions, such as Møller-Plesset perturbation theory or coupled cluster theory, tend to rely on matrix-algebraic manipulations of large, high-dimensional arrays of anharmonic force constants, tasks which require large amounts of computer storage space and which are very difficult to implement in a parallel-scalable fashion. On the other hand, existing quantum Monte Carlo (QMC) methods for vibrational wave functions tend to lack robust techniques for obtaining excited-state energies, especially for large systems. By exploiting analytical identities for matrix elements of position operators in a harmonic oscillator basis, we have developed stochastic implementations of the size-extensive vibrational self-consistent field (MC-XVSCF) and size-extensive vibrational Møller-Plesset second-order perturbation (MC-XVMP2) theories which do not require storing the potential energy surface (PES). The programmable equations of MC-XVSCF and MC-XVMP2 take the form of a small number of high-dimensional integrals evaluated using Metropolis Monte Carlo techniques. The associated integrands require independent evaluations of only the value, not the derivatives, of the PES at many points, a task which is trivial to parallelize. However, unlike existing vibrational QMC methods, MC-XVSCF and MC-XVMP2 can calculate anharmonic frequencies directly, rather than as a small difference between two noisy total energies, and do not require user-selected coordinates or nodal surfaces. MC-XVSCF and MC-XVMP2 can also directly sample the PES in a given approximation without analytical or grid-based approximations, enabling us to quantify the errors induced by such approximations.

RG05 **2:38 – 2:53**

DIAGRAMMATIC VIBRATIONAL COUPLED-CLUSTER

JACOB A FAUCHEAUX, SO HIRATA, *Department of Chemistry, University of Illinois at Urbana-Champaign, Urbana, IL, USA.*

A diagrammatic vibrational coupled-cluster method for calculation of zero-point energies and an equation-of-motion coupled-cluster method for calculation of anharmonic vibrational frequencies are developed. The methods, which we refer to as XVCC and EOM-XVCC respectively, rely on the size-extensive vibrational self-consistient field (XVSCF) method for reference wave functions. The methods retain the efficiency advantages of XVSCF making them suitable for applications to large molecules and solids, while they are numerically shown to accurately predict zero-point energies and frequencies of small molecules as well. In particular, EOM-XVCC is shown to perform well for modes which undergo Fermi resonance where traditional perturbative methods fail. Rules for the systematic generation and interpretation of the XVCC and EOM-XVCC diagrams to any order are presented.

RG06 **2:55 – 3:10**

VIBRATIONAL JAHN-TELLER EFFECT IN NON-DEGENERATE ELECTRONIC STATES

MAHESH B. DAWADI, BISHNU P THAPALIYA, *Department of Chemistry, The University of Akron, Akron, OH, USA*; RAM BHATTA, *Polymer Science, The University of Akron, Akron, OH, USA*; DAVID S. PERRY, *Department of Chemistry, The University of Akron, Akron, OH, USA.*

The Jahn-Teller theorem [a] states that "All non-linear nuclear configurations are therefore unstable for an orbitally degenerate electronic state." In 1982, Kellman [b] realized that the Jahn-Teller theorem also applies to nonlinear molecular species in non-degenerate electronic states when there are high-frequency vibrations that are degenerate at a symmetrical reference geometry. When those high frequencies can be considered as adiabatic functions of degenerate low-frequency coordinates, there is a spontaneous Jahn-Teller distortion that lifts the degeneracy of the high-frequency vibrations. Kellman applied the vibrational Jahn-Teller (vJT) concept to the Van der Waals dimer $(SF_6)_2$.

In this talk, the vJT concept is applied to $E \otimes e$ systems that are small bound molecules in non-degenerate electronic states. The first case considered in systems for which the global minimum of the electronic potential has C_{3v} symmetry.For such systems, including $(C_6H_6)Cr(CO)_3$ and CH_3CN, the vJT effect leads to a significant splitting of the degenerate high-frequency vibrations (CH or CO stretches), but the spontaneous vJT distortion is exceptionally small. The second case in systems for which the global minimum of the electronic potential is substantially distorted from the C_{3v} reference geometry. For the second case systems, including CH_3OH and CH_3SH, the vJT splitting of the degenerate CH stretches is much larger, on the order of several 10Äôs of cm^{-1}). For both cases, there is the symmetry-required vibrational conical intersection at the C_{3v} reference geometry. For the second case systems, there are additional symmetry-allowed vibrational conical intersections far from the C_{3v} geometry but energetically accessible to the molecule at thermal energies. For both cases, the vibrationally adiabatic surfaces, including the multiple conical intersections, are well described by modest extensions to a high-order Hamiltonian that was developed for the electronic Jahn-Teller problem.[c]

[a] H. A. Jahn, and E. Teller, *Proc. R. Soc. Lond. A.* **161**, 220, (1937).
[b] M. E. Kellman, *Chem. Phys. Lett.* **87**, 171, (1982).
[c] A. Viel, and W. Eisfeld, *J. Chem. Phys.* **120**, 4603, (2004).

RG07 **3:12 – 3:27**

SPECTRAL SIGNATURES AND STRUCTURAL MOTIFS IN NEUTRAL AND PROTONATED HISTAMINE: A COMPUTATIONAL STUDY

SANTOSH KUMAR SRIVASTAVA, VIPIN BAHADUR SINGH, *Department of Physics, Udai Pratap Autonomous College, Varanasi, India.*

Histamine is an important neurotransmitter that acts as a chemical messenger to exhibit various functions in central and peripheral tissues. The knowledge of its most favored forms is thus of great interest because its shape plays an important role in the key-and-hole recognition process that occurs at the receptor sites. In the present work the conformational landscapes of neutral and protonated histamine have been investigated by MP2 and DFT (employing the various density functionals M06-2X, B97X-D, B3LYP etc) methods. The ground state geometry optimization of the four experimentally observed lowest energy structures of the neutral histamine were performed at the MP2/aug-cc-pVDZ level of theory, for the first time. The conformers 1G-IVa (A)1,2 and 3G-Ib (C)1,2 are predicted to be the most stable and the second most stable isolated structures of the neutral histamine, respectively. Theoretical IR and Raman spectra of the above four conformers were investigated and it was found that the alkyl CH stretch frequencies provides the most useful diagnostic of the ethylamine side chain conformation and are most sensitive to conformational changes in histamine. The computed IR frequencies and intensities of the experimentally observed most stable structure of protonated histamine are used to assign the observed vibrational fundamentals. NH-stretching vibration in the vicinity of the amino group is predicted to be red shifted by 594 cm-1 due to the strong intra-molecular hydrogen bonding (N-H...N). In the alkyl CH- stretching region the CH2(alpha) symmetric stretching vibration attributed to the most intense IR band whereas the CH2(beta) symmetric stretching vibration belongs to the most intense Raman band. The ring CC/CN stretch vibrational fundamental observed in the IRMPD spectra of histamineH+ is found to be in remarkable agreement with the predicted frequency of 1597 cm-1 computed at B3LYP/6-311++G(2d,2p) level. We investigated the low-lying excited states of each experimentally observed conformer of neutral histamine by means of coupled cluster singles and approximate doubles (CC2) method and TDDFT methods and a satisfactory interpretation of the electronic absorption spectra is obtained. 1. P. D. Godfrey and R. D. Brown, J. Am. Chem. Soc. 1998, 120, 10724-10732.

2. E. G. Robertson and J. P. Simons, Phys. Chem. Chem. Phys. 2001, 3, 1-18.

RG08 **3:29 – 3:39**

ANALOG OF DUSCHINSKY MATRIX AND CO-ASSIGNMENT OF FREQUENCIES IN DIFFERENT ELECTRONIC STATES

YURII PANCHENKO, ALEXANDER ABRAMENKOV, *Department of Chemistry, Lomonosov Moscow State University, Moscow, Russia.*

The analog of the Duschinsky matrix[a] D is defined as $D=(L_I)^{-1}L_{II}$ where L_I and L_{II} are the matrices of the vibrational modes of the molecule under investigation. They are obtained by solving the vibrational problems in the I and II electronic states, respectively. Choosing the dominant elements in columns of the D matrix and permuting these columns to arrange these elements along the diagonal of the transformed matrix D^* makes it possible to establish the correct co-assignment of the calculated frequencies in the I and II electronic states. The rows of D^* are for the vibrations in the I electronic state, whereas the columns are for vibrations in the II electronic state. The results obtained may be tested by analogous calculations of D^* for isotopologues.

[a]F. Duschinsky, Acta Physicochim. URSS, 7 (4), 551-566 (1937).

Intermission

RG09 **3:58 – 4:13**

HIGH RESOLUTION INFRARED SPECTRA OF TRIACETYLENE

KIRSTIN D DONEY, DONGFENG ZHAO, HAROLD LINNARTZ, *Leiden Observatory, Sackler Laboratory for Astrophysics, Universiteit Leiden, Leiden, Netherlands.*

Triacetylene, HC_6H, is the longest poly-acetylene chain found in space, and is believed to be involved in the formation of longer chain molecules and polycyclic aromatic hydrocarbons (PAHs). However, abundances are expected to be low, and observational confirmation requires knowledge of the gas-phase spectra, which up to now has been incomplete with only the weak, low lying bending modes being known. We present new infrared (IR) spectra in the C-H stretch region obtained using ultra-sensitive and highly precise IR continuous wave cavity ring-down spectroscopy (cw-CRDS), combined with supersonic plasma expansions[a]. The talk reviews the accurate determination of the rotational constants of the asymmetric fundamental mode, ν_5, including discussion on the perturber state, and associated hot bands[b]. The determined molecular parameters are accurate enough to aid astronomical searches with such facilities as ALMA (Atacama Large Millimeter Array) or the upcoming JWST (James Webb Space Telecscope), which can now probe even trace molecules (abundances of $\sim 10^{-6}$ - 10^{-10} with respect to H_2).

[a]D. Zhao, J. Guss, A. Walsh, H. Linnartz, Chem. Phys. Lett., 565, 132 (2013)
[b]K.D. Doney, D. Zhao, H. Linnartz, in preparation

RG10 **4:15 – 4:30**

INFRARED AND ULTRAVIOLET SPECTROSCOPY OF GAS-PHASE IMIDAZOLIUM AND PYRIDINIUM IONIC LIQUIDS.

JUSTIN W. YOUNG, RYAN S BOOTH, CHRISTOPHER ANNESLEY, JAIME A. STEARNS, *Space Vehicles Directorate, Air Force Research Lab, Kirtland AFB, NM, USA.*

Ionic liquids (ILs) are a highly variable and potentially game-changing class of molecules for a number of Air Force applications such as satellite propulsion, but the complex nature of IL structure and intermolecular interactions makes it difficult to adequately predict structure-property relationships in order to make new IL-based technology a reality. For example, methylation of imidazolium ionic liquids leads to a substantial increase in viscosity but the underlying physical mechanism is not understood. In addition, the role of hydrogen bonding in ILs, and especially its relationship to macroscopic properties, is a matter of ongoing research. Here we describe the gas-phase spectroscopy of a series of imidazolium- and pyridinium-based ILs, using a combination of infrared spectroscopy and density functional theory to establish the intermolecular interactions present in various ILs, to assess how well they are described by theory, and to relate microscopic structure to macroscopic properties.

RG11 **4:32 – 4:47**

GROUND AND EXCITED STATE ALKYL CH STRETCH IR SPECTRA OF STRAIGHT-CHAIN ALKYLBENZENES

DANIEL M. HEWETT, JOSEPH A. KORN, TIMOTHY S. ZWIER, *Department of Chemistry, Purdue University, West Lafayette, IN, USA.*

Vibrational spectra of alkanes in the CH stretch region are often complicated by Fermi resonance with the overtone of the CH bends. This complication has made the CH stretch region difficult to use as a spectroscopic tool for assigning structures to experimental infrared spectra. A first-principles model accounting for Fermi resonance has been developed by Sibert and co-workers, and has been successfully implemented to predict the CH stretch region of alkyl groups in a variety of settings (both -CH_2- and -CH_3). We have recorded jet-cooled, single-conformation infrared spectra of a series of straight chain alkylbenzenes having chain lengths of two carbons and longer, serving as a foundation for further tests and refinement of the theoretical model. Ground and excited state IR spectra of these alkylbenzenes were acquired using fluorescence dip infrared spectroscopy. A novel approach for taking the excited state spectra that utilizes the gain of a second, infrared-induced fluorescence peak will be discussed and compared to the typical depletion spectra, using ethylbenzene as a prototypical system.

RG12 **4:49 – 5:04**

ASYMMETRY OF $M^+(H_2O)RG$ COMPLEXES, (M=V, Nb) REVEALED WITH INFRARED SPECTROSCOPY

TIMOTHY B WARD, *Department of Chemistry, University of Georgia, Athens, GA, USA*; EVANGELOS MILIORDOS, SOTIRIS XANTHEAS, *Physical Sciences Division, Pacific Northwest National Laboratory, Richland, WA, USA*; MICHAEL A DUNCAN, *Department of Chemistry, University of Georgia, Athens, GA, USA.*

$M^+(H_2O)Ar$ and $M^+(H_2O)Ne$ clusters (M=V, Nb) were produced in a laser vaporization/pulsed nozzle source. The clusters were then mass selected in a time-of-flight mass spectrometer and studied with infrared photodissociation spectroscopy in the OH stretching region. Spectra showed two bands, with the asymmetric band showing k-type rotational structure. Previous work has shown that most metal-water rare gas-tagged systems adopt C_{2v} geometry and exhibit the well-known 3:1 ortho:para ratio in the k-type rotational structure in asymmetric stretch band. However these two metals display a pattern that indicates a breaking of the C_{2v} symmetry. Computational work confirms the breaking of C_{2v} symmetry giving an Ar-M^+-O angle of 163.7 degrees for V and 172.1 degrees for Nb. In the ground state we obtain rotational constants that match up well with obtained spectra using 166 degrees for V and 175 degrees for Nb.

RG13 **5:06 – 5:21**

INFRARED SPECTROSCOPY OF PROTONATED ACETYLACETONE AND MIXED ACETYLACETONE/WATER CLUSTERS

DANIEL MAUNEY, JON MANER, *Department of Chemistry, University of Georgia, Athens, GA, USA*; DAVID C McDONALD, *Chemistry, University of Georgia, Athens, GA, USA*; MICHAEL A DUNCAN, *Department of Chemistry, University of Georgia, Athens, GA, USA.*

Acetylacetone (acac) is the simplest of the beta-diketones. which have both keto and enol tautomers with multiple protonation sites. We readily produce the protonated forms in the gas phase and the current investigation uses vibrational spectroscopy coupled with argon tagging to determine which protonated isomers are present in clusters of acac and the effects of solvation on the isomers observed.

RG14 **5:23 – 5:38**

HEAVY ATOM VIBRATIONAL MODES AND LOW-ENERGY VIBRATIONAL AUTODETACHMENT IN NITROMETHANE ANIONS

MICHAEL C THOMPSON, *JILA and the Department of Chemistry and Biochemistry, University of Colorado-Boulder, Boulder, CO, USA*; JOSHUA H BARABAN, *Department of Chemistry, University of Colorado, Boulder, CO, USA*; JOHN F. STANTON, *Department of Chemistry, The University of Texas, Austin, TX, USA*; J. MATHIAS WEBER, *JILA and the Department of Chemistry and Biochemistry, University of Colorado-Boulder, Boulder, CO, USA.*

We use Ar predissociation and vibrational autodetachment below 2100 $\mathrm{cm}^{-1}$to obtain vibrational spectra of the low-energy modes of nitromethane anion. We interpret the spectra using anharmonic calculations, which reveal strong mode coupling and Fermi resonances. Not surprisingly, the number of evaporated Ar atoms varies with photon energy, and we follow the propensity of evaporating two versus one Ar atoms as photon energy increases. The photodetachment spectrum is discussed in the context of threshold effects and the importance of hot bands.

RG15 ***Post-Deadline Abstract*** **5:40 – 5:55**

OBSERVATION OF DIPOLE-BOUND STATE AND HIGH-RESOLUTION PHOTOELECTRON IMAGING OF COLD ACETATE ANIONS

GUO-ZHU ZHU, DAO-LING HUANG, LAI-SHENG WANG, *Department of Chemistry, Brown University, Providence, RI, USA.*

We report the observation of a dipole-bound state and a high-resolution photoelectron imaging study of cryogenically cooled acetate anions (CH_3COO^-). Both high-resolution non-resonant and resonant photoelectron spectra via the dipole-bound state of CH_3COO^- are obtained. The binding energy of the dipole-bound state relative to the detachment threshold is determined to be 53 ± 8 cm^{-1}. The electron affinity of the $CH_3COO\bullet$ neutral radical is measured accurately as 26 236 ± 8 cm^{-1}(3.2528 ± 0.0010 eV) using high-resolution photoelectron imaging. This accurate electron affinity is validated by observation of autodetachment from two vibrational levels of the dipole-bound state of CH_3COO^-. Excitation spectra to the dipole-bound states yield rotational profiles, allowing the rotational temperature of the trapped CH_3COO^- anions to be evaluated [1].

[1] D. L. Huang, G. Z. Zhu and L. S. Wang, *J.Chem.Phys.*, 2015, **142**, 091103

RH. Clusters/Complexes

Thursday, June 25, 2015 – 1:30 PM

Room: B102 Chemical and Life Sciences

Chair: Galen Sedo, University of Virginia's College at Wise, Wise, VA, USA

RH01 **1:30 – 1:45**

CHIRPED PULSE AND CAVITY FT MICROWAVE SPECTROSCOPY OF THE HCCH-2,6-DIFLUOROPYRIDINE WEAKLY BOUND COMPLEX

CHRIS DEWBERRY, BECCA MACKENZIE, KEN LEOPOLD, *Chemistry Department, University of Minnesota, Minneapolis, MN, USA.*

The microwave spectrum of the HCCH-2,6-difluoropyrine complex has been observed using a chirped pulse and conventional cavity-type Fourier transform microwave spectroscopy. The acetylene moiety forms a hydrogen bond to the nitrogen of the 2,6-difluoropyridine, and this structure is contrasted with several systems involving HCCH or CO_2 bound to pyridine or 2,6-difluoropyridine. The results of DFT calculations support the experimental observations and are reported as well. The chirped pulse spectrometer is new in our laboratory and is built in tandem with our cavity-type spectrometer with a design that allows for switching between the two modes of operation without having to break vacuum. Pertinent details of the spectrometer will also be given.

RH02 **1:47 – 1:57**

MICROWAVE SPECTRUM, VAN DER WAALS BOND LENGTH, AND ^{131}Xe QUADRUPOLE COUPLING CONSTANT OF Xe-SO_3

CHRIS DEWBERRY, *Chemistry Department, University of Minnesota, Minneapolis, MN, USA*; ANNA HUFF, *Chemistry Deparment, Gustavus Adolphus College, St. Peter, MN, USA*; BECCA MACKENZIE, KEN LEOPOLD, *Chemistry Department, University of Minnesota, Minneapolis, MN, USA.*

Nine isotopologues of Xe-SO_3 have been observed by pulsed-nozzle Fourier transform microwave spectroscopy. The complex is a symmetric top with a Xe-S van der Waals distance of 3.577(2) Å. The increase in rare gas distance relative to that in Kr-SO_3 is equal to the difference in van der Waals radii between Xe and Kr. The ^{131}Xe nuclear quadrupole coupling constant indicates that the electric field gradient at the xenon nucleus is 78% larger than that at the Kr nucleus in Kr-SO_3.

RH03 **1:59 – 2:14**

DIMETHYL SULFIDE-DIMETHYL ETHER AND ETHYLENE OXIDE-ETHYLENE SULFIDE COMPLEXES INVESTIGATED BY FOURIER TRANSFORM MICROWAVE SPECTROSCOPY AND AB INITIO CALCULATION

YOSHIYUKI KAWASHIMA, YOSHIO TATAMITANI, TAKAYUKI MASE, *Applied Chemistry, Kanagawa Institute of Technology, Atsugi, Japan*; EIZI HIROTA, *The Central Office, The Graduate University for Advanced Studies, Hayama, Kanagawa, Japan.*

The ground-state rotational spectra of the dimethyl sulfide-dimethyl ether (DMS-DME) and the ethylene oxide and ethylene sulfide (EO-ES) complexes were observed by Fourier transform microwave spectroscopy, and a-type and c-type transitions were assigned for the normal, ^{34}S, and three ^{13}C species of the DMS-DME and a-type and b-type rotational transitions for the normal, ^{34}S, and two ^{13}C species of the EO-ES. The observed transitions were analyzed by using an S-reduced asymmetric-top rotational Hamiltonian. The rotational parameters thus derived for the DMS-DME were found consistent with a structure of C_s symmetry with the DMS bound to the DME by two C-H(DMS)—O and one S—H-C(DME) hydrogen bonds. The barrier height V_3 to internal rotation of the "free" methyl group in the DME was determined to be 915.4 (23) cm^{-1}, which is smaller than that of the DME monomer, 951.72 (70) cm^{-1},[a] and larger than that of the DME dimer, 785.4 (52) cm^{-1}.[b] For the EO-ES complex the observed data were interpreted in the terms of an antiparallel C_s geometry with the EO bound to the ES by two C-H(ES)—O and two S—H-C(EO) hydrogen bonds. We have applied a natural bond orbital (NBO) analysis to the DMS-DME and EO-ES to calculate the stabilization energy CT (= $\Delta E\sigma\sigma^*$), which were closely correlated with the binding energy E_B, as found for other related complexes.

[a] Y. Niide and M. Hayashi, *J. Mol. Spectrosc.* **220**, 65-79 (2003).

[b] Y. Tatamitani, B. Liu, J. Shimada, T. Ogata, P. Ottaviani, A. Maris, W. Caminati, and J. L. Alonso, *J. Am. Chem. Soc.* **124**, 2739-2743 (2002).

RH04 **2:16 – 2:31**

INTERNAL DYNAMICS IN $SF_6 \cdots NH_3$ OBSERVED BY BROADBAND ROTATIONAL SPECTROSCOPY

DROR M. BITTNER, DANIEL P. ZALESKI, SUSANNA L. STEPHENS, NICK WALKER, *School of Chemistry, Newcastle University, Newcastle-upon-Tyne, United Kingdom*; ANTHONY LEGON, *School of Chemistry, University of Bristol, Bristol, United Kingdom.*

The rotational spectra of $SF_6 \cdots NH_3$ isotopologues have been observed in a pulsed nozzle chirped pulse Fourier-transform microwave spectrometer in the frequency range 6.5-18.5 GHz. The spectrum of $SF_6 \cdots {}^{14}NH_3$ has been fitted to a Hamiltonian describing a symmetric top complex in which two symmetric top subunits undergo free internal rotation about a common symmetry axis. The distance between the centers of mass of the two monomers was found to be 4.15776(7) Å. Challenges associated with fitting $|m|=1$ transitions (correlating with K of free NH_3) for $SF_6 \cdots {}^{14}ND_3$ imply complicated internal dynamics occurs within the complex.

RH05 **2:33 – 2:48**

EVIDENCE FOR A COMPLEX BETWEEN THF AND ACETIC ACID FROM BROADBAND ROTATIONAL SPECTROSCOPY

DANIEL P. ZALESKI, DROR M. BITTNER, JOHN CONNOR MULLANEY, SUSANNA L. STEPHENS, *School of Chemistry, Newcastle University, Newcastle-upon-Tyne, United Kingdom*; ADRIAN KING, MATTHEW HABGOOD, *Sensors and Spectroscopy, Atomic Weapons Establishment, Aldermaston, United Kingdom*; NICK WALKER, *School of Chemistry, Newcastle University, Newcastle-upon-Tyne, United Kingdom.*

Evidence for a complex between tetrahydrofuran (THF) and acetic acid from broadband rotational spectroscopy will be presented. Transitions believed to belong to the complex were first identified in a gas mixture containing small amounts of THF, triethyl borane, and acetic acid balanced in argon. Ab initio calculations suggest a complex between THF and acetic acid is more likely to form compared to the analogous acetic acid complex with triethyl borane, the initial target. The observed rotational constants are also more similar to those predicted for a complex formed between THF and acetic acid, than for those of a complex formed between triethyl borane and acetic acid. Subsequently, multiple isotopologues of acetic acid have been measured, confirming its presence in the structure. No information has yet been obtained through isotopic substitution within the THF sub-unit. Ab initio calculations predict the most likely structure is one where the acetic acid subunit coordinates over the ring creating a "bridge" between the THF oxygen, the carboxylic O-H, and the carbonyl oxygen to a hydrogen atom on the back of the ring.

RH06 **2:50 – 3:00**

THE ROTATIONAL SPECTRUM OF PYRIDINE-FORMIC ACID

LORENZO SPADA, QIAN GOU, *Dep. Chemistry 'Giacomo Ciamician', University of Bologna, Bologna, Italy*; BARBARA MICHELA GIULIANO, *Department of Chemistry, University of Bologna, Bologna, Italy*; WALTHER CAMINATI, *Dep. Chemistry 'Giacomo Ciamician', University of Bologna, Bologna, Italy.*

The rotational spectrum of three 1:1 complexes of pyridine with formic acid has been observed and assigned using pulsed jet Fourier transform microwave technique. The two subunits are held together through one O-H⋯N hydrogen bond and one C-H⋯O weak hydrogen bond, forming a seven-membering cyclic structure. The rotational spectrum of the pyridine-HCOOD isotopologue is considerably shifted towards lower frequencies, with respect to the "rigid" model, suggesting a considerable Ubbelohde effect, similar in nature to that observed in the bi-molecules of carboxylic acids.

Intermission

RH07 **3:19 – 3:34**

FOURIER-TRANSFORM MICROWAVE AND MILLIMETERWAVE SPECTROSCOPY OF THE H_2-HCN MOLECULAR COMPLEX

KEIICHI TANAKA, KENSUKE HARADA, *Department of Chemistry, Kyushu University, Fukuoka, Japan*; YOSHIHIRO SUMIYOSHI, *Division of Pure and Applied Science, Faculty of Science and Technology, Gunma University, Maebashi, Japan*; MASAKAZU NAKAJIMA, YASUKI ENDO, *Department of Basic Science, The University of Tokyo, Tokyo, Japan.*

Fourier-Transform microwave (FTMW) spectroscopy has been applied to observe the J = 1 - 0 rotational transitions of the H_2-HCN/DCN complexes containing both the $para$-H_2 (I_{H2}=0) and $ortho$-H_2 (I_{H2}=1) molecule[a]. Rotational spectra of H_2-HCN/DCN up to J = 5 - 4 were also observed in the millimeter-wave (MMW) region below 180 GHz[b]. Observed FTMW lines for H_2-HCN/DCN split into hyperfine components due to the nuclear quadrupole interaction of N and D nuclei. For the $ortho$-H_2 species, the hyperfine splitting due to the magnetic interaction between the hydrogen nuclear spin of $ortho$-H_2 part (j_{H2}=1, I_{H2}=1) was also observed, but not for the $para$-H_2 species (j_{H2}=0, I_{H2}=0). From the observed nuclear spin-spin coupling constants of $ortho$-H_2 species, d = 21.90(47) and 24.66(68) kHz for HCN and DCN complexes, respectively, the average values of $< P_2(\cos\theta) >$ = 0.380(8) and 0.439(10) were derived indicating the nearly free rotation of H_2 in the complex with j_{H2}= 1 and k_{H2}= 0.

The nuclear quadrupole interaction constants due to N and D nuclei show that the HCN/DCN part executes a floppy motion with a large mean square amplitude of about 29/25 and 33/30 degree in the $para$ and $ortho$ species, respectively. From the observed rotational constants, the center-of-mass distances of H_2 and HCN/DCN were derived to be 3.9617(5)/4.00356 (43) Å for the $ortho$ species and 4.1589(13)/4.1596 (36) Å for the $para$ species. The isotope effect on rotational constants confirmed the totally different configurations in the $ortho$ and $para$ species: H_2 is attached to the H/D end of HCN/DCN for the $para$ species, while to the N end for the $ortho$ species, as suggested by IR spectroscopy and theoretical study.

[a]M. Ishiguro et al., *Chem. Phys. Lett.* **554**, 33 (2012).
[b]M. Ishiguro et al., *J. Chem. Phys.* **115**, 5155 (2001).

RH08 **3:36 – 3:51**

MICROWAVE SPECTROSCOPY OF THE CYCLOPENTANOL - WATER DIMER

BRANDON CARROLL, IAN A FINNERAN, GEOFFREY BLAKE, *Division of Chemistry and Chemical Engineering, California Institute of Technology, Pasadena, CA, USA.*

Observations of gas-phase dimers are one of the simplest methods for studying bimolecular interactions. These dimers are excellent model systems for molecular interactions, providing a benchmark for theoretical studies, and a basis for understanding and modeling more complex interactions. Of particular interest are studies of strong (O—H···O—H) and weak (C—H···O—H) long-range interactions of water. We have recently recorded the pure rotational spectrum of the cyclopentanol-water dimer with chirped-pulse Fourier transform microwave spectroscopy (CP-FTMW). We will present the spectrum of this dimer and discuss its structure in the context of C—H···O—H and O—H···O—H bonding.

RH09 **3:53 – 4:08**

HYDROGEN-BONDING AND HYDROPHOBIC INTERACTIONS IN THE ETHANOL-WATER DIMER

IAN A FINNERAN, BRANDON CARROLL, MARCO A. ALLODI, GEOFFREY BLAKE, *Division of Chemistry and Chemical Engineering, California Institute of Technology, Pasadena, CA, USA.*

The conformational energy landscape of the ethanol-water dimer is determined by the relative hydrogen-bond donor and acceptor strengths of the two molecules, as well as weaker hydrophobic interactions between the water and the ethyl group. Using a combination of *ab initio* calculations and chirped-pulse Fourier transform microwave spectroscopy, we have recorded the first rotationally-resolved, jet-cooled spectrum of the ethanol-water dimer between 8-18.5 GHz and identified two water-donor ethanol-acceptor conformers. The lowest energy conformer is chiral, has ethanol in the gauche configuration, and is consistent with previous raman and infrared results.[a] The second conformer corresponds to the trans-ethanol configuration, and exhibits a significant splitting.

[a]Nedić, Marija, et al. PCCP 13.31 (2011): 14050-14063.

RH10 **4:10 – 4:25**

THE INFLUENCE OF FLUORINATION ON STRUCTURE OF THE TRIFLUOROACETONITRILE WATER COMPLEX

WEI LIN, *Department of Chemistry, University of Texas, Brownsville, TX, USA*; ANAN WU, XIN LU, *Department of Chemistry, Xiamen University, Xiamen, China*; DANIEL A. OBENCHAIN, STEWART E. NOVICK, *Department of Chemistry, Wesleyan University, Middletown, CT, USA*.

Acetonitrile, CH_3CN, and trifluoroacetonitrile, CF_3CN, are symmetric tops. In a recent study of the rotational spectrum of the acetonitrile and water complex, it was observed that the structure was also an effective symmetric top[a], with the external hydrogen freely rotating about the O – H bond aligned towards the nitrogen of the cyanide of CH_3CN. Unlike the $CH_3CN - H_2O$ complex, the $CH_3CN - Ar$ and $CF_3CN - Ar$ complexes were observed to be asymmetric tops. Having a series of symmetric and asymmetric top complexes of acetonitrile and trifluoracetonitrile for comparison, we report the rotational spectrum of the weakly bound complex between trifluoroacetonitrile and water. Rotational constants and quadrupole coupling constants will be presented, and the structure of $CF_3CN - H_2O$ will be revealed.

SPOILER ALERT: It's an asymmetric top.

[a]Lovas, F.J.; Sobhanadri, J. Microwave rotational spectral study of $CH_3CN - H_2O$ and $Ar - CH_3CN$. *J. Mol. Spetrosc.* **2015**, *307*, 59-64.

RH11 **4:27 – 4:42**

THE POSITION OF DEUTERIUM IN THE HOD – N_2O AS DETERMINED BY STRUCTURAL AND NUCLEAR QUADRUPOLE COUPLING CONSTANTS

DANIEL A. OBENCHAIN, DEREK S. FRANK, STEWART E. NOVICK, *Department of Chemistry, Wesleyan University, Middletown, CT, USA*; WILLIAM KLEMPERER, *Department of Chemistry and Chemical Biology, Harvard University, Cambridge, MA, USA*.

A recent investigation of the HOD – N_2O complex measuring the OH + OD excited band in the near-IR was completed by Foldes *et al.*[a] During this study, one of us (WAK) was contacted about the position of deuterium in the HOD – N_2O complex, as his group completed the original microwave study of $H_2O - N_2O$ and its deuterated isotopologues[b] in 1992. The results of this microwave study did not give the orientation of HOD in the complex, however, we present here a supplementary study to the original microwave work using a Balle-Flygare cavity instrument, attempting to determine the orientation of HOD relative to the N_2O. In addition to a Kraitchman and a least-squares inertial structure fit of the molecule, we present the nuclear quadrupole coupling tensor of deuterium to determine the position of HOD in the complex.

[a]Földes, T; Lauzin, C.; Vanfleteren, T.; Herman, M.; Lièvin, J.; Didriche. K. High-resolution, near-infrared CW-CRDS and ab initio investigations of N_2O – HDO.*Mol. Phys.* **2015**, *113(5)*,473-482.

[b]Zolandz, D.; Yaron, D.; Peterson, K.I.; Klemperer, W. Water in weak interactions: The structure of the water-nitrous oxide complex. *J. Chem. Phys.* **1992**, 97,2861.

RH12 **4:44 – 4:59**

THE CP-FTMW SPECTROSCOPY AND ASSIGNMENT OF THE MONO- AND DIHYDRATE COMPLEXES OF PERFLUOROPROPIONIC ACID

G. S. GRUBBS II, *Department of Chemistry, Missouri University of Science and Technology, Rolla, MO, USA*; DANIEL A. OBENCHAIN, DEREK S. FRANK, STEWART E. NOVICK, *Department of Chemistry, Wesleyan University, Middletown, CT, USA*; S. A. COOKE, *Natural and Social Science, Purchase College SUNY, Purchase, NY, USA*; AGAPITO SERRATO III, WEI LIN, *Department of Chemistry, University of Texas, Brownsville, TX, USA*.

While searching for the chirped pulse spectra of allyl phenyl ether, the authors used current rotational spectroscopic fitting tools to assign multiple sets of spectra of unknown origin. Previous chirped pulse experiments searching for hydrate complexes of perfluoropropionic acid had not been successful but, through theoretical agreement, it was determined that at least one of the sets of unknown spectra observed belonged to the perfluoropropionic acid-water complex. Further determination showed that the dihydrate had also been observed. The determination process and spectral assignment will be discussed. Structural determinations of the complexes will also be discussed.

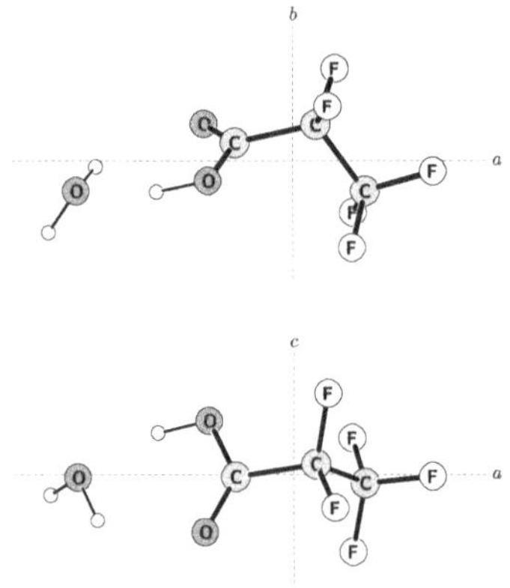

RH13 **5:01 – 5:16**

HYDROGEN BONDING IN 4-AMINOPHENYL ETHANOL: A COMBINED IR-UV DOUBLE RESONANCE AND MICROWAVE STUDY

CAITLIN BRAY, CARA RAE RIVERA, E. A. ARSENAULT, DANIEL A. OBENCHAIN, STEWART E. NOVICK, JOSEPH L. KNEE, *Department of Chemistry, Wesleyan University, Middletown, CT, USA*.

Both amine and hydroxyl functional groups are present in 4-aminophenyl ethanol (4-AE), and each functional group can form hydrogen bonds with carboxylic acids, such as formic acid and acetic acid. Predicting the structures of such complexes involving 4-AE is rather complex, given the many possible conformations and their similar (and method and basis-dependent) energies. In particular, the carboxyl group, -COOH, can act as both as a hydrogen bond donor or acceptor, or both at once. In this study we report the formic acid – 4-AE hydrogen bonded complex. An infrared-ultraviolet double resonance spectrometer is used to examine the shifts in IR frequencies of 4-AE from the monomer to the complex. Fourier transform microwave spectroscopy is used to determine structures of the species. Results from both experiments are compared to DFT and *ab initio* results. Time permitting, results of the water complex with 4-AE will also be presented.

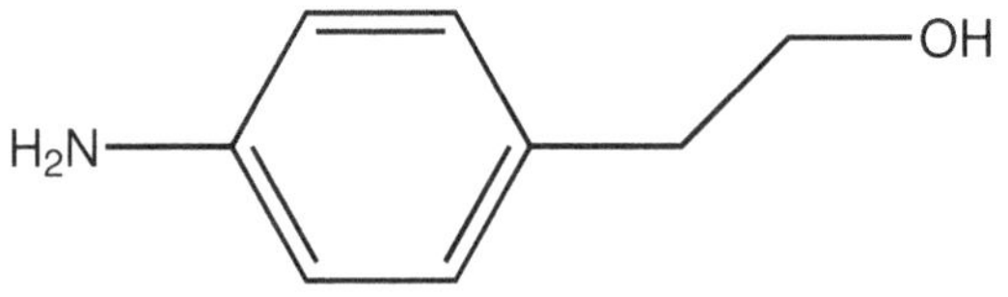

RH14 **5:18 – 5:28**

THEORETICAL STUDY OF THE EFFECT OF DOPING CLUSTERS (ZNO) 6 BY THE SELENIUM USING THE DFT

NOUR EL HOUDA BENSIRADJ, OURIDA OUAMERALI, *Laboratory lctcp, University USTHB, Algiers, Algeria.*

Nano structures (ZnO) 6 have a great interest in the creation of new materials used in energy technologies. We have used the technique of doping ie introducing impurities at the geometry of these clusters; replacing each time by an oxygen atom a selenium atom. This implies a change in the electronic and energetic properties. Clusters obtained (containing selenium) have interesting characteristics in the development of solar energy systems and the field of radiology in medicine. The clusters studied are shown in a 3D geometry (crystal form b) .The geometric parameters of these systems are calculated using the theory of density functional (DFT) . The optimisation the first excited state is performed at the Hartree-Fock method, Single Configuration interaction (HF / CIS), the transmission-absorption spectra are given by the TDDFT method. The results for the excited states have a good process for materials for application in solar cells. The emission spectra of these clusters are located in the tera-hertz region (between the far infrared and microwave), which is less than the ionizing X-ray spectral region and could soon replace for applications medicine.

RH15 ***Post-Deadline Abstract*** **5:30 – 5:45**

BORONYL MIMICS GOLD: A PHOTOELECTRON SPECTROSCOPY STUDY

TIAN JIAN, GARY LOPEZ, LAI-SHENG WANG, *Department of Chemistry, Brown University, Providence, RI, USA.*

Previous studies have found that gold atom and boronyl bear similarities in bonding in many gas phase clusters.[a b c d] $B_{10}(BO)$, $B_{12}(BO)$, $B_3(BO)_n$ (n=1, 2) were found to possess similar bonding and structures to $B_{10}Au$, $B_{12}Au$, B_3Au_n (n=1, 2), respectively. During the recent photoelectron spectroscopy experiments, the spectra of $BiBO^-$ and $BiAu^-$ clusters are found to exhibit similar patterns, hinting that they possess similar geometric structures. While $BiAu^-$ is a linear molecule, $BiBO^-$ is also linear. The similarity in bonding between $BiBO^-$ and $BiAu^-$ is owing to the fact that Au and BO are monovalent σ ligands. The electron affinities are measured to be 1.79$\pm$0.04eV for $BiBO^-$ and 1.36$\pm$0.02eV for $BiAu^-$. The current results provide new examples for the BO/Au isolobal analogy and enrich the chemistry of boronyl and gold.

[a]H.-J. Zhai, C.-Q. Miao, S.-D. Li, L.-S. Wang, J. Phys. Chem. A 2010, 114, 12155–1216

[b]Q. Chen, H. Bai, H.-J. Zhai, S.-D. Li, L.-S. Wang, J. Chem. Phys. 2013, 139, 044308

[c]H. Bai, H.-J. Zhai, S.-D. Li, L.-S. Wang, Phys. Chem. Chem. Phys., 2013, 15, 9646–9653

[d]H.-J. Zhai, Q. Chen, H. Bai, S.-D. Li, L.-S. Wang, Acc. Chem. Res. 2014, 47, 2435-2445

RI. Astronomy

Thursday, June 25, 2015 – 1:30 PM

Room: 274 Medical Sciences Building

Chair: Harshal Gupta, California Institute of Technology, Pasadena, CA, USA

RI01 **1:30 – 1:45**

THE COMPLETE, TEMPERATURE RESOLVED SPECTRUM OF METHYL FORMATE BETWEEN 214 AND 265 GHZ

JAMES P. McMILLAN, SARAH FORTMAN, CHRISTOPHER F. NEESE, FRANK C. DE LUCIA, *Department of Physics, The Ohio State University, Columbus, OH, USA.*

We have studied methyl formate, one of the so-called 'astronomical weeds', in the 214–265 GHz band. We have experimentally gathered a set of intensity calibrated, complete, and temperature resolved spectra from across the astronomically significant temperature range of 248–406 K. Using our previously reported method of analysis[a], the point by point method, we are capable of generating the complete spectrum at an arbitrary temperature. Thousands of lines, of nontrivial intensity, which were previously not included in the available astrophysical catalogs have been found.

The sensitivity of the point by point analysis is such that we are able to identify lines which would not have manifest in a single scan across the band. The consequence has been to reveal not only a number of new methyl formate lines, but also trace amounts of contaminants. We show how the intensities from the contaminants can be removed with indiscernible impact on the signal from methyl formate. To do this we use the point by point results from our previous studies of these contaminants. The efficacy of this process serves as strong proof of concept for usage of our point by point results on the problem of the 'weeds'. The success of this approach for dealing with the weeds has also previously been reported.[b]

[a] J. McMillan, S. Fortman, C. Neese, F. DeLucia, ApJ. 795, 56 (2014)
[b] S. Fortman, J. McMillan, C. Neese, S. Randall, and A. Remijan, J. Mol. Spectrosc. 280, 11 (2012).

RI02 **1:47 – 2:02**

ROTATIONAL SPECTROSCOPY OF 4-HYDROXY-2-BUTYNENITRILE

R. A. MOTIYENKO, L. MARGULÈS, *Laboratoire PhLAM, UMR 8523 CNRS - Université Lille 1, Villeneuve d'Ascq, France*; J.-C. GUILLEMIN, *Institut des Sciences Chimiques de Rennes, UMR 6226 CNRS - Université de Rennes 1, Rennes, France.*

Recently we studied the rotational spectrum of hydroxyacetonitrile ($HOCH_2CN$, HAN) in order to provide a firm basis for its possible detection in the interstellar medium[a]. Different plausible pathways of the formation of HAN in the interstellar conditions were proposed;[b] however, up to now, the searches for this molecule were unsuccessful. To continue the study of nitriles that represent an astrophysical interest we present in this talk the analysis of the rotational spectrum of 4-hydroxy-2-butynenitrile ($HOCH_2CC$-CN, HBN), the next molecule in the series of hydroxymethyl nitriles. Using the Lille spectrometer the spectrum of HBN was measured in the frequency range 50 – 500 GHz. From the spectroscopic point of view HBN molecule is rather similar to HAN, because of -OH group tunnelling in *gauche* conformation. As it was previously observed for HAN, due to this large amplitude motion, the splittings in the rotational spectra of HBN are easily resolved making the spectral analysis more difficult. Additional difficulties arise from the near symmetric top character of HBN ($\kappa = -0.996$), and very dense spectrum because of relatively small values of rotational constants and a number of low-lying excited vibrational states. The analysis carried out in the frame of reduced axis system approach of Pickett[c] allows to fit within experimental accuracy all the rotational transitions in the ground vibrational state. Thus, the results of the present study provide a reliable catalog of frequency predictions for HBN.

The support of the Action sur Projets de l'INSU PCMI, and ANR-13-BS05-0008-02 IMOLABS is gratefully acknowledged

[a] Margulès L., Motiyenko R.A., Guillemin J.-C. *68th ISMS*, **2013**, TI12.
[b] Danger G. et al. *Phys. Chem. Chem. Phys.* **2014**, 16, 3360.
[c] Pickett H.M. *J. Chem. Phys.* **1972**, 56, 1715.

RI03 **2:04 – 2:19**

TIME-DOMAIN TERAHERTZ SPECTROSCOPY OF ISOLATED PAHS

BRANDON CARROLL, MARCO A. ALLODI, *Division of Chemistry and Chemical Engineering, California Institute of Technology, Pasadena, CA, USA*; BRETT A. McGUIRE, *NAASC, National Radio Astronomy Observatory, Charlottesville, VA, USA*; SERGIO IOPPOLO, *Department of Physical Sciences, The Open University, Milton Keynes, UK*; GEOFFREY BLAKE, *Division of Chemistry and Chemical Engineering, California Institute of Technology, Pasadena, CA, USA.*

Polycyclic aromatic hydrocarbons (PAHs) are a strong candidate as carriers of the unidentified infrared features (UIRs). As UIR carriers, PAHs may account for up to 20% of the interstellar carbon budget and may play key roles in many chemical and physical processes in the ISM. Laboratory and astronomical observations in the TeraHertz (THz) regime offer a unique method to study these species through observations of low frequency vibrational modes of individual molecules as well as bulk phonon modes. Embedding PAHs in matrices enables the differentiation of bulk and single-molecule modes, as well as the investigation of the interaction between PAHs and both polar and apolar matrices. Such work provides a basis for further studies of the effects of PAHs on interstellar ices. We will present the THz time-domain spectra (0.3 - 7 THz) of PAHs embedded in N_2 and H_2O matrices, and discuss the importance of these spectra for future laboratory and astronomical studies.

RI04 **2:21 – 2:36**

HIGH-RESOLUTION IR ABSORPTION SPECTROSCOPY OF POLYCYCLIC AROMATIC HYDROCARBONS: SHINING LIGHT ON THE INTERSTELLAR 3 MICRON EMISSION BANDS

ELENA MALTSEVA, *Van' t Hoff Institute for Molecular Sciences, University of Amsterdam, Amsterdam, Netherlands*; ALESSANDRA CANDIAN, XANDER TIELENS, *Leiden Observatory, University of Leiden, Leiden, Netherlands*; ANNEMIEKE PETRIGNANI, J. OOMENS, *Institute for Molecules and Materials (IMM), Radboud University Nijmegen, Nijmegen, Netherlands*; WYBREN JAN BUMA, *Van' t Hoff Institute for Molecular Sciences, University of Amsterdam, Amsterdam, Netherlands.*

Various astronomical objects show distinctive series of IR emission bands indicated as unidentified infrared emission bands. These features are nowadays mainly attributed to the IR fluorescence of Polycyclic Aromatic Hydrocarbons (PAHs) even though an unambiguous identification of which PAHs are involved has not been possible yet. We present here a high-resolution IR absorption study of a number of jet-cooled polycyclic aromatic hydrocarbons in the 3.3 μm region obtained by IR-UV ion depletion techniques. The experimental spectra display many more bands than expected, and lead to the conclusion that the appearance of the spectrum is dominated by fourth-order vibrational coupling terms. This has far-reaching consequences since up till now the assignment of infrared emission features observed in different types of space objects in this wavelength region -and the conclusions drawn from these assignments on the evolution of interstellar gas- has relied heavily on harmonic quantum chemical calculations. We also observe that the presence of bay-hydrogen sites in a PAH leads to a shift of the overall spectrum to the high-energy side and to a broadening of the 3 μm band. This observation provides an appealing explanation for previous speculations that the emission of 3 μm band consists of two components. Moreover, it paves for using this structure to derive the composition of different objects.

RI05 **2:38 – 2:53**

EXPLORING MOLECULAR COMPLEXITY WITH ALMA (EMoCA): HIGH-ANGULAR-RESOLUTION OBSERVATIONS OF SAGITTARIUS B2(N) AT 3 mm

HOLGER S. P. MÜLLER, *I. Physikalisches Institut, Universität zu Köln, Köln, Germany*; ARNAUD BELLOCHE, KARL M. MENTEN, *Millimeter- und Submillimeter-Astronomie, Max-Planck-Institut für Radioastronomie, Bonn, NRW, Germany*; ROBIN T. GARROD, *Departments of Chemistry and Astronomy, The University of Virginia, Charlottesville, VA, USA.*

Sagittarius (Sgr for short) B2 is the most massive and luminous star-forming region in our Galaxy, located close to the Galactic Center. We have carried out a molecular line survey with the IRAM 30 m telescope toward its two major sites of star-formation, Sgr B2(M) and (N).[a] Toward the latter source, which is particularly rich in Complex Organic Molecules (COMs), we detected three molecules for the first time in space, aminoacetonitrile, ethyl formate, and *n*-propyl cyanide.

We have recently obtained ALMA data of Sgr B2(N) between ∼84 and ∼111 GHz within Cycle 0 and one additional setup up to 114.4 GHz within Cycle 1. At angular resolutions of 1.8″ and 1.4″, respectively, the two main hot cores, the prolific Sgr B2(N-LMH) (or Sgr B2(N)-SMA1) and the likely less evolved Sgr B2(N)-SMA2 are well separated, and line confusion is reduced greatly for the latter. As a consequence, we have been able to identify the first branched alkyl molecule in space, *iso*-propyl cyanide, toward Sgr B2(N)-SMA2.[b] Our ongoing analyses include investigations of cyanides and isocyanides, alkanols and thioalkanols, and deuterated molecules among others. We will present some of our results.

[a] A. Belloche et al., *A&A* **559** (2013) Art. No. A47.
[b] A. Belloche et al., *Science* **345** (2014) 1584.

RI06 **2:55 – 3:10**

FIRST SPECTROSCOPIC STUDIES AND DETECTION IN SgrB2 OF ^{13}C-DOUBLY SUBSTITUED ETHYL CYANIDE

L. MARGULÈS, R. A. MOTIYENKO, *Laboratoire PhLAM, UMR 8523 CNRS - Université Lille 1, Villeneuve d'Ascq, France*; J.-C. GUILLEMIN, *Institut des Sciences Chimiques de Rennes, UMR 6226 CNRS - Université de Rennes 1, Rennes, France*; HOLGER S. P. MÜLLER, *I. Physikalisches Institut, Universität zu Köln, Köln, Germany*; ARNAUD BELLOCHE, *Millimeter- und Submillimeter-Astronomie, Max-Planck-Institut für Radioastronomie, Bonn, NRW, Germany.*

Ethyl cyanide (CH_3CH_2CN) is one of the most abundant complex organic molecules in the interstellar medium firstly detected in OMC-1 and Sgr B2 in 1977[a]. The vibrationally excited states are enough populated under ISM conditions and could be detected[b],[c]. Apart from the deuterated ones, all mono-substituted isotopologues of ethyl cyanide (^{13}C[d] and ^{15}N[e]) have been detected in the ISM. The detection of isotopologues in the ISM is important: it can give information about the formation process of complex organic molecules, and it is essential to clean the ISM spectra from the lines of known molecules in order to detect new ones. The ^{12}C/^{13}C ratio found in SgrB2: 20-30 suggests that the doubly ^{13}C could be present in the spectral line survey recently obtained with ALMA (EMoCA)[f], but no spectroscopic studies exist up to now. We measured and analyzed the spectra of the ^{13}C-doubly-substitued species up to 1 THz with the Lille solid-state based spectrometer. The spectroscopic results and and the detection of the doubly ^{13}C species in SgrB2 will be presented.
This work was supported by the CNES and the Action sur Projets de l'INSU, PCMI. This work was also done under ANR-13-BS05-0008-02 IMOLABS. Support by the Deutsche Forschungsgemeinschaft via SFB 956, project B3 is acknowledged

[a] D. R. Johnson, et al., *Astrophys. J.* 1977, **218**, L370
[b] A. Belloche, et al., *A&A* 2013, **559**, A47
[c] A.M. Daly, et al., *Astrophys. J.* 2013, **768**, 81
[d] K. Demyk, et al. *A&A* 2007 **466**, 255
[e] Margulès, et al. *A&A* 2009, **493**, 565
[f] Belloche et al. 2014, *Science*, **345**, 1584

RI07 **3:12 – 3:27**

MILLIMETERWAVE SPECTROSCOPY OF ETHANIMINE AND PROPANIMINE AND THEIR SEARCH IN ORION

L. MARGULÈS, R. A. MOTIYENKO, *Laboratoire PhLAM, UMR 8523 CNRS - Université Lille 1, Villeneuve d'Ascq, France*; J.-C. GUILLEMIN, *Institut des Sciences Chimiques de Rennes, UMR 6226 CNRS - Université de Rennes 1, Rennes, France*; JOSE CERNICHARO, *Departamento de Astrofísica, Centro de Astrobiología CAB, CSIC-INTA, Madrid, Spain.*

The aldimines are important to understand amino acids formation process as they appear in reaction scheme of Strecker-type synthesis. Following the detection in the ISM of methanimine (CH_2NH) in 1973[a] and the more recent one of ethanimine (CH_3CHNH)[b], we decided to investigate the next molecule in the series: propanimine (CH_3CH_2CHNH). For this molecule no spectroscopic information was available up to now. We measured the rotational spectrum of propanimine in the frequency range up to 500 GHz. Since the spectroscopic studies of ethanimine were limited to 130 GHz[c], we also extended the measurements up to 300 GHz. The spectra of both E- and Z- isomers are analyzed for the two molecules. Usually aldimines, which are unstable molecules, are obtained by discharge or pyrolysis methods, here pure sample were obtained by synthesis process.

For ethanimine, the methyl top internal rotation should be taken into account, therefore the analysis is performed using new version of RAM36 code[d] which includes the treatment of the nuclear quadrupole hyperfine structure.

The spectroscopic results and their searches in Orion will be presented.

This work was supported by the CNES and the Action sur Projets de l'INSU, PCMI. This work was also done under ANR-13-BS05-0008-02 IMOLABS

[a] Godfrey, P. D.; *et al. Astrophys. Lett.* **13**, (1973) 119
[b] Loomis, R. A.; *et al. ApJ. Lett.* **765**, (2013) L9
[c] Lovas, F. J.; *et al. J. Chem. Phys.* **72**, (1980) 4964
[d] Ilyushin, V.V. et al; *J. Mol. Spectrosc.* **259**, (2010) 26

Intermission

RI08 **3:46 – 4:01**

FURTHER STUDIES OF λ 5797.1 DIFFUSE INTERSTELLAR BAND

TAKESHI OKA, L. M. HOBBS, DANIEL E. WELTY, DONALD G. YORK, *Department of Astronomy and Astrophysics, The University of Chicago, Chicago, IL, USA*; JULIE DAHLSTROM, *Department of Physics and Astronomy, Carthage College, Kenosha, WI, USA*; ADOLF N. WITT, *Department of Physics and Astronomy, University of Toledo, Toledo, OH, USA.*

The λ 5797.1 DIB is unique with its sharp central feature.[a] We simulated the spectrum based on three premises: (1) Its carrier molecule is polar as concluded from the anomalous spectrum toward the star Herschel 36.[b] (2) The central feature is Q-branch of a parallel band of a prolate top. (3) The radiative temperature of the environment is T_r = 2.73 K. A comparison with observed spectrum indicated that the carrier contains 5-7 heavy atoms.[c]

To further strengthen this hypothesis, we have looked for vibronic satellites of the λ 5797.1 DIB. Since its anomaly toward Her 36 was ascribed to the lengthening of bonds upon the electronic excitation, vibronic satellites involving stretch vibrations are expected. Among the 73 DIBs observed toward HD 183143 to the blue of 5797.1 Å, two DIBs, λ 5545.1 and λ 5494.2 stand out as highly correlated with λ 5797.1 DIB. Their correlation coefficients 0.941 and 0.943, respectively, are not sufficiently high to establish the vibronic relation by themselves but can be explained as due to high uncertainties due to their weakness and their stellar blends. They are above the λ 5797.1 DIB by 784.0 cm^{-1} and 951.2 cm^{-1}, respectively, approximately expected for stretching vibrations.

Another observations which may possibly be explained by our hypothesis is the emission at 5800 Å from the Red Rectangle Nebula called RR 5800.[d] Our analysis suggests that λ 5797.1 DIB and RR 5800 are consistently explained as caused by the same molecule.

[a] T.H. Kerr, R.E. Hibbins, S.J. Fossey, J.R. Miles, P.J. Sarre, ApJ 495, 941 (1998)
[b] T. Oka, D.E. Welty, S. Johnson, D.G. York, J. Dahlstrom, L.M. Hobbs, ApJ 773, 42 (2013)
[c] J. Huang, T. Oka, Mol. Phys. J.P. Maier Special Issue in press.
[d] G.D. Schmidt, A.N. Witt, ApJ 383, 698 (1991)

RI09 **4:03 – 4:18**

LABORATORY OPTICAL SPECTROSCOPY OF THE PHENOXY RADICAL AS A DIFFUSE INTERSTELLAR BANDS CANDIDATE

MITSUNORI ARAKI, YUKI MATSUSHITA, *Faculty of Science Division I, Tokyo University of Science, Shinjuku-ku, Tokyo, Japan*; KOICHI TSUKIYAMA, *Faculty of Science Division I, Tokyo University of Science, Shinjuku-ku, Tokyo, Japan.*

Diffuse Interstellar Bands (DIBs) are optical absorption lines observed in diffuse clouds in interstellar space. They still remain the longest standing unsolved problem in spectroscopy and astrochemistry, although several hundreds of DIBs have been already detected. Aromatic radicals in a gas phase are potential DIB candidate molecules. The electronic transitions of aromatic radicals result in optical absorption. Last year we reported the gas-phase optical absorption spectrum of the $^2A_2 \leftarrow X^2B_1$ transition of the thiophenoxy radical C_6H_5S using a cavity ringdown spectrometer.[a,b] As the next step, we observed the $B^2A_2 \leftarrow X^2B_1$ transition of the phenoxy radical C_6H_5O in the discharge of anisole. The four broad and asymmetric peaks making a progression of 500 cm^{-1} were detected in the 5700–6450 Åregion. The progression was assigned to the 6a mode, and the broad and asymmetric peak profiles were accounted for by the sequences of the 10b mode. Each vibrational component has a broad structure of 23 Å, which can be explained by lifetime broadening. Based on the assignment of the progression and the sequences, the vibronic components from $v = 0$ in the X^2B_1 ground state can be extracted from the broad and asymmetric peak profiles to compare the laboratory bands with DIBs. Although the components did not agree with the reported DIBs, the upper limit of the column density for the phenoxy radical in the diffuse clouds toward HD 204827 was evaluated to be 4×10^{14} cm^{-2}. Therefore the most fundamental aromatic radicals, the thiophenoxy and phenoxy radicals, could not explain DIBs observed at present.

[a]MF14
[b]Araki, Niwayama, and Tsukiyama, *Astronomical Journal*, 148, 87 (5pp), 2014

RI10 **4:20 – 4:35**

INVESTIGATION OF CARBONACEOUS INTERSTELLAR DUST ANALOGUES BY INFRARED SPECTROSCOPY: EFFECTS OF ENERGETIC PROCESSING

BELÉN MATÉ, MIGUEL JIMÉNEZ-REDONDO, ISABEL TANARRO, VICTOR JOSE HERRERO, *Molecular Physics, Instituto de Estructura de la Materia (IEM-CSIC), Madrid, Spain.*

Carbonaceous compounds, both solids and gas-phase molecules, are found in very diverse astronomical media[a]. A significant amount of the elemental carbon is found in small dust grains. This carbonaceous dust, mostly formed in the last stages of evolution of C-rich stars, is the carrier of characteristic IR absorption bands revealing the presence of aliphatic, aromatic and olefinic functional groups in variable proportions[b]. Among the various candidate materials investigated as possible carriers of these bands, hydrogenated amorphous carbon (a-C:H) has led to the best agreement with the observations. Carbonaceous grains are processed by H atoms, UV radiation, cosmic rays and interstellar shocks in their passage from asymptotic giant branch stars to planetary nebulae and to the diffuse interstellar medium. The mechanisms of a-C:H production and evolution in astronomical media are presently a subject of intensive investigation. In this work we present a study of the stability of carbonaceous dust analogues generated in He+CH_4 radiofrequency discharges. In order to simulate the processing of dust in the interstellar environments, the samples have been subjected to electron bombardment, UV irradiation, and both He and H_2 plasma processing. IR spectroscopy is employed to monitor the changes in the structure and composition of the carbonaceous films.

[a]A.G.G.M. Tielens. Rev. Mod. Phys., 85, 1021 (2013)
[b]J.E. Chiar, A.G.G.M. Tielens, A.J. Adamson and A. Ricca. Astrophys. J., 770, 78 (2013)

RI11 **4:37 – 4:52**

REACTIONS OF GROUND STATE NITROGEN ATOMS $N(^4S)$ WITH ASTROCHEMICALLY-RELEVANT MOLECULES ON INTERSTELLAR DUSTS

LAHOUARI KRIM, SENDRES NOURRY, *Department of Chemistry, MONARIS, CNRS, UMR 8233, Sorbonne Universités, UPMC Univ Paris 06, Paris, France.*

In the last few years, ambitious programs were launched to probe the interstellar medium always more accurately. One of the major challenges of these missions remains the detection of prebiotic compounds and the understanding of reaction pathways leading to their formation. These complex heterogeneous reactions mainly occur on icy dust grains, and their studies require the coupling of laboratory experiments mimicking the extreme conditions of extreme cold and dilute media. For that purpose, we have developed an original experimental approach that combine the study of heterogeneous reactions (by exposing neutral molecules adsorbed on ice to non-energetic radicals H, OH, N...) and a neon matrix isolation study at very low temperatures, which is of paramount importance to isolate and characterize highly reactive reaction intermediates. Such experimental approach has already provided answers to many questions raised about some astrochemically-relevant reactions occurring in the ground state on the surface of dust grain ices in dense molecular clouds. The aim of this new present work is to show the implication of ground state atomic nitrogen on hydrogen atom abstraction reactions from some astrochemically-relevant species, at very low temperatures (3K-20K), without providing any external energy. Under cryogenic temperatures and with high barrier heights, such reactions involving $N(^4S)$ nitrogen atoms should not occur spontaneously and require an initiating energy. However, the detection of some radicals species as byproducts, in our solid samples left in the dark for hours at 10K, proves that hydrogen abstraction reactions involving ground state $N(^4S)$ nitrogen atoms may occur in solid phase at cryogenic temperatures. Our results show the efficiency of radical species formation stemming from non-energetic N-atoms and astrochemically-relevant molecules. We will then discuss how such reactions, involving nitrogen atoms in their ground states, might be the first key step towards complex organic molecules production in the interstellar medium.

RI12 **4:54 – 5:09**

STABILITY OF GLYCINE TO ENERGETIC PROCESSING UNDER ASTROPHYSICAL CONDITIONS INVESTIGATED VIA INFRARED SPECTROSCOPY

BELÉN MATÉ, VICTOR JOSE HERRERO, ISABEL TANARRO, RAFAEL ESCRIBANO, *Molecular Physics, Instituto de Estructura de la Materia (IEM-CSIC), Madrid, Spain.*

Glycine, the simplest aminoacid, has been detected in comets and meteorites in our Solar System. Its detection in the interstellar medium is not improbable since other organic molecules of comparable complexity have been observed[a]. Information of how complex organic molecules resist the energetic processing that they may suffer in different regions of space is of great interest for astrochemists and astrobiologists.

Further to previous investigations[b] we have studied in this work, via infrared spectroscopy, the effect of 2 keV electron bombardment on amorphous and crystalline glycine layers at low temperatures, to determine its destruction cross section under astrophysical conditions. Energetic electrons are known to be present in the solar wind and in planetary magnetospheres, and are also formed in the interaction of cosmic rays with matter. Moreover, we have probed the shielding effect of water ice layers grown on top of the glycine samples at 90 K. These experiment aim to mimic the conditions of the aminoacid in ice mantles on dust grains in the interstellar medium or in some outer Solar System objects, with a water ice surface crust. A residual material, product of glycine decomposition, was found at the end of the processing. A tentative assignment of the infrared spectra of the residue will be discussed in the presentation.

[a] E. Herbst and E. F. van Dishoeck, Annu. Rev. Astro. Astrophys. 2009, 47:427-480

[b] B. Maté, Y. Rodriguez-Lazcano, O. Gálvez, I. Tanarro and R. Escribano, Phys Chem Chem Phys, 2011, 13, 12268. B. Maté, I. Tanarro, M.A. Moreno, M. Jiménez-Redondo, R. Escribano, and V. J. Herrero, Faraday Discussions, 2014, DOI: 10.1039/c3fd00132f.

RI13 **5:11 – 5:21**

MILLIMETER AND SUBMILLIMETER STUDIES OF INTERSTELLAR ICE ANALOGUES

AJ MESKO, IAN C WAGNER, HOUSTON HARTWELL SMITH, *Department of Chemistry, Emory University, Atlanta, GA, USA*; STEFANIE N MILAM, *Astrochemistry, NASA Goddard Space Flight Center, Greenbelt, MD, USA*; SUSANNA L. WIDICUS WEAVER, *Department of Chemistry, Emory University, Atlanta, GA, USA.*

The chemistry of interstellar ice analogues has been a topic of great interest to astrochemists over the last 20 years. Currently, the models of interstellar chemistry feature icy-grain reactions as a primary mechanism for the formation of many astrochemical species as well as potentially astrobiologically-relevant complex organic molecules. This talk presents new spectral results collected by a millimeter and submillimeter spectrometer coupled to a vacuum chamber designed to study the sublimation or sputtered products of icy-grain reactions initiated by thermal-processing or photo-processing of interstellar ice analogues. Initial results from thermal desorption and UV photoprocessing experiments of pure water ice and water + methanol ice mixtures will be presented.

RI14 **5:23 – 5:38**

UNTANGLING MOLECULAR SIGNALS OF ASTROCHEMICAL ICES IN THE THz: DISTINGUISHING AMORPHOUS, CRYSTALLINE, AND INTRAMOLECULAR MODES WITH BROADBAND THz SPECTROSCOPY

BRETT A. McGUIRE, *NAASC, National Radio Astronomy Observatory, Charlottesville, VA, USA*; SERGIO IOPPOLO, *Department of Physical Sciences, The Open University, Milton Keynes, UK*; XANDER DE VRIES, *Theoretical Chemistry, University of Nijmegen, Nijmegen, Netherlands*; MARCO A. ALLODI, BRANDON CARROLL, GEOFFREY BLAKE, *Division of Chemistry and Chemical Engineering, California Institute of Technology, Pasadena, CA, USA.*

We have previous reported at this meeting on the initial construction of a broadband (0.3 – 7.5 THz) TeraHertz time-domain spectrometer to study condensed-phase samples of astrophysically-relevant species. Here, we present the latest results from this instrument, focusing on the intersection of theory with experiment in the interpretation of our spectra. We will present both simple (CO_2) and more complex (CH_3OH and beyond) species, in their purely-crystalline and purely-amorphous states, at varying levels of matrix isolation, and as mixtures of these species. We will discuss the relative contributions of individual molecular motions (i.e. torsional modes) and bulk motions within the ice to the observed laboratory spectra. We will also touch upon the feasibility of direct interstellar detection of species from these spectra, and the results of proof-of-concept observations with the FIFI-LS instrument on the SOFIA telescope, currently scheduled for Spring 2015.

RI15 **5:40 – 5:55**

QUANTUM CHEMICAL STUDY OF THE REACTION OF C^+ WITH INTERSTELLAR ICE: PREDICTIONS OF VIBRATIONAL AND ELECTRONIC SPECTRA OF REACTION PRODUCTS

DAVID E. WOON, *Department of Chemistry, University of Illinois at Urbana-Champaign, Urbana, IL, USA.*

The C^+ cation (CII) is the dominant form of carbon in diffuse clouds and an important tracer for star formation in molecular clouds. We studied the low energy deposition of C^+ on ice using density functional theory calculations on water clusters as large as 18 H_2O. Barrierless reactions occur with water to form two dominant sets of products: HOC + H_3O^+ and CO^- + $2H_3O^+$. In order to provide testable predictions, we have computed both vibrational and electronic spectra for pure ice and processed ice clusters. While vibrational spectroscopy is expected to be able to discern that C^+ has reacted with ice by the addition of H_3O^+ features not present in pure ice, it does not provided characteristic bands that would discern between HOC and CO^-. On the other hand, predictions of electronic spectra suggest that low energy absorptions may occur for CO^- and not HOC, making it possible to distinguish one product from the other.

RJ. Cold/Ultracold/Matrices/Droplets

Thursday, June 25, 2015 – 1:30 PM

Room: 217 Noyes Laboratory

Chair: Gary E. Douberly, The University of Georgia, Athens, GA, USA

RJ01 **1:30 – 1:45**

IR SPECTRA OF COLD PROTONATED METHANE

OSKAR ASVANY, *I. Physikalisches Institut, Universität zu Köln, Köln, Germany*; KOICHI MT YAMADA, *EMTech, National Institute of Advanced Industrial Science and Technology (AIST), Tsukuba, Japan*; SANDRA BRÜNKEN, ALEXEY POTAPOV, STEPHAN SCHLEMMER, *I. Physikalisches Institut, Universität zu Köln, Köln, Germany*.

High-resolution infrared spectra of mass selected protonated methane, CH_5^+, have been recorded in the C-H stretching region in a 22-pole ion trap experiment at low temperatures. The frequencies of the infrared OPO system (pump and signal) have been calibrated using a NIR frequency comb. As a result the ro-vibrational IR transition frequencies of CH_5^+ could be determined to an accuracy in the MHz regime.[a] In this contribution we discuss different techniques of laser induced reactions which enabled recording spectra at different temperatures.[b] The spectra simplify dramatically at a nominal trap temperature of 4 K. Nevertheless an assignment of these spectra is very difficult. We apply the idea of the Rydberg-Ritz combination principle to the complex spectra of protonated methane in order to get first hints at the energy level structure of this enigmatic molecule.

[a] O. Asvany, J. Krieg, and S. Schlemmer, Frequency comb assisted mid-infrared spectroscopy of cold molecular ions, Review of Scientific Instruments, 83 (2012), 076102.

[b] O. Asvany, S. Brünken, L. Kluge, and S. Schlemmer, COLTRAP: a 22-pole ion trapping machine for spectroscopy at 4 K, Applied Physics B: Lasers and Optics, 114 (2014), 203-211

RJ02 **1:47 – 2:02**

PROGRESS ON OPTICAL ROTATIONAL COOLING OF SiO+

PATRICK R STOLLENWERK, YEN-WEI LIN, BRIAN C. ODOM, *Department of Physics and Astronomy, Northwestern University, Evanston, IL, USA*.

Producing ultracold molecules is the first step in precision molecular spectroscopy. Here we present some of the challenges and advantages of SiO+ as well as some of our progress toward meeting those challenges. To demonstrate ground state SiO+, we first load about 100 SiO+ via 2+1 REMPI into an ion trap. Translational motion of SiO+ is then sympathetically cooled by co-trapped Ba+, which is laser cooled. To prepare the population into the ground state, we optically pump the P-branch (rotational cooling transitions) in the B:Σ(v'=0) ← X:Σ(v=0) band with broadband radiation. Because the band is highly diagonal, population can be effectively driven into the rotational ground state before falling into other manifolds. The broadband source, a fs laser, is spectrally filtered using an ultrashort pulse shaping technique to drive only the P-branch. Attention must be paid when aligning the optics to obtain sufficient masking resolution. We have achieved 3 cm^{-1} resolution, which is sufficient to modify a broadband source for rotationally cooling SiO+.

RJ03 **2:04 – 2:19**

THE OPTICAL BICHROMATIC FORCE IN MOLECULAR SYSTEMS[a]

LELAND M. ALDRIDGE, SCOTT E. GALICA, EDWARD E. EYLER, *Department of Physics, University of Connecticut, Storrs, CT, USA.*

The bichromatic optical force (BCF), which can greatly exceed radiative forces, seems ideal for laser slowing and cooling of molecules because it minimizes the effects of radiative decay. However, it relies on sustained coherences between optically coupled states, and molecules, with their many sublevels and decay pathways, present new challenges in maintaining these coherences compared with simple atoms. We have conducted extensive numerical simulations of BCFs in model molecular systems based on the $B \leftrightarrow X$ transition in CaF, and have begun experimental tests in a molecular beam.

In our modeling, the effects of fine and hyperfine structure are examined using a simplified level scheme that is still sufficiently complete to include the major pathways leading to loss or decoherence. To circumvent optical pumping into coherent dark states we explore two possible schemes: (1) a skewed dc magnetic field, and (2) rapid optical polarization switching. The effects of repumping to compensate for out-of-system radiative decay are also examined. Our results verify that the BCF is a promising method for creating large forces in molecular beams while minimizing out-of-system radiative losses, and provide detailed guidance for experimental designs. Compared to a two-level atom, the peak force is reduced by about an order of magnitude, but there is little reduction in the velocity range over which the force is effective. Our experiments on deflection and slowing using the CaF $B \leftrightarrow X$, (0-0) transition, still at an early stage, include studies of both the $P_{11}(1.5)/^{P}Q_{12}(0.5)$ branch, a quasi-cycling configuration with extensive hfs, and the $R_{11}(0.5)/^{R}Q_{21}(0.5)$ branch, which has a much simpler hfs but requires rotational repumping.

[a]Supported by the National Science Foundation

RJ04 **2:21 – 2:36**

A NEW EQUATION OF STATE FOR SOLID *para*-HYDROGEN

LECHENG WANG, ROBERT J. LE ROY, PIERRE-NICHOLAS ROY, *Centre for Graduate Work in Chemistry and Biochemistry, University of Waterloo, Waterloo, Ontario, Canada.*

Solid *para*-H_2 is a popular accommodating host for impurity spectroscopy due to its unique softness and the spherical symmetry of *para*-H_2 in its *J*=0 rotational level.[a,b] To simulate the properties of impurity-doped solid *para*-H_2, a reliable model for the 'soft' pure solid *para*-H_2 at different pressures is highly desirable. While a couple of experimental[c] and theoretical[d] studies aimed at elucidating the equation of state (EOS) of solid *para*-H_2 have been reported, the calculated EOS was shown to be heavily dependent on the potential energy surface (PES) between two *para*-H_2 that was used in the simulations.[e] The current study also demonstrates that different choices of the parameters governing the Quantum Monte Carlo simulation could produce different EOS curves.

To obtain a reliable model for pure solid *para*-H_2, we used a new 1-D *para*-H_2 PES reported by Faruk *et al.*[f] that was obtained by averaging over Hinde's highly accurate 6-D H_2–H_2 PES.[g] The EOS of pure solid *para*-H_2 was calculated using the PIMC algorithm with periodic boundary conditions (PBC). To precisely determine the equilibrium density of solid *para*-H_2, both the value of the PIMC time step τ and the number of particles in the PBC cell were extrapolated to convergence. The resulting EOS agreed well with experimental observations, and the *hcp* structured solid *para*-H_2 was found to be more stable than the *fcc* one at 4.2K, in agreement with experiment. The vibrational frequency shift of *para*-H_2 as a function of the density of the pure solid was also calculated, and the value of the shift at the equilibrium density is found to agree well with experiment.

[a] T. Momose, H. Honshina, M. Fushitani and H. Katsuki, *Vib. Spectrosc.* **34**, 95(2004).
[b] M. E. Fajardo, *J. Phys. Chem. A* **117**, 13504 (2013).
[c] I. F. Silvera, *Rev. Mod. Phys.* **52**, 393(1980).
[d] F. Operetto and F. Pederiva, *Rhys. Rev. B* **73**, 184124(2006).
[e] T. Omiyinka and M. Boninsegni, *Rhys. Rev. B* **88**, 024112(2013).
[f] N. Faruk, M. Schmidt, H. Li, R. J. Le Roy, and P.-N. Roy,*J. Chem. Phys.* bf 141, 014310(2014).
[g] R. J. Hinde, *J. Chem. Phys.* **128**, 154308(2008).

RJ05 **2:38 – 2:53**

INFRARED SPECTROSCOPY OF NOH SUSPENDED IN SOLID PARAHYDROGEN: PART TWO

MORGAN E. BALABANOFF, FREDRICK M. MUTUNGA, DAVID T. ANDERSON, *Department of Chemistry, University of Wyoming, Laramie, WY, USA.*

The only report in the literature on the infrared spectroscopy of the parent oxynitrene NOH was performed using Ar matrix isolation spectroscopy at 10 K.[a] In this previous study, they performed detailed isotopic studies to make definitive vibrational assignments. NOH is predicted by high-level calculations to be in a triplet ground electronic state,[b] but the Ar matrix isolation spectra cannot be used to verify this triplet assignment. In our 2013 preliminary report,[c] we showed that 193 nm in situ photolysis of NO trapped in solid parahydrogen can also be used to prepare the NOH molecule. Over the ensuing two years we have been studying the infrared spectroscopy of this species in more detail. The spectra reveal that NOH can undergo hindered rotation in solid parahydrogen such that we can observe both a-type and b-type rovibrational transitions for the O-H stretch vibrational mode, but only a-type for the mode assigned to the bend. In addition, both observed a-type infrared absorption features (bend and OH stretch) display fine structure; an intense central peak with weaker peaks spaced symmetrically to both lower and higher wavenumbers. The spacing between the peaks is nearly identical for both vibrational modes. We now believe this fine structure is due to spin-rotation interactions and we will present a detailed analysis of this fine structure. Currently, we are performing additional experiments aimed at making ^{15}NOH to test these preliminary assignments. The most recent data and up-to-date analysis will be presented in this talk.

[a] G. Maier, H. P. Reisenauer, M. De Marco, *Angew. Chem. Int. Ed.* **38**, 108-110 (1999).
[b] U. Bozkaya, J. M. Turney, Y. Yamaguchi, and H. F. Schaefer III, *J. Chem. Phys.* **136**, 164303 (2012).
[c] David T. Anderson and Mahmut Ruzi, *68th Ohio State University International Symposium on Molecular Spectroscopy*, talk TE01 (2013).

RJ06 **2:55 – 3:10**

HIGH RESOLUTION INFRARED SPECTROSCOPY OF CH_3F-(*ortho*-H_2)$_n$ CLUSTER IN SOLID *para*-H_2

HIROYUKI KAWASAKI, ASAO MIZOGUCHI, HIDETO KANAMORI, *Department of Physics, Tokyo Institute of Technology, Tokyo, Japan.*

The absorption spectrum of the ν_3 (C-F stretching) mode of CH_3F in solid *para*-H_2 by FTIR showed a series of equal interval peaks[a]. Their interpretation was that the *n*-th peak of this series was due to CH_3F-(*ortho*-H_2)$_n$ clusters which were formed CH_3F and *n*'s *ortho*-H_2 in first nearest neighbor sites of the *para*-H_2 crystal with *hcp* structure. In order to understand this system in more detail, we have studied these peaks, especially $n = 0 - 3$ corresponding to 1037 - 1041 cm^{-1}, by using high-resolution and high-sensitive infrared quantum cascade (QC) laser spectroscopy. Before now, we found many peaks around each *n*-th peak of the cluster, which we didn't know their origins[b]. We observed photochromic phenomenon of these peaks by taking an advantage of the high brightness of the laser[c]. In this study, we focus on satellite series consisting of six peaks which locate at the lower energy side of each main peak. All the peaks showed a common red shouldered line profile, which corresponds to partly resolved transitions of *ortho*- and *para*- CH_3F. The spectral pattern and time behavior of the peaks may suggest that these satellite series originate from a family of CH_3F clusters involving *ortho*-H_2 in second nearest neighbor sites. A model function assuming this idea is used to resolve the observed spectrum into each Lorentzian component, and then some common features of the satellite peaks are extracted and the physical meanings of them will be discussed.

[a] K. Yoshioka and D. T. Anderson, J. Chem. Phys. 119 (2003) 4731-4742
[b] A. R. W. McKellar, A. Mizoguchi, and H. Kanamori, J. Chem. Phys. 135 (2011) 124511
[c] A. R. W. McKellar, A. Mizoguchi, and H. Kanamori, Phys. Chem. Chem. Phys. 13 (2011) 11587-11589

Intermission

RJ07 **3:29 – 3:44**

REACTIVE INTERMEDIATES IN ^{4}He NANODROPLETS: INFRARED LASER STARK SPECTROSCOPY OF DIHYDROXYCARBENE

BERNADETTE M. BRODERICK, CHRISTOPHER P. MORADI, GARY E. DOUBERLY, *Department of Chemistry, University of Georgia, Athens, GA, USA*; LAURA McCASLIN, JOHN F. STANTON, *Department of Chemistry, The University of Texas, Austin, TX, USA.*

Singlet dihydroxycarbene (HOCOH) is produced via pyrolytic decomposition of oxalic acid, captured by helium nanodroplets, and probed with infrared laser Stark spectroscopy. Rovibrational bands in the OH stretch region are assigned to either trans,trans- or *trans,cis-* rotamers on the basis of symmetry type, nuclear spin statistical weights, and comparisons to electronic structure theory calculations. Stark spectroscopy provides the inertial components of the permanent electric dipole moments for these rotamers. The dipole components for *trans,trans-* and *trans,cis-* rotamers are (μ_a, μ_b) = (0.00, 0.68(6)) and (1.63(3), 1.50(5)), respectively. The infrared spectra lack evidence for the higher energy *cis,cis-* rotamer, which is consistent with a previously proposed pyrolytic decomposition mechanism of oxalic acid and computations of HOCOH torsional interconversion and tautomerization barriers.

RJ08 **3:46 – 4:01**

INFRARED LASER STARK SPECTROSCOPY OF THE PRE-REACTIVE Cl···HCl COMPLEX FORMED IN SUPERFLUID ^{4}He DROPLETS

CHRISTOPHER P. MORADI, GARY E. DOUBERLY, *Department of Chemistry, University of Georgia, Athens, GA, USA.*

Chlorine atoms, generated through the thermal decomposition of Cl_2, are solvated in superfluid helium nanodroplets and clustered with HCl molecules. The H–Cl stretching modes of these clusters are probed via infrared laser spectroscopy. A broad band centered at $\approx$2880.9 cm^{-1} is assigned to the binary Cl···HCl complex. The band center is red shifted by only 7.4 cm^{-1} from the "free" HCl stretch (ν_1) of $(HCl)_2$ and, as such, is consistent with an assignment to a similarly "free" HCl stretch. Also, the breadth of the band ($\approx$2 cm^{-1} FWHM) is consistent with assignment to a mostly *b*-type component of the H–Cl stretch; the band is lifetime broadened to a similar extent as the predominantly *b*-type ν_1 stretch of $(HCl)_2$, due to fast rotational relaxation facilitated by the helium droplet environment. Despite the lack of rotational structure, which would verify our assignment, the spectrum is consistent with stabilization of a weakly-bound complex having an L-shaped geometry. Computations reveal that the projection of the transition dipole moment onto the *a*-axis results in a dramatic decrease ($\approx$700 times) in the intensity of the *a*-type band relative to the *b*-type band intensity; indeed, the signal-to-noise ratio in our experiment precluded observation of an *a*-type band for this complex. No bands were observed that could derive from a strongly H-bonded Cl···HCl complex. Additionally, we located two bands at 2764.0 and 2798.5 cm^{-1} that are consistent with the pick-up of two HCl molecules and are therefore assigned to vibrations of the Cl···$(HCl)_2$ complex.

RJ09 **4:03 – 4:13**

HELIUM NANODROPLET INFRARED SPECTROSCOPY OF THE TROPYL RADICAL

MATIN KAUFMANN, *Physikalische Chemie II, Ruhr University Bochum, Bochum, Germany*; BERNADETTE M. BRODERICK, GARY E. DOUBERLY, *Department of Chemistry, University of Georgia, Athens, GA, USA.*

Helium nanodroplet spectroscopy is a well-established experimental technique to study weakly bound complexes and reactive species. The superfluid helium interacts weakly with the embedded species, leading to only small matrix-induced shifts in vibrational spectra. This technique has been applied for the spectroscopic study of the resonance-stabilized allyl radical and its reactions and complexes.[a,b] The tropyl radical is another example of a π-conjugated radical, being of importance as a reaction intermediate in organic chemistry. Having an electron in a pair of degenerate orbitals, its geometry is subject to the Jahn-Teller effect.[c] The Jahn-Teller distortion of the ground electronic state is probed with IR laser spectroscopy.

[a] C. M. Leavitt, C. P. Moradi, B. W. Acrey, G. E. Douberly; J. Chem. Phys. 2013, 139, 234301.
[b] D. Leicht, D. Habig, G. Schwaab, M. Havenith; J. Phys. Chem. A 2015, 119, 1007.
[c] E. P. F. Lee, T. G. Wright; J. Phys. Chem. A 1998, 102, 4007.

RJ10 **4:15 – 4:30**

MICROSOLVATION STUDIES IN HELIUM NANODROPLETS

GERHARD SCHWAAB, MATIN KAUFMANN, DANIEL LEICHT, RAFFAEL SCHWAN, THEO FISCHER, DEVENDRA MANI, MARTINA HAVENITH, *Physikalische Chemie II, Ruhr University Bochum, Bochum, Germany.*

In bulk aqueous solutions the interactions between solute and solvent are still not fully understood. We apply spectroscopy in Helium nanodroplets to investigate solvation processes step by step (bottom up approach). Recently, the Bochum helium nanodroplet spectrometer has been equipped with a quantum cascade laser spanning the frequency ranges from 1000–1400, 1600–1700, and 2500–2600 cm^{-1}. First results with the extended setup will be presented.

RJ11 **4:32 – 4:47**

INFRARED SPECTRA OF THE CO_2-H_2O, CO_2-$(H_2O)_2$, and $(CO_2)_2$-H_2O COMPLEXES ISOLATED IN SOLID NEON BETWEEN 90 AND 5300 cm^{-1}

BENOÎT TREMBLAY, *Chemistry/ MONARIS, CNRS, UMR 8233, Sorbonne Universités, UPMC Univ Paris 06, Paris, France*; PASCALE SOULARD, *MONARIS UMR8233, CNRS - UNiversité Paris 6 UPMC, Paris, France.*

The van der Waals complex of H_2O with CO_2 has attracted considerable theoretical interest since it is a typical example of a weak binding complex (less than 3 kcal/mol), but a very few IR data are available in gas. For these reasons, we have studied in solid neon hydrogen bonded complexes involving carbon dioxide and water molecules. Evidence for the existence of at least three $(CO_2)_m(H_2O)_n$, or m:n, complexes has been obtained from the appearance of many new absorptions near the well-know monomers fundamental transitions. Concentration effects and detailed vibrational analysis allowed identification of fifteen, eleven and four transitions for the 1:1, 1:2, and 2:1 complexes, respectively. Careful examination of the far infrared allows the assignment of several 1:1 and 1:2 intermolecular modes, confirmed by the observation of combinations of intra+intermolecular transitions. All of these results significantly increase the number of one and, especially, two quanta vibrational transitions observed for these complexes, and anharmonic coupling constants have been derived. This study shows the high sensibility of the solid neon isolation for the spectroscopy of the hydrogen-bonded complexes since two quanta transitions can't be easily observed in gas phase.

RJ12 **4:49 – 4:59**

MATRIX ISOLATION AND COMPUTATIONAL STUDY OF [2C, 2N, X] (X=S, SE) ISOMERS

TAMAS VOROS, GYORGY TARCZAY, *Institute of Chemistry, Eotvos University, Budapest, Hungary.*

The [2C, 2N, S] and the [2C, 2N, Se] systems were investigated by quantum chemical computations and matrix isolation IR spectroscopy. For both systems nine isomers were computationally investigated, for which harmonic and anharmonic vibrational wavenumbers and infrared (IR) intensities were calculated using the CCSD(T)/aug-cc-pVTZ level of theory. The results show that each of the isomers have two or more detectable bands in the mid IR region, which have one or two orders of magnitude larger intensity compared to the IR intensity of the most intense bands of the most stable NCSCN and NCSeCN isomers'. It follows that if the most stable isomers can be detected, then the other previously unobserved isomers generated from NCSCN or NCSeCN should also be detectable with IR spectroscopy. UV spectra were also computed for each isomer at the TD-DFT B3LYP/aug-cc-pVTZ level of theory. These computations showed that the most stable isomers (NCSCN and NCSeCN) can absorb the UV radiation around 250 nm, and the irradiation may promote photoisomerization. This means that if the initial isomers are irradiated by narrow-band UV radiation, new isomers may be generated, which likely decompose by irradiating broad-band UV radiation.

The two most stable isomers, sulphur dicyanide (NCSCN) and selenium dicyanide (NCSeCN), were prepared following literature methods. The matrix isolation IR spectra of these molecules in Ar and Kr were measured for the first time. As a result of a selective 254 nm-irradiation of the deposited matrices some new bands appeared in the IR spectra, while the intensity of the bands of NCSCN or NCSeCN were decreased at the same time. Irradiation of the matrices with broad-band UV light decreased the intensity of the bands corresponding to the deposited isomers and some of the bands appeared on the 254 nm-irradiation. On the basis of the analysis of the formation rates of the different bands upon 254 nm photolysis and by comparison with the results of the quantum chemical calculations these bands could be assigned to new isomers. In the case of sulphur analogue NCSNC and NCNCS were unambiguously identified, and for selenium analogue the formation of NCSeNC and NCNCSe isomers were observed.

RJ13 **5:01 – 5:16**

MATRIX ISOLATION SPECTROSCOPY AND PHOTOCHEMISTRY OF TRIPLET 1,3-DIMETHYLPROPYNYLIDENE (MeC_3Me)

STEPHANIE N. KNEZZ, *Department of Chemistry, The Univeristy of Wisconsin, Madison, WI, USA*; TERESE A WALTZ, *Department of Chemistry, Geoscience, and Physics, Edgewood College, Madison, WI, USA*; BENJAMIN C. HAENNI, *Department of Chemistry, University of Wisconsin–Madison, Madison, WI, USA*; NICOLA J. BURRMANN, *Department of Chemistry, Heartland Community College, Normal, IL, USA*; ROBERT J. McMAHON, *Department of Chemistry, The Univeristy of Wisconsin, Madison, WI, USA.*

Acetylenic carbenes and conjugated carbon chain molecules of the HC_nH family are relevant to the study of combustion and chemistry in the interstellar medium (ISM). Propynylidene (HC_3H) has been thoroughly studied and its structure and photochemistry determined.[a] Here, we produce triplet diradical 1,3-dimethylpropynylidene (MeC_3Me) photochemically from a precursor diazo compound in a cryogenic matrix (N_2 or Ar) at 10 K, and spectroscopic analysis is carried out. The infrared, electronic absorption, and electron paramagnetic resonance spectra were examined in light of the parent (HC_3H) system to ascertain the effect of alkyl substituents on delocalized carbon chains of this type. Computational analysis, EPR, and infrared analysis indicate a triplet ground state with a quasilinear structure. Infrared experiments reveal photochemical reaction to penten-3-yne upon UV irradiation. Further experimental and computational results pertaining to the structure and photochemistry will be presented.

[a]Seburg, R. A.; Patterson, E. V.; McMahon, R. J., Structure of Triplet Propynylidene (HCCCH) as Probed by IR, UV/vis, and EPR Spectroscopy of Isotopomers. Journal of the American Chemical Society 2009, 131 (26), 9442-9455.

RJ14 **5:18 – 5:33**

EVIDENCE OF INTERNAL ROTATION IN THE O-H STRETCHING REGION OF THE 1:1 METHANOL-BENZENE COMPLEX IN AN ARGON MATRIX

JAY AMICANGELO, IAN CAMPBELL, JOSHUA WILKINS, *School of Science (Chemistry), Penn State Erie, Erie, PA, USA.*

Co-depositions of methanol (CH_3OH) and benzene (C_6H_6) in an argon matrix at 20 K result in the formation of a 1:1 methanol-benzene complex (CH_3OH-C_6H_6) as evidenced by the observation of distinct infrared bands attributable to the complex near the O-H, C-H, and C-O stretching fundamental vibrations of CH_3OH and the hydrogen out-of-plane bending fundamental vibration of C_6H_6. Co-deposition experiments were also performed using isotopically labeled methanol (CD_3OD) and benzene (C_6D_6) and the corresponding deuterated complexes were also observed. Based on ab initio and density functional theory calculations, the structure of the complex is thought to be an H-π complex in which the CH_3OH is above the C_6H_6 ring with the OH hydrogen atom interacting with the π cloud of the ring. Close inspection of the O-H and O-D stretching peaks of the complexes reveals small, distinct satellite peaks that are approximately 3 – 4 cm^{-1}lower than the primary peak. A series of experiments have been performed to ascertain the nature of the satellite peaks. These consist of co-depositions in which the concentrations of both monomers were varied over a large range (1:200 to 1:1600 S/M ratios), annealing experiments (20 K to 35 K), and lower temperature cycling experiments (20 K to 8 K). Based on the results of these experiments, it is concluded that the satellite peaks are due to rotational structure and not due to matrix site effects, higher aggregation or distinct complex geometries. Given the rigidity of a low temperature argon matrix, it is proposed that the rotational motion responsible for the satellite peaks is internal rotation within the methanol subunit of the complex rather than overall molecular rotation of the complex.

FA. Electronic structure, potential energy surfaces

Friday, June 26, 2015 – 8:30 AM

Room: 116 Roger Adams Lab

Chair: Timothy Steimle, Arizona State University, Tempe, AZ, USA

FA01 **8:30 – 8:45**

CHARACTERIZATION OF THE $1\,^5\Pi_u$ - $1\,^5\Pi_g$ BAND OF C_2 BY TWO-COLOR RESONANT FOUR-WAVE MIXING AND LIF

PETER RADI, *General Energy, Paul Scherrer Institute, Villigen, Switzerland.*

The application of two-color resonant four-wave mixing (TC-RFWM) in combination with a discharge slit-source in a molecular beam environment is advantageous for the study of perturbations in C_2. Initial investigations have shown the potential of the method by a detailed deperturbation of the d $^3\Pi_g$, $v = 4$ state.[a]

The deperturbation of the d $^3\Pi_g$, $v = 6$ state unveiled the presence of the energetically lowest high-spin state of C_2. This dark state gains transition strength through the perturbation process with the d $^3\Pi_g$, $v = 6$ state yielding weak spectral features that are observable by the high sensitivity of the TC-RFWM technique. The successful deperturbation study of the d $^3\Pi_g$, $v = 6$ state resulted in the spectroscopic characterization of the quintet ($1\,^5\Pi_g$) and an additional triplet state (b $^3\Sigma_g^-$, $v = 19$). [b]

More recently, investigations have been performed by applying unfolded TC-RFWM to obtain further information on the quintet manifold. The first high-spin transition ($1\,^5\Pi_u$- $1\,^5\Pi_g$) has been observed *via* an intermediate "gateway" state exhibiting both substantial triplet and quintet character owing to the perturbation between the $1\,^5\Pi_g$, $v = 0$ and the d $^3\Pi_g$, $v = 6$ states. The high-lying quintet state is found to be predissociative and displays a shallow potential that accommodates three vibrational levels only. [c]

Further studies of the high-spin system will be presented in this contribution. By applying TC-RFWM and laser-induced fluorescence, data on the vibrational structure of the $1\,^5\Pi_u$- $1\,^5\Pi_g$ system is obtained. The results are combined with high-level ab initio computations at the multi-reference configuration interaction (MRCI) level of theory and the largest possible basis currently implemented in the 2012 version of MOLPRO.

[a] P. Bornhauser, G. Knopp, T. Gerber, and P.P. Radi, Journal of Molecular Spectroscopy 262, 69 (2010)
[b] P. Bornhauser, Y. Sych, G. Knopp, T. Gerber, and P.P. Radi, Journal of Chemical Physics 134, 044302 (2011)
[c] Bornhauser, P., Marquardt, R., Gourlaouen, C., Knopp, G., Beck, M., Gerber, T., van Bokhoven, JA, and Radi, P. P., Journal of Physical Chemistry, submitted

FA02 **8:47 – 9:02**

SIGN CHANGES IN THE ELECTRIC DIPOLE MOMENT OF EXCITED STATES IN RUBIDIUM-ALKALINE EARTH DIATOMIC MOLECULES

JOHANN V. POTOTSCHNIG, *Institute of Experimental Physics, Graz University of Technology, Graz, Austria*; FLORIAN LACKNER, *UXSL, Chemical Sciences Division, Lawrence Berkeley National Laboratory, Berkeley, CA, USA*; ANDREAS W. HAUSER, WOLFGANG E. ERNST, *Institute of Experimental Physics, Graz University of Technology, Graz, Austria.*

In a recent series of combined experimental and theoretical studies we investigated the ground state and several excited states of the Rb-alkaline earth molecules RbSr[a] and RbCa.[b] The group of alkali-alkaline earth (AK-AKE) molecules has drawn attention for applications in ultracold molecular physics and the measurement of fundamental constants[c] due to their large permanent electric and magnetic dipole moments in the ground state. These properties should allow for an easy manipulation of the molecules and simulations of spin models in optical lattices.[d]

In our studies we found that the permanent electric dipole moment points in different directions for certain electronically excited states, and changes the sign in some cases as a function of bond length. We summarize our results, give possible causes for the measured trends in terms of molecular orbital theory and extrapolate the tendencies to other combinations of AK and AKE - elements.

[a] F. Lackner, G. Krois, T. Buchsteiner, J. V. Pototschnig, and W. E. Ernst, Phys. Rev. Lett., 2014, 113, 153001; G. Krois, F. Lackner, J. V. Pototschnig, T. Buchsteiner, and W. E. Ernst, Phys. Chem. Chem. Phys., 2014, 16, 22373; J. V. Pototschnig, G. Krois, F. Lackner, and W. E. Ernst, J. Chem. Phys., 2014, 141, 234309
[b] J. V. Pototschnig, G. Krois, F. Lackner, and W. E. Ernst, J. Mol. Spectrosc., in Press (2015), doi:10.1016/j.jms.2015.01.006
[c] M. Kajita, G. Gopakumar, M. Abe, and M. Hada, J. Mol. Spectrosc., 2014, 300, 99-107
[d] A. Micheli, G. K. Brennen, and P. Zoller, Nature Physics, 2006, 2, 341-347

FA03 **9:04 – 9:19**

HIGH RESOLUTION VELOCITY MAP IMAGING PHOTOELECTRON SPECTROSCOPY OF THE BERYLLIUM OXIDE ANION, BEO-

AMANDA REED, KYLE MASCARITOLO, MICHAEL HEAVEN, *Department of Chemistry, Emory University, Atlanta, GA, USA.*

The photodetachment spectrum of BeO^- has been studied for the first time using high resolution velocity map imaging photoelectron spectroscopy. Vibrational contours were imaged and compared with Franck-Condon simulations for the ground and excited states of the neutral. The first measured electron affinity of BeO, and anisotropies of several transitions were also measured. Experimental findings are compared to high level *ab initio* calculations.

FA04 **9:21 – 9:36**

ELECTRONIC AUTODETACHMENT SPECTROSCOPY AND IMAGING OF THE ALUMINUM MONOXIDE ANION, ALO-

AMANDA REED, KYLE MASCARITOLO, ADRIAN GARDNER, MICHAEL HEAVEN, *Department of Chemistry, Emory University, Atlanta, GA, USA.*

The $^1\Sigma^+ \leftarrow ^1\Sigma^+$ ground state to dipole bound state electronic transition of AlO^- has been studied with both rotationally resolved autodetachment spectroscopy and high resolution velocity map imaging photoelectron spectroscopy in a newly constructed apparatus. Vibrational and rotational molecular constants have been determined for both the ground state ($\nu'' =$ 0,1) and excited dipole bound state ($\nu' = 0,1$) of the aluminum monoxide anion. The spectra yield the electron binding energy of the dipole bound state, and a more accurate electron affinity for AlO. The photoelectron anisotropies of several transitions were measured. Experimental findings are compared to high level *ab initio* calculations. Additionally, high resolution photodetachment imaging of AlO^- $^1\Sigma^+$ within energy ranges well above the detachment threshold were measured and compared to previous, low resolution photodetachment results.

FA05 **9:38 – 9:53**

SPECTROSCOPY OF THE LOW-ENERGY STATES OF BaO^+

JOSHUA BARTLETT, ROBERT A. VANGUNDY, MICHAEL HEAVEN, *Department of Chemistry, Emory University, Atlanta, GA, USA.*

The BaO^+ cation is a promising candidate for studies conducted at ultra-cold temperatures. It is known that the ion can be formed by the reaction of laser-cooled Ba^+ with N_2O or O_2. Spectroscopic data are now needed for the BaO^+ cation, for both characterization of the internal state population distributions and the design of population transfer schemes. We have obtained the first spectroscopic data for BaO^+ using the pulsed-field ionization, zero kinetic energy (PFI-ZEKE) photoelectron technique. Two-color ionization was carried out via the $A^1\Sigma^+$-$X^1\Sigma^+$ transition of BaO. Vibronic levels of the $X^2\Sigma^+$, $A^2\Pi_{3/2}$ and $A^2\Pi_{1/2}$ states of BaO^+ have been characterized. The results are compared with the predictions of high-level electronic structure calculations.

Intermission

FA06 **10:12 – 10:27**

THE OPTICAL SPECTRUM OF SrOH RE-VISITED: ZEEMAN EFFECT, HIGH-RESOLUTION SPECTROSCOPY AND FRANCK-CONDON FACTORS.

TRUNG NGUYEN, DAMIAN L KOKKIN, TIMOTHY STEIMLE[a], *Department of Chemistry and Biochemistry, Arizona State University, Tempe, AZ, USA*; IVAN KOZYRYEV, JOHN M. DOYLE, *Department of Physics, Harvard University, Cambridge, MA, USA.*

Motivated by a diverse range of applications in physics and chemistry, currently there is great interest in the cooling of molecules to very low temperatures ($\leq$1 mK). Direct laser cooling has been previously demonstrated for the diatomic radicals SrF[b,c], YO[d,e], and CaF[f], and most recently a three-dimensional magneto-optical trap (MOT) of SrF molecules was achieved[g,h]. To determine the possibility of laser cooling for polyatomic molecules containing three or more atoms, detailed information is required about their Franck-Condon factors (FCFs) for emission from the excited states of interest. Here we report on the high-resolution laser excitation spectra, recorded field-free and in the presence of a static magnetic field, and on the dispersed fluorescence (DF) spectra for the $A^2\Pi_{1/2} \leftarrow X^2\Sigma^+$ and $B^2\Sigma^+ \leftarrow X^2\Sigma^+$ electronic transitions of SrOH. The DF spectra were analyzed to precisely determine FCFs and compared with values predicted using a normal coordinate GF matrix approach. The recorded Zeeman spectra were analyzed to determine the magnetic moments. Implication for proposed laser cooling and trapping experiments for SrOH will be presented.

---

[a]NSF CHE-1265885

[b]E.S. Shuman, J.F. Barry and D. DeMille, Nature 467, 820 (2010)

[c]J.F. Barry, E.S. Shuman, E.B. Norrgard and D. DeMille, Phys. Rev. Lett. 108, 103002 (2012)

[d]M.T. Hummon, M. Yeo, B.K. Stuhl, A.L. Collopy, Y. Xia, and J. Ye, Phys. Rev. Lett. 110, 143001 (2013)

[e]M. Yeo, M.T. Hummon, A.L. Collopy, B. Yan, B. Hemmerling, E. Chae, J.M. Doyle, and J. Ye, arXiv:1501.04683 (2015)

[f]V. Zhelyazkova, A. Cournol, T.E. Wall, A. Matsushima, J.J. Hudson, E.A. Hinds, M.R. Tarbutt, and B.E. Sauer, Phys. Rev. A 89, 053416 (2014)

[g]J.F. Barry, D.J. McCarron, E.B. Norrgard, M.H. Steinecker and D. DeMille, Nature 512, 286 (2014)

[h]D.J. McCarron, E.B. Norrgard, M.H. Steinecker and D. DeMille, arXiv:1412.8220 (2014)

FA07 **10:29 – 10:44**

SPECTROSCOPIC ACCURACY IN QUANTUM CHEMISTRY: A BENCHMARK STUDY ON Na_3

ANDREAS W. HAUSER, JOHANN V. POTOTSCHNIG, WOLFGANG E. ERNST, *Institute of Experimental Physics, Graz University of Technology, Graz, Austria.*

Modern techniques of quantum chemistry allow the prediction of molecular properties to good accuracy, provided the systems are small and their electronic structure is not too complex. For most users of common program packages, ‘chemical’ accuracy in the order of a few kJ/mol for relative energies between different geometries is sufficient. The demands of molecular spectroscopists are typically much more stringent, and often include a detailed topographical survey of multi-dimensional potential energy surfaces with an accuracy in the range of wavenumbers. In a benchmark study of current predictive capabilities we pick the slightly sophisticated, but conceptually simple and well studied case of the Na_3 ground state, and present a thorough investigation of the interplay between Jahn-Teller-, spin-orbit-, rovibrational- and hyperfine-interactions based only on ab initio calculations. The necessary parameters for the effective Hamiltonian are derived from the potential energy surface of the $1^2E'$ ground state and from spin density evaluations at selected geometries, without any fitting adjustments to experimental data. We compare our results to highly resolved microwave spectra.[a]

---

[a]L. H. Coudert, W. E. Ernst and O. Golonzka, J. Chem. Phys. 117, 7102–7116 (2002)

FA08 **10:46 – 11:01**

ACCURATE FIRST-PRINCIPLES SPECTRA PREDICTIONS FOR ETHYLENE AND ITS ISOTOPOLOGUES FROM FULL 12D AB INITIO SURFACES

THIBAULT DELAHAYE, *CNRS et Universités Paris Est et Paris Diderot, Laboratoire Interuniversitaire des Systèmes Atmosphériques (LISA), Créteil, France*; MICHAEL REY, VLADIMIR TYUTEREV, *Groupe de Spectrométrie Moléculaire et Atmosphérique, UMR CNRS 7331, Université de Reims, Reims Cedex 2, France*; ANDREI V. NIKITIN, *Atmospheric Spectroscopy Div., Institute of Atmospheric Optics, Tomsk, Russia*; PETER SZALAY, *Institute of Chemistry, Eotvos University, Budapest, Hungary*.

Hydrocarbons such as ethylene (C_2H_4) and methane (CH_4) are of considerable interest for the modeling of planetary atmospheres and other astrophysical applications. Knowledge of rovibrational transitions of hydrocarbons is of primary importance in many fields but remains a formidable challenge for the theory and spectral analysis. Essentially two theoretical approaches for the computation and prediction of spectra exist. The first one is based on empirically-fitted effective spectroscopic models. Several databases aim at collecting the corresponding data but the information about C_2H_4 spectrum present in these databases remains limited, only some spectral ranges around 1000, 3000 and 6000 cm^{-1} being available. Another way for computing energies, line positions and intensities is based on global variational calculations using *ab initio* surfaces. Although they do not yet reach the spectroscopic accuracy, they could provide reliable predictions which could be quantitatively accurate with respect to the precision of available observations and as complete as possible. All this thus requires extensive first-principles quantum mechanical calculations essentially based on two necessary ingredients: (i) accurate intramolecular potential energy surface and dipole moment surface components and (ii) efficient computational methods to achieve a good numerical convergence. We report predictions of vibrational and rovibrational energy levels of C_2H_4 using our new ground state potential energy surface obtained from extended ab initio calculations[a]. Additionally we will introduce line positions and line intensities predictions based on a new dipole moment surface for ethylene. These results will be compared with previous works on ethylene and its isotopologues.

[a] T. Delahaye, A. V. Nikitin, M. Rey, P. G. Szalay, and Vl. G. Tyuterev, J. Chem. Phys. 2014, 141, 104301

FA09 **11:03 – 11:18**

HIGH-RESOLUTION LASER SPECTROSCOPY OF S_1-S_0 TRANSITION OF NAPHTHALENE: MEASUREMENT OF VIBRATIONALLY EXCITED STATES

TAKUMI NAKANO, RYO YAMAMOTO, *Graduate School of Science, Kobe University, Kobe, Japan*; SHUNJI KASAHARA, *Molecular Photoscience Research Center, Kobe University, Kobe, Japan*.

Naphthalene is one of the simple polycyclic aromatic molecule, and it is interesting that the excited state dynamics take place. To understand the excited state dynamics, rotationally resolved fluorescence excitation spectra of several vibronic bands were measured.[a] [b] In this work, we have measured high-resolution fluorescence excitation spectra across a single mode laser and molecular beam at light angle. Vibronic bands, which lies 2866 cm^{-1} and 3068 cm^{-1} above the 0-0 band ($0^0{}_0$ + 2866 cm^{-1} band and $0^0{}_0$ + 3068 cm^{-1} band), were measured. Absolute wavenumber was calibrated with accuracy 0.0002 cm^{-1} by the measurement of Doppler-free absorption spectrum of I_2 molecule and transmitting light intensity of the stabilized etalon. Rotational lines of the $0^0{}_0$ + 2866 cm^{-1} band were almost resolved. A part of the rotational lines were assigned, and several energy shifts were found. On the other hand, rotational lines were not completely resolved for the $0^0{}_0$ + 3068 cm^{-1} band.

[a] K. Yoshida, Y. Semba, S. Kasahara, T. Yamanaka, and M. Baba, *J. Chem. Phys.* **130,** 19304 (2009)
[b] H. Katô, M. Baba, and S. Kasahara, *Bull. Chem. Soc. Jpn.* **80,** 456 (2007)

FA10 **11:20 – 11:35**

HIGH-RESOLUTION LASER SPECTROSCOPY OF THE $S_1 \leftarrow S_0$ TRANSITION OF Cl-NAPHTHALENES

SHUNJI KASAHARA, *Molecular Photoscience Research Center, Kobe University, Kobe, Japan*; RYO YAMAMOTO, *Graduate School of Science, Kobe University, Kobe, Japan.*

High-resolution fluorescence excitation spectra of the $S_1 \leftarrow S_0$ electronic transition have been observed for 1-Cl naphthalene (1-ClN) and 2-Cl naphthalene (2-ClN). Sub-Doppler excitation spectra were measured by crossing a single-mode UV laser beam perpendicular to a collimated molecular beam. The absolute wavenumber was calibrated with accuracy 0.0002 cm^{-1} by measurement of the Doppler-free saturation spectrum of iodine molecule and fringe pattern of the stabilized etalon. For 2-ClN, the rotationally resolved high-resolution spectra were obtained for the 0^0_0 and $0^0_0 + 1042$ cm^{-1} bands, and these molecular constants were determined in high accuracy. The obtained molecular constants of the 0^0_0 band are good agreement with the ones reported by Plusquellic *et. al.* [a] For the $0^0_0 + 1042$ cm^{-1} band, the local energy shifts were found. On the other hand, for 1-ClN, the rotational lines were not fully resolved because the fluorescence lifetime is shorter than the one of 2-ClN. Then we determined the molecular constants of 1-ClN from the comparison the observed spectrum with calculated one.

[a] D. F. Plusquellic, S. R. Davis, and F. Jahanmir, *J. Chem. Phys.*, **115**, 225 (2001).

FB. Spectroscopy as an analytical tool

Friday, June 26, 2015 – 8:30 AM

Room: 100 Noyes Laboratory

Chair: Christopher F. Neese, The Ohio State University, Columbus, OH, USA

FB01 **8:30 – 8:45**

CONTINUOUS MONITORING OF PHOTOLYSIS PRODUCTS BY THZ SPECTROSCOPY

ABDELAZIZ OMAR, ARNAUD CUISSET, GAËL MOURET, FRANCIS HINDLE, SOPHIE ELIET, ROBIN BOCQUET, *Laboratoire de Physico-Chimie de l'Atmosphère, Université du Littoral Côte d'Opale, Dunkerque, France.*

We demonstrate the potential of THz spectroscopy to monitor the real time evolution of the gas phase concentration of photolysis products and determine the kinetic reaction rate constant[a]. In the primary work, we have chosen to examine the photolysis of formaldehyde (H_2CO) [b]. Exposure of H_2CO to a UVB light (250 to 360 nm) in a single pass of 135 cm length cell leads to decomposition via two mechanisms: the radical channel with production of HCO and the molecular channel with production of CO. A commercial THz source [c] (frequency multiplication chain) operating in the range 600-900 GHz was used to detect and quantify the various chemical species as a function of time. Monitoring the concentrations of CO and H_2CO via rotational transitions, allowed the kinetic rate of H_2CO consummation to be obtained, and an estimation of the rate constants for both the molecular and radical photolysis mechanisms.

We have modified our experimental setup to increase the sensitivity of the spectrometer and changed sample preparation protocol specifically to quantify the HCO concentration. Acetaldehyde was used as the precursor for photolysis by UVC resulting in the decompositon mechanism can be described by:

$$CH_3CHO + h\nu \rightarrow CH_3 + HCO \rightarrow CH_4 + CO$$

Frequency modulation of the source and Zeeman modulation is used to achieve the high sensitivity required. Particular attention has been paid to the mercury photosensitization effect that allowed us to increase the HCO production enabling quantification of the monitored radical. We quantify the HCO radical and start a spectroscopic study of the line positions.

[a] H. M. Pickett and T. L. Boyd, Chem. Phys. Lett, Vol 58, 446-449, (1978)

[b] S. Eliet, A. Cuisset, M Guinet, F. Hindle, G. Mouret, R. Bocquet, and J. Demaison, Journal of Molecular Spectroscopy, Vol 279, 12-15 (2012).

[c] G. Mouret, M. Guinet, A. Cuisset, L. Croizé, S. Eliet, R. Bocquet and F. Hindle, Sensors Journal. IEEE, Vol 13, 133 – 138, (2013)

FB02 **8:47 – 9:02**

MEDIUM RESOLUTION CAVITY SPECTROSCOPY FOR THE STUDY OF LARGE MOLECULES

SATYAKUMAR NAGARAJAN, CHRISTOPHER F. NEESE, FRANK C. DE LUCIA, *Department of Physics, The Ohio State University, Columbus, OH, USA.*

It is well known that as molecules become larger the spectral lines of their high-resolution rotational spectra begin to merge, first into modest blends, then into clusters of many lines, and finally into continua. In addition to impacting specificity, the usual signal processing strategies used to separate spectral information from background become ineffective. Medium Resolution Cavity Spectroscopy trades the usual excess of specificity of rotational spectroscopy for a means of obtaining spectra of large molecules with congested or semi-continua spectra. The chief scientific question to be answered is how large (according the several definitions of 'large') can a molecule be and still have structure at medium resolution. The chief technical question to be answered is how to develop strategies to approach white noise sensitivity limits. Experimental details and results, and theoretical results will be presented.

FB03 **9:04 – 9:19**

SUBMILLIMETER/INFRARED DOUBLE RESONANCE: REGIMES FOR MOLECULAR SENSORS

SREE SRIKANTAIAH, *Department of Physics, The Ohio State University, Columbus, OH, USA*; IVAN MEDVEDEV, *Department of Physics, Wright State University, Dayton, OH, USA*; CHRISTOPHER F. NEESE, *Department of Physics, The Ohio State University, Columbus, OH, USA*; DANE PHILLIPS, *IERUS Technologies, Huntsville, AL, USA*; HENRY O. EVERITT, *Army Aviation and Missile Research Development and Engineering Center, Redstone Arsenal, AL, USA*; FRANK C. DE LUCIA, *Department of Physics, The Ohio State University, Columbus, OH, USA.*

Submillimeter/Infrared Double Resonance is a well-established technique. It has been used for spectroscopy, studies of collisional energy transfer, and diagnostics. The high level of molecule specific spectroscopic specificity achieved through this technique makes it an attractive candidate for sensor application. Here we will discuss its application to sensor development, with emphasis on regimes of applicability that range from mTorr to atmospheric pressure. System requirements and development as well as theoretical and experimental results will be discussed.

FB04 **9:21 – 9:36**

ROTATIONAL SPECTROSCOPY AS A TOOL TO INVESTIGATE INTERACTIONS BETWEEN VIBRATIONAL POLYADS IN SYMMETRIC TOP MOLECULES: LOW-LYING STATES $v_8 \leq 2$ OF METHYL CYANIDE

HOLGER S. P. MÜLLER, MATTHIAS H. ORDU, FRANK LEWEN, *I. Physikalisches Institut, Universität zu Köln, Köln, Germany*; LINDA BROWN, BRIAN DROUIN, JOHN PEARSON, KEEYOON SUNG, *Jet Propulsion Laboratory, California Institute of Technology, Pasadena, CA, USA*; ISABELLE KLEINER, *Laboratoire Interuniversitaire des Systèmes Atmosphériques (LISA), CNRS et Universités Paris Est et Paris Diderot, Créteil, France*; ROBERT SAMS, *Chemical Physics, Pacific Northwest National Laboratory, Richland, WA, USA.*

Rotational and rovibrational spectra of methyl cyanide were recorded to analyze interactions in low-lying vibrational states and to construct line lists for radio astronomical observations as well as for infrared spectroscopic investigations of planetary atmospheres. The rotational spectra cover large portions of the 36–1627 GHz region.[a] In the infrared (IR), a spectrum was recorded for this study in the region of $2\nu_8$ around 717 cm^{-1} with assignments covering 684–765 cm^{-1}. Additional spectra in the ν_8 region were used to validate the analysis.

Using ν_8 data[b] as well as spectroscopic parameters for $v_4 = 1$, $v_7 = 1$, and $v_8 = 3$ from previous studies,[c] we analyzed rotational data involving $v = 0$, $v_8 = 1$, and $v_8 = 2$ up to high J and K quantum numbers. We analyzed a strong $\Delta v_8 = \pm 1$, $\Delta K = 0$, $\Delta l = \pm 3$ Fermi resonance between $v_8 = 1^{-1}$ and $v_8 = 2^{+2}$ at $K = 14$ and obtained preliminary results for two further Fermi resonances between $v_8 = 2$ and 3. We also found resonant $\Delta v_8 = \pm 1$, $\Delta K = \mp 2$, $\Delta l = \pm 1$ interactions between $v_8 = 1$ and 2 and present the first detailed analysis of such a resonance between $v_8 = 0$ and 1.

We discuss the impact of this analysis on the $v_8 = 1$ and 2 as well as on the axial $v = 0$ parameters and compare selected CH_3CN parameters with those of CH_3CCH and CH_3NC.

We evaluated transition dipole moments of ν_8, $2\nu_8 - \nu_8$, and $2\nu_8$ for remote sensing in the IR.

[a] Part of this work was carried out at the Jet Propulsion Laboratory under contract with the National Aeronautics and Space Administration.

[b] M. Koivusaari et al., *J. Mol. Spectrosc.* **152** (1992) 377–388.

[c] A.-M. Tolonen et al., *J. Mol. Spectrosc.* **160** (1993) 554–565.

FB05 **9:38 – 9:53**

VIBRATIONAL SUM FREQUENCY STUDY OF THE INFLUENCE OF WATER-IONIC LIQUID MIXTURES IN THE CO_2 ELECTROREDUCTION ON SILVER ELECTRODES

NATALIA GARCIA REY, DANA DLOTT, *Department of Chemistry, University of Illinois at Urbana-Champaign, Urbana, IL, USA.*

Understand the molecular dynamics on buried electrodes under electrochemical transformations is of significant interest. There is a big gap of knowledge in the CO_2 electroreduction mechanism due to the limitations to access and probe the liquid-metal interfaces [1,2]. Vibrational Sum Frequency Spectroscopy (VSFS) is a non-invasive and surface sensitive technique, with molecular level detection that can be used to probe electrochemical reactions occurring on the electrolyte-electrode interface [2]. We observed the CO_2 electroreduction to CO in ionic liquids (ILs) on poly Ag using VSFS synchronized with cyclic voltammetry. In order to follow the CO_2 reaction in situ on the ionic liquid-Ag interface; the CO, CO_2 and imidazolium vibrational modes (resonant SFS) were monitored as a function of potential. We identified at which potential the CO was produced and how the EMIM-BF_4 played an important role in the electron transfer to the CO_2, lowering the CO_2^- energy barrier. A new approach to reveal the double layer dynamics to the electrostatic environment is presented by the study of the nonresonant sum frequency intensity as a function of the applied potential. By this method, we studied the influence of water-ionic liquid mixtures in the CO_2 electroreduction on Ag electrode. We observed a shift to lower potentials in the CO_2 electroreduction in water-ILs electrolyte. Previous studies in gas diffusion fuel cells have shown the CO_2 electroreduction in a water-imidazolium–based ILs on Ag nanoparticles at lower overpotential [3]. Our VSFS study helps to understand the fundamental electrochemical mechanism, showing how the ILs structural transition influences the CO_2 electroreduction.

[1] Polyansky, D. E.; Electroreduction of Carbon Dioxide, 2014, Encyclopedia of Applied Electrochemistry, Springer New York, pag 431-437. [2] Bain, C. D.; J. Chem. Soc., Faraday Trans., 1995, 91, 1281. [3] Rosen, B. A. et al; Science, 2011, 334 (6056), 643. Rosen, B. A. et al.; J. electrochem. Soc., 2013, 160 (2), H138.

Intermission

FB06 **10:12 – 10:27**

ELUCIDATING THE COMPLEX LINESHAPES RESULTING FROM THE HIGHLY SENSITIVE, ION SELECTIVE, TECHNIQUE NICE-OHVMS

JAMES N. HODGES, BRIAN SILLER[a], *Department of Chemistry, University of Illinois at Urbana-Champaign, Urbana, IL, USA*; BENJAMIN J. McCALL, *Departments of Chemistry and Astronomy, University of Illinois at Urbana-Champaign, Urbana, IL, USA.*

The technique Noise Immune Cavity Enhanced Optical Heterodyne Velocity Modulation Spectroscopy, or NICE-OHVMS, has been used to great effect to precisely and accurately measure a variety of molecular ion transitions from species such as H_3^+, CH_5^+, HeH^+, and HCO^+, achieving MHz or in some cases sub-MHz uncertainty.[bc] It is a powerful technique, but a complete theoretical understanding of the complex NICE-OHVMS lineshape is needed to fully unlock its potential.

NICE-OHVMS is the direct result of the combination of the highly sensitive spectroscopic technique Noise Immune Cavity Enhanced Optical Heterodyne Molecular Spectroscopy(NICE-OHMS) with Velocity Modulation Spectroscopy(VMS), applying the most sensitive optical detection method with ion species selectivity.[d] The theoretical underpinnings of NICE-OHMS lineshapes are well established,[e] as are those of VMS.[f] This presentation is the logical extension of those two preceding bodies of work. Simulations of NICE-OHVMS lineshapes under a variety of conditions and fits of experimental data to the model are presented. The significance and accuracy of the various inferred parameters, along with the prospect of using them to extract additional information from observed transitions, are discussed.

[a]Present Address: Tiger Optics, LLC, Warrington, PA 18976, USA
[b]J. N. Hodges, *et al. J. Chem. Phys.* (2013), **139**, 164201.
[c]A. J. Perry, *et al. J. Chem. Phys.* (2014), **141**, 101101.
[d]K. N. Crabtree, *et al. Chem. Phys. Lett.* (2012), **551**, 1-6.
[e]F. M. Schmidt, *et al. J. Opt. Soc. Amer. A* (2008), **24**, 1392–1405.
[f]J. W. Farley, *J. Chem. Phys.* (1991), **95**, 5590–5602.

FB07 **10:29 – 10:44**

CHARACTERIZATION AND INFRARED EMISSION SPECTROSCOPY OF BALL PLASMOID DISCHARGES

SCOTT E. DUBOWSKY, *Department of Chemistry, University of Illinois at Urbana-Champaign, Urbana, IL, USA*; BENJAMIN J. McCALL, *Departments of Chemistry and Astronomy, University of Illinois at Urbana-Champaign, Urbana, IL, USA.*

Plasmas at atmospheric pressure serve many purposes, from ionization sources for ambient mass spectrometry (AMS) to plasma-assisted wound healing. Of the many naturally occurring ambient plasmas, ball lightning is one of the least understood; there is currently no solid explanation in the literature for the formation and lifetime of natural ball lightning. With the first measurements of naturally occurring ball lightning being reported last year,[a] we have worked to replicate the natural phenomenon in order to elucidate the physical and chemical processes by which the plasma is sustained at ambient conditions.

We are able to generate ball-shaped plasmoids (self-sustaining plasmas) that are analogous to natural ball lightning using a high-voltage, high-current, pulsed DC system.[b] Improvements to the discharge electronics used in our laboratory and characterization of the plasmoids that are generated from this system will be described. Infrared emission spectroscopy of these plasmoids reveals emission from water and hydroxyl radical – fitting methods for these molecular species in the complex experimental spectra will be presented. Rotational temperatures for the stretching and bending modes of H_2O along with that of OH will be presented, and the non-equilibrium nature of the plasmoid will be discussed in this context.

[a]Cen, J.; Yuan, P.; Xue, S. *Phys. Rev. Lett.* **2014**, *112*, 035001.

[b]Dubowsky, S.E.; Friday, D.M.; Peters, K.C.; Zhao, Z.; Perry, R.H.; McCall, B.J. *Int. J. Mass Spectrom.* **2015**, *376*, 39-45.

FB08 **10:46 – 11:01**

VUV FLUORESCENCE OF WATER & AMMONIA FOR SATELLITE THRUSTER PLUME CHARACTERIZATION.

JUSTIN W. YOUNG, CHRISTOPHER ANNESLEY, RYAN S BOOTH, JAIME A. STEARNS, *Space Vehicles Directorate, Air Force Research Lab, Kirtland AFB, NM, USA.*

A quantified description of photoemission from thruster plume species, such as water and ammonia, is necessary for complete characterization of a thruster plume. Photoemission in a plume is due to excitation of molecular species from solar photons. For instance, electronic excitation of water with Lyman-alpha (121.6 nm) causes dissociation to the OH radical by following one of several possible pathways. One pathway leads to an electronically excited OH radical which fluoresces near 300 nm. Here, four-wave mixing is used to generate vacuum ultraviolet photons to excite a plume species seeded in a jet expansion. The resulting fluorescence is analyzed and used to describe the temperature dependence of the fluorescence signature.

FB09 **11:03 – 11:18**

REACTIONS OF 3-OXETANONE AT HIGH TEMPERATURES

EMILY WRIGHT, BRIAN WARNER, HANNAH FOREMAN, *Department of Chemistry, Marshall University, Huntington, WV, USA*; KIMBERLY N. URNESS, *Department of Mechanical Engineering, University of Colorado Boulder, Boulder, CO, USA*; LAURA R. McCUNN, *Department of Chemistry, Marshall University, Huntington, WV, USA.*

The pyrolysis of 3-oxetanone, $O(CH_2)_2CO$, has been studied in a resistively heated SiC tubular reactor at 400-1200°C. Products of pyrolysis were identified via matrix-isolation FTIR spectroscopy and photoionization mass spectrometry in separate experiments. While 3-oxetanone is expected to dissociate into ketene and formaldehyde, these experiments show that ethylene oxide and carbon monoxide are also produced. Methyl radical and ethylene were observed as additional products and are thought to be the result of reactions involving ethylene oxide.

FC. Comparing theory and experiment

Friday, June 26, 2015 – 8:30 AM

Room: B102 Chemical and Life Sciences

Chair: Edwin Sibert, The Univeristy of Wisconsin, Madison, WI, USA

FC01 — 8:30 – 8:45

VIBRATIONAL COUPLING IN SOLVATED FORM OF EIGEN PROTON

JHENG-WEI LI, KAITO TAKAHASHI, JER-LAI KUO, *Institute of Atomic and Molecular Sciences, Academia Sinica, Taipei, Taiwan.*

The most simple solvated proton, the hydronium ion H_3O^+ has been studied experimentally in its bare case as well as with the messenger techniques. Recent studies have shown that features in the vibrational spectra can be modulated not only by the different messengers, but also by the number of messengers. Theoretical molecular dynamics simulations have shed some light on the $H_3O^+(H_2)_n$ clusters, but understanding on the effect of microsolvaton by the messengers toward the spectra is still far from complete. We compare the experimental $H_3O^+Ar_m$ m=1-3 spectra with accurate theoretical simulations and oobtain the peak position and absorption intensity by solving the quantum vibrational Schrodinger equation using the potential and dipole moment obtained from DFT methods. One of the main goals of the study is to glean into the vibrational couplings induced by the microsolvation by the argon on the spectra region of 1500-3800 cm^{-1}, and to provide assignment on the peaks observed in these regions.

FC02 — 8:47 – 9:02

BINDING BETWEEN NOBEL GAS ATOMS AND PROTONATED WATER MONOMER AND DIMER

YING-CHENG LI, JER-LAI KUO, *Institute of Atomic and Molecular Sciences, Academia Sinica, Taipei, Taiwan.*

H_3O^+ and $H_5O_2^+$, Eigen and Zundel forms of the excess proton, are the basic moieties of hydrated proton in aqueous media. Using vibrational pre-dissotion spectra, vibrational spectra of messenager-tagged species are often measured; however, only neat species have been studied in detail by theoretical and computational means. To bridge this gap, we carry out extensive CCSD(T)/aug-cc-pvTZ calculations to investigate the binding between commonly used noble gas (NG) messenagers (He, Ne and Ar) with H_3O^+ and $H_5O_2^+$ to get an accurate estimate on the binding energy which yields the upper limits of vibrational temperature of NG-tagged clusters. The binding sites of NG and low-lying transition states have also been searched to give a better description on the energy landscape. In addition, a few exchange/correlation functionals have been tested to access the accuracy of these methods for future and more sophisticated theoretical studies.

FC03 — 9:04 – 9:19

ANALYSIS OF HYDROGEN BONDING IN THE OH STRETCH REGION OF PROTONATED WATER CLUSTERS

LAURA C. DZUGAN, ANNE B McCOY, *Department of Chemistry and Biochemistry, The Ohio State University, Columbus, OH, USA.*

There are two types of bands in the OH stretch region of the vibrational spectra of hydrogen-bonded complexes; narrow peaks due to isolated OH stretches and a broadened feature reflecting the OH stretches involved in strong hydrogen bonding. This second region can be as wide as several hundred wavenumbers and is shifted to the red of the narrow peaks. In this work we focus on $H^+(H_2O)_n$, where n = 3 or 4.[a] Both of these systems exhibit a very intense, broad H-bonded band. This breadth arises from coupling between the OH stretches and the low frequency modes. To understand the broadening observed in the spectra, we have developed a computational scheme in which we sample displacement geometries from the equilibrium structure based on the ground state harmonic wavefunction.[b] Then we combine the harmonic spectra in the OH stretch region for each computed geometry to generate the spectrum for each protonated water structure. Based on the large anharmonicities at play in these modes, we extend the approach using second-order perturbation theory to solve the reduced-dimensional Hamiltonian that involves only the HOH bends and the OH stretches. This is done by expressing the normal modes used to expand the Hamiltonian as linear combinations of internal coordinates. In this talk we will describe the approach used for these anharmonic calculations and report preliminary results for these protonated water clusters.

[a] Relph, R. A.; Guasco, T. L.; Elliot, B. M.; Kamrath, M. Z.; McCoy, A. B.; Steele, R. P.; Schofield, D. P.; Jordan, K. D.; Viggiano, A. A.; Ferguson, E. E.; Johnson, M. A. *Science*, **2010**, *327(5963)*, 308-312.

[b] Johnson, C. J.; Dzugan, L. C.; Wolk, A. B.; Leavitt, C. M.; Fournier, J. A.; McCoy, A. B.; Johnson, M. A. *J. Phys. Chem. A*, **2014**, *118*, 7590-7597.

FC04 **9:21 – 9:36**

SEMIEXPERIMENTAL STRUCTURE OF THE NON-RIGID BF_2OH MOLECULE BY COMBINING HIGH RESOLUTION INFRARED SPECTROSCOPY AND AB INITIO CALCULATIONS.

NATALJA VOGT, JEAN DEMAISON, *Section of Chemical Information Systems, Universität Ulm, Ulm, Germany*; AGNES PERRIN, *LISA, CNRS, Universités Paris Est Créteil et Paris Diderot, Créteil, France*; HANS BÜRGER, *Anorganische Chemie, Bergische Universität Wuppertal, Wuppertal, Germany.*

In BF_2OH, difluoroboric acid, the OH group is the subject of a large amplitude torsion motion which induces a splitting in the rotational spectrum as well as in the high-resolution infrared spectrum. It is interesting to check whether it is still posible to determine a semiexperimental equilibrium structure for such a molecule. For this goal, the rotation-vibration interactions constants have been experimentally determined by analyzing all the fondamental bands. They have also been computed ab initio using two different levels of theory. The results of the analysis as well as the determination of the structure will be reported.

FC05 **9:38 – 9:48**

CONFORMATIONAL, VIBRATIONAL AND ELECTRONIC PROPERTIES OF C5H3XOS (X = H, F, Cl OR Br): HALOGEN AND SOLVENT EFFECTS

MUSTAFA SENYEL, *Department of Physics, Anadolu University, Eskisehir, Turkey*; GUNES ESMA, *Physics, Anadolu University, Eskisehir, TURKEY*; CEMAL PARLAK, *Physics, Dumlupinar University, Kutahya, TURKEY.*

The effects of halogen and solvent on the conformer, vibrational and electronic properties of thiophene-2-carbaldehyde (C5H4OS) and thiophene-2-carbonyl-halogens [C5H3XOS; X = F, Cl or Br] were investigated employing the DFT and TD-DFT methods. The B3LYP functional was used with the 6-31++G(d,p) basis set. Computations were focused on the two conformational isomers of the compounds in the gas phase and both in a non-polar solvent and in a polar solvent. The present work explores the effects of both the halogen and the medium on the conformational preference, geometrical parameter, dipole moment, vibrational spectra, UV spectrum and HOMO-LUMO orbital. The findings of this work can be useful to those systems involving changes in the conformations analogous to the compounds studied.

FC06 **9:50 – 10:05**

COMBINED EXPERIMENTAL AND THEORETICAL STUDIES ON THE VIBRATIONAL AND ELECTRONIC SPECTRA OF 5-QUINOLINECARBOXALDEHYDE

MUSTAFA KUMRU, MUSTAFA KOCADEMIR, TAYYIBE BARDAKCI, *Department of Physics, Fatih University, Istanbul, Turkey.*

Experimental and theoretical investigations have been performed on the structure, vibrational and electronic spectra of 5-quinolinecarboxaldehyde (5QC). The 4000-50 cm^{-1}region FT-IR and FT-Raman and the 190-1100 nm region UV–Vis spectra of 5QC were recorded at the room temperature. Structural and spectroscopic properties of the cis and trans conformers of 5QC were calculated by Hartree-Fock (HF) and B3LYP density functional methods using the 6-311++G(d,p) basis set. Although calculated B3LYP frequencies are found to be closer to the experimental frequencies than the HF calculation results, scaled frequencies of both HF and B3LYP levels are in good agreement with the experimental spectra. The time-dependent density functional theory (TDDFT) is also used to find excitation energies, absorption wavelength, oscillator strengths and HOMO and LUMO energies of the title molecule.

Keywords: FT-IR, FT-Raman, and UV-vis spectra; HF; DFT, HOMO-LUMO.

1. V. Kucuk, A. Altun, M. Kumru, Spectrochim. Acta Part A 85(2012)9298.
2. M. Kumru, V. Kucuk, T. Bardakci, Spectrochim. Acta Part A 90(2012)2834.
3. M. Kumru, V. Kucuk, M. Kocademir, Spectrochim. Acta Part A, 96 (2012) 242251.
4. M. Kumru, V. Kucuk, P. Akyürek, Spectrochim. Acta Part A, 113 (2013) 72–79.
5. M. Kumru, et al., Spectrochimica Acta Part A, 134 (2015) 81–89.

This work was supported by the Scientific Research Fund of Fatih University under the project number P50011001G (1457).

Intermission

FC07 **10:24 – 10:39**

COMBINED COMPUTATIONAL AND EXPERIMENTAL STUDIES OF THE DUAL FLUORESCENCE IN DIMETHYLAMINOBENZONITRILE (DMABN)

ANASTASIA EDSELL, STEVEN SHIPMAN, *Department of Chemistry, New College of Florida, Sarasota, FL, USA.*

The dual florescence of dimethylaminobenzonitrile (DMABN) has been investigated since the 1960s. Despite more than 50 years of previous research, the spatial configuration of the excited state causing the dual fluorescence is still controversial. We have performed excited state calculations of DMABN in a variety of solvents of varying hydrogen-bonding affinity and polarity using implicit solvation (COSMO-PCM) at the M06-HF level of theory, and we have also collected steady-state absorption and fluorescence spectra of DMABN in these solvents. Our experimental spectra are broadly consistent with previous work, and the computational results show a significant solvent dependence.

FC08 **10:41 – 10:56**

MODELING SPIN-ORBIT COUPLING IN THE HALOCARBENES

PHALGUN LOLUR, RICHARD DAWES, *Department of Chemistry, Missouri University of Science and Technology, Rolla, MO, USA*; SCOTT REID, SILVER NYAMBO, *Department of Chemistry, Marquette University, Milwaukee, WI, USA.*

Halocarbenes are organic reactive intermediates with a neutral divalent carbon atom that is covalently bonded with a halogen and another substituent. Being the smallest carbenes that exhibit closed shell ground states, they have contributed greatly to our understanding of the reactivity of singlet carbene species and the factors that contribute to singlet-triplet energy gaps. We report an analysis of spin-orbit coupling in the mono-halocarbenes, CH(D)X, where X = Cl, Br, I. Single Vibronic Level (SVL) emission spectroscopy and Stimulated Emission Pumping (SEP) spectroscopy have been used to probe the ground vibrational level structures in these carbenes which have indicated the presence of perturbations involving the low-lying triplet state. In this talk, we present two approaches to model these interactions. Anharmonic constants, singlet-triplet gaps and geometry-dependent spin-orbit (SO) coupling surfaces were computed using high-level explicitly correlated methods such as CCSD(T)-F12b and MRCI-F12. These were used to evaluate SO coupling matrix elements and hence predict/fit mixed-perturbed singlet-triplet experimental levels. Results are also compared to those from a simpler model using a geometry-independent SO-constant.

FC09 **10:58 – 11:13**

GAS-PHASE CONFORMATIONS AND ENERGETICS OF PROTONATED 2′-DEOXYADENOSINE-5′-MONOPHOSPHATE AND ADENOSINE-5′-MONOPHOSPHATE: IRMPD ACTION SPECTROSCOPY AND THEORETICAL STUDIES

RANRAN WU, Y-W NEI, CHENCHEN HE, LUCAS HAMLOW, *Department of Chemistry, Wayne State University, Detroit, MI, USA*; GIEL BERDEN, J. OOMENS, *Institute for Molecules and Materials (IMM), Radboud University Nijmegen, Nijmegen, Netherlands*; M T RODGERS, *Department of Chemistry, Wayne State University, Detroit, MI, USA*.

Nature uses protonation to alter the structures and reactivities of molecules to facilitate various biological functions and chemical transformations. For example, in nucleobase repair and salvage processes, protonation facilitates nucleobase removal by lowering the activation barrier for glycosidic bond cleavage. Systematic studies of the structures of protonated 2’-deoxyribonucleotides and ribonucleotides may provide insight into the roles protonation plays in altering the nucleobase orientation relative to the glycosidic bond and sugar puckering. In this study, infrared multiple photon dissociation (IRMPD) action spectroscopy experiments in conjunction with electronic structure calculations are performed to probe the effects of protonation on the structures and stabilities of 2′-deoxyadenosine-5′-monophosphate (pdAdo) and adenosine-5′-monophosphate (pAdo). Photodissociation as a function of IR wavelength is measured to generate the IRMPD action spectra. Geometry optimizations and frequency analyses performed at the B3LYP/6-311+G(d,p) level of theory are used to characterize the stable low-energy structures and to generate their linear IR spectra. Single point energy calculations performed at the B3LYP/6-311+G(2d,2p) and MP2(full)/6-311+G(2d,2p) levels of theory provide relative stabilities of the optimized conformations. The structures accessed in the experiments are determined by comparing the calculated linear IR spectra for the stable low-energy conformers computed to the measured IRMPD action spectra. The effects of the 2′-hydroxyl moiety are elucidated by comparing the structures and IRMPD spectra of $[pAdo+H]^+$ to those of its DNA analogue. Comparisons are also made to the deprotonated forms of these nucleotides and the protonated forms of the analogous nucleosides to elucidate the effects of protonation and the phosphate group on the structures.

FD. Atmospheric science

Friday, June 26, 2015 – 8:30 AM

Room: 274 Medical Sciences Building

Chair: Kyle N Crabtree, University of California, Davis, Davis, CA, USA

FD01 **8:30 – 8:45**

OBSERVATION OF THE SIMPLEST CRIEGEE INTERMEDIATE CH_2OO IN THE GAS-PHASE OZONOLYSIS OF ETHYLENE

CARRIE WOMACK, *Department of Chemistry, MIT, Cambridge, MA, USA*; MARIE-ALINE MARTIN-DRUMEL, *Spectroscopy Lab, Harvard-Smithsonian Center for Astrophysics, Cambridge, MA, USA*; GORDON G BROWN, *Department of Science and Mathematics, Coker College, Hartsville, SC, USA*; ROBERT W FIELD, *Department of Chemistry, MIT, Cambridge, MA, USA*; MICHAEL C McCARTHY, *Atomic and Molecular Physics, Harvard-Smithsonian Center for Astrophysics, Cambridge, MA, USA.*

Criegee intermediates (R_1R_2COO) are understood to be critical intermediates in the ozonolysis of alkenes, but their high reactivity has traditionally made them very difficult to study directly. Although the smallest Criegee intermediates have now been generated in the laboratory using a diiodomethane photolysis scheme, numerous questions still remain about the product branching ratios of Criegee intermediates formed directly from ozonolysis. This talk will discuss our recent detection of the simplest Criegee intermediate, CH_2OO, in the ozonolysis of ethylene, using Fourier transform microwave spectroscopy and a modified pulsed nozzle. Nine other product species of the reaction were also detected, in abundances that qualitatively support the published mechanisms and rate constants.

FD02 **8:47 – 9:02**

HIGH-RESOLUTION SPECTRA OF CH_2OO : ASSIGNMENTS OF ν_5 AND $2\nu_9$ BANDS AND OVERLAPPED BANDS OF ICH_2OO

YU-HSUAN HUANG, LI-WEI CHEN, YUAN-PERN LEE, *Applied Chemistry, National Chiao Tung University, Hsinchu, Taiwan.*

The simplest Criegee intermediate CH_2OO, important in atmospheric chemistry, has recently been detected with infrared (IR) absorption in the reaction of $CH_2I + O_2$.[a] We have recorded high-resolution infrared spectrum of CH_2OO with rotational lines partially resolved. In additional to derivation of some critical spectral parameters to confirm the previous assignments of ν_3 at 1434.1 cm^{-1}, ν_4 at 1285.7 cm^{-1}, ν_6 at 909.2 cm^{-1}, and ν_8 at 847.4 cm^{-1}, the high-resolution spectra enable us to assign with confidence the $2\nu_9$ at 1233.5 cm^{-1} and ν_5 at 1213.0 cm^{-1}. Observed vibrational wavenumbers, relative intensities, and rotational structures agree well with those predicted by high-level quantum calculations. Some additional hot bands and combination bands are also observed. We also recorded the IR spectrum of ICH_2OO under high-pressure conditions. Observed IR intensities and vibrational wavenumbers of 1233.8 (ν_4), 1221 (ν_5), 1087 (ν_6), and 923 (ν_7) cm^{-1} agree with those simulated according to theoretical predictions and those observed in solid p-H_2.[b] The ν_4 band of ICH_2OO interferes with the $2\nu_9$ band of CH_2OO even at pressure as low as 100 Torr. With direct detection of both CH_2OO and ICH_2OO, we determined the pressure dependence of the yield of CH_2OO. The yield of CH_2OO near one atmosphere is greater than previous reports.

[a] Y.-T. Su, Y.-H. Huang, H. A. Witek, and Y.-P. Lee, Science **340**, 174 (2013).
[b] Y.-F. Lee and Y.-P. Lee, Chem. Phys. DOI: 10.1080/00268976.2015.1012129

FD03 **9:04 – 9:19**

DIRECT INFRARED IDENTIFICATION OF THE CRIEGEE INTERMEDIATES *syn*- and *anti*-CH_3CHOO AND THEIR DISTINCT CONFORMATION-DEPENDENT REACTIVITY

HUI-YU LIN, YU-HSUAN HUANG, *Applied Chemistry, National Chiao Tung University, Hsinchu, Taiwan*; XIAOHONG WANG, JOEL BOWMAN, *Department of Chemistry, Emory University, Atlanta, GA, USA*; YOSHIFUMI NISHIMURA, HENRY A WITEK, YUAN-PERN LEE, *Applied Chemistry, National Chiao Tung University, Hsinchu, Taiwan.*

The Criegee intermediates are carbonyl oxides that play critical roles in ozonolysis of alkenes in the atmosphere. Su et al. reported the mid-infrared spectrum of the simplest Criegee intermediate CH_2OO.[a] Methyl substitution of CH_2OO produces two conformers of CH_3CHOO and consequently complicates the infrared spectrum. We report the transient infrared spectrum of both *syn*- and *anti*-CH_3CHOO, produced from $CH_3CHI + O_2$ in a flow reactor, using a step-scan Fourier-transform spectrometer. Guided and supported by high-level full-dimensional quantum calculations, rotational contours of the four observed bands are simulated successfully and provide definitive identification of both conformers. Although nearly all observed bands of *anti*-CH_3CHOO overlapped with *syn*-CH_3CHOO, the Q-branch of ν_8 near 1090.6 cm^{-1} is contributed solely by *syn*-CH_3CHOO, and that of ν_7 near 1280.8 cm^{-1} is also dominated by *syn*-CH_3CHOO. Furthermore, *anti*-CH_3CHOO shows a reactivity greater than *syn*-CH_3CHOO toward NO/NO_2; at the later period of reaction, the spectrum can be simulated with only *syn*-CH_3CHOO. Without NO/NO_2, *anti*-CH_3CHOO also decays much faster than *syn*-CH_3CHOO. The direct infrared detection of *syn*- and *anti*-CH_3CHOO should prove useful for field measurements and laboratory investigations of the Criegee mechanism.

[a] Y.-T. Su, Y.-H. Huang, H. A. Witek, Y.-P. Lee, *Science* **340**, 174 (2013).

FD04 **9:21 – 9:36**

THE $\tilde{A}$-$\tilde{X}$ ELECTRONIC TRANSITIONS OF THE CH_2BrOO AND CH_2ClOO RADICALS IN THE NEAR INFRARED REGION

NEAL KLINE, MENG HUANG, TERRY A. MILLER, *Department of Chemistry and Biochemistry, The Ohio State University, Columbus, OH, USA.*

Moderate resolution cavity ring-down spectroscopy(CRDS) is used to obtain the $\tilde{A}$-$\tilde{X}$ electronic transition of the CH_2BrOO and CH_2ClOO radicals in the near-infrared region at room temperature. The CH_2BrOO radical was generated by 248nm excimer laser photolysis of a gas mixture of CH_2Br_2, O_2 and inert gas. The CH_2ClOO radical was generated similarly except for using CH_2ClI as the precursor. In both spectra, the first strong transition is located near 6800 cm^{-1}, and is assigned as the origin band. Several transitions are observed in the region between the origin and 9000 cm^{-1}. A strong vibrational transition is observed around 800 cm^{-1} to the blue of the origin and attributed to the OO stretch which is characteristic of the peroxy radical spectra. Our analysis of the vibrational structure is conducted using frequencies and Franck-Condon factors based on electronic structure calculations. Rotational structure analyses with ab-initio calculated rotational constants and dipole moments show good agreement with the contour of the origin band. Numerous transitions around the origin band in the CH_2BrOO radical spectrum can be explained by excitation from low-lying torsional levels in the $\tilde{X}$ state that are populated at room temperature.

FD05 **9:38 – 9:53**

THE $\tilde{A}$-$\tilde{X}$ ELECTRONIC TRANSITION OF CH_2IOO RADICAL IN THE NEAR INFRARED REGION

NEAL KLINE, MENG HUANG, TERRY A. MILLER, *Department of Chemistry and Biochemistry, The Ohio State University, Columbus, OH, USA*; PHALGUN LOLUR, RICHARD DAWES, *Department of Chemistry, Missouri University of Science and Technology, Rolla, MO, USA.*

In the past few years, the photolysis of CH_2I_2 in the presence of O_2 has received much attention. It has been shown to be an attractive method for producing the Criegee intermediate, CH_2O_2. Under certain conditions the reaction is also expected to produce the iodomethyl peroxy radical, CH_2IO_2. Interestingly both species are expected to have electronic transitions in the near infrared (NIR). The transition in CH_2O_2 would be analogous to the $\tilde{a} - \tilde{X}$ singlet-triplet transition in O_3 and a NIR $\tilde{A} - \tilde{X}$ transition in well-known to be characteristic of peroxy radicals. Notwithstanding the above, NIR spectra have not been reported for either CH_2O_2 or CH_2IO_2.

Based upon these considerations, we have performed the CH_2I_2 photolysis with O_2 in the optical cavity of our room temperature cavity ringdown spectrometer and have discovered a spectrum in the NIR. Our recorded spectrum stretches from a complex origin structure at $\approx$6800 $\mathrm{cm}^{-1}$to beyond 9000 cm^{-1}. Aside from the origin its strongest feature is a similar, complex band $\approx$870 $\mathrm{cm}^{-1}$to the blue of it, which is likely an O-O stretch vibrational transition, which is present in peroxy radicals but might also be expected for CH_2O_2. With the aid of high-level *ab initio* calculations (described in detail in the subsequent talk) we have undertaken the analysis of the spectrum. We find that a spectral analysis, including a number of hot bands arising from populated torsional levels, is consistent with the electronic structure calculations for the $\tilde{A}$ and $\tilde{X}$ states of CH_2IO_2.

FD06 **9:55 – 10:10**

A THEORETICAL CHARACTERIZATION OF ELECTRONIC STATES OF CH_2IOO AND CH_2OO RADICALS RELEVANT TO THE NEAR IR REGION

RICHARD DAWES, PHALGUN LOLUR, *Department of Chemistry, Missouri University of Science and Technology, Rolla, MO, USA*; MENG HUANG, NEAL KLINE, TERRY A. MILLER, *Department of Chemistry and Biochemistry, The Ohio State University, Columbus, OH, USA.*

Criegee intermediates (R_1R_2COO or CIs) arise from ozonolysis of biogenic and anthropogenic alkenes, which is an important process in the atmosphere. Recent breakthroughs in producing them in the gas phase have resulted in a flurry of experimental and theoretical studies. Producing the simplest CI (CH_2OO) in the lab via photolysis of CH_2I_2 in the presence of O_2 yields both CH_2OO and CH_2IOO with pressure dependent branching.

As discussed in the preceding talk, both species might be expected to have electronic transitions in the near IR (NIR). Here we discuss electronic structure calculations used to characterize the electronic states of both systems in the relevant energy range. Using explicitly-correlated multireference configuration interaction (MRCI-F12) and coupled-cluster (UCCSD(T)-F12b) calculations we were first able to exclude CH_2OO as the carrier of the observed NIR spectrum. Next, by computing frequencies and relaxed full torsional scans for the Ã and X̃ states, we were able to aid in analysis and assignment of the NIR spectrum attributed to CH_2IOO.

Intermission

FD07 **10:29 – 10:44**

JET-COOLED LASER-INDUCED FLUORESCENCE SPECTROSCOPY OF T-BUTOXY

NEIL J REILLY[a], *Department of Chemistry, University of Louisville, Louisville, KY, USA*; LAN CHENG, JOHN F. STANTON, *Department of Chemistry, The University of Texas, Austin, TX, USA*; TERRY A. MILLER, *Department of Chemistry and Biochemistry, The Ohio State University, Columbus, OH, USA*; JINJUN LIU, *Department of Chemistry, University of Louisville, Louisville, KY, USA*.

The vibrational structures of the $\tilde{A}^2A_1$ and $\tilde{X}^2E$ states of t-butoxy were obtained in jet-cooled laser-induced fluorescence (LIF) and dispersed fluorescence (DF) spectroscopic measurements. The observed transitions are assigned based on vibrational frequencies calculated using Complete Active Space Self-Consistent Field (CASSCF) method and the predicted Franck-Condon factors. The spin-orbit (SO) splitting was measured to be 35(5) cm^{-1} for the lowest vibrational level of the ground ($\tilde{X}^2E$) state and increases with increasing vibrational quantum number of the CO stretch mode. Vibronic analysis of the DF spectra suggests that Jahn-Teller (JT)-active modes of the ground-state t-butoxy radical are similar to those of methoxy and would be the same if methyl groups were replaced by hydrogen atoms. Coupled-cluster calculations show that electron delocalization, introduced by the substitution of hydrogens with methyl groups, reduces the electronic contribution of the SO splittings by only around ten percent, and a calculation on the vibronic levels based on quasidiabatic model Hamiltonian clearly attributes the relatively small SO splitting of the $\tilde{X}^2E$ state of t-butoxy mainly to stronger reduction of orbital angular momentum by the JT-active modes when compared to methoxy. The rotational and fine structure of the LIF transition to the first CO stretch overtone level of the $\tilde{A}^2A_1$ state has been simulated using a spectroscopic model first proposed for methoxy, yielding an accurate determination of the rotational constants of both $\tilde{A}$ and $\tilde{X}$ states.

[a]Current address: Department of Chemistry, Marquette University.

FD08 **10:46 – 11:01**

NITROSYL IODIDE, INO: MILLIMETER-WAVE SPECTROSCOPY GUIDED BY *AB INITIO* QUANTUM CHEMICAL COMPUTATION

STEPHANE BAILLEUX, DENIS DUFLOT, *Laboratoire PhLAM, Université de Lille - Sciences et Technologies, Villeneuve d'Ascq, France*; SHOHEI AIBA, HIROYUKI OZEKI, *Department of Environmental Science, Toho University, Funabashi, Japan*.

In the series of the nitrosyl halides, XNO (where X = F, Cl, Br, I), the millimeter-wave spectrum of INO remains so far unknown. We report our investigation on the first high-resolution rotational spectroscopy of nitrosyl iodide, INO.

One of the motivation for this work comes from the growing need in developing a more complete understanding of atmospheric chemistry, especially halogen and nitrogen oxides chemistry that adversely impacts ozone levels. In the family of the nitrogen oxyhalides such as nitrosyl (XNO), nitryl (XNO), nitrite (XONO), and nitrate ($X$$ONO_2$) halides, those with X = F, Cl, Br have been well studied, both theoretically and experimentally. However, relatively little is known about the iodine-containing analogues, although they also are of potential importance in tropospheric chemistry. In 1991, the Fourier-transform IR spectroscopic detection of INO, INO_2 and $IONO_2$ in the gas phase has been reported[a].

The INO molecule was generated by *in situ* mixing continuously I_2 and NO in a 50-cm long reaction glass tube whose outlet was connected to the absorption cell using a teflon tube. At the time of writing this abstract, 68 μ_a-type transitions ($K_a = 0 - 10$), all weak, have been successfully assigned. The hyperfine structures due to both I and N nuclei will also be presented.

S.B. and D.D. acknowledge support from the Laboratoire d'Excellence CaPPA (Chemical and Physical Properties of the Atmosphere) through contract ANR-10-LABX-005 of the Programme d'Investissement d'Avenir.

[a]I. Barnes, K. H. Becker and J. Starcke, J. Phys. Chem. 1991, 95, 9736-9740.

FD09 **11:03 – 11:18**

DISPERSED FLUORESCENCE SPECTROSCOPY OF JET-COOLED ISOBUTOXY, 2-METHYL-1-BUTOXY, AND ISOPENTOXY RADICALS

MD ASMAUL REZA, NEIL J REILLY[a], JAHANGIR ALAM, AMY MASON, JINJUN LIU, *Department of Chemistry, University of Louisville, Louisville, KY, USA.*

It is well known that rate constants of certain reactions of alkoxy radicals, e.g., unimolecular dissociation (decomposition by C-C bond fission) and isomerization via 1,5 H-shift, are highly sensitive to the molecular structure. In the present and the next talks, we report dispersed fluorescence (DF) spectra of various alkoxy radicals obtained under supersonic jet-cooled conditions by pumping different vibronic bands of their $\tilde{B} \leftarrow \tilde{X}$ laser induced fluorescence (LIF) excitation spectra.[b,c,d] This talk focuses on the DF spectra of 2-methyl-1-propoxy (isobutoxy), 2-methyl-1-butoxy, and 3-methyl-1-butoxy (isopentoxy). In all cases, strong CO-stretch progressions were observed, as well as transitions to other vibrational levels, including low-frequency ones. Quantum chemical calculations were carried out to aid the assignment of the DF spectra. Franck-Condon factors were calculated using the ezSpectrum program.[e]

[a]Current address: Department of Chemistry, Marquette University
[b]Wu, Q.; Liang, G.; Zu, L.; Fang, W. *J. Phys. Chem A* **2012**, *116*, 3156-3162.
[c]Lin, J.; Wu, Q.; Liang, G.; Zu, L.; Fang, W. *RSC Adv.* **2012**, *2*, 583-589.
[d]Liang, G.; Liu , C.; Hao, H.; Zu, L.; Fang, W. *J. Phys. Chem. A* **2013**, *117*, 13229- 13235.
[e]V. Mozhayskiy and A. I. Krylov, http://iopenshell.usc.edu/

FD10 **11:20 – 11:35**

PHOTODISSOCIATION OF METHYL ISOTHIOCYANATE STUDIED USING CHIRPED PULSE UNIFORM FLOW SPECTROSCOPY

NUWANDI M ARIYASINGHA, LINDSAY N. ZACK, CHAMARA ABEYSEKERA, BAPTISTE JOALLAND, ARTHUR SUITS, *Department of Chemistry, Wayne State University, Detroit, MI, USA.*

Chirped-Pulse Fourier-transform microwave spectroscopy has been applied in a uniform supersonic flow (Chirped-pulse/Uniform flow, CPUF) to study the 193 nm photodissociation of methyl isothiocyanate (MITC). Several products (CH_3NC, NCS, H_2CS, HCN and HNC) were identified via their pure rotational spectra. Observation of CH_3NC and NCS are consistent with previous studies of this system, however it is the first detection of H_2CS and HCN/HNC. Branching ratios were obtained from these data and will be discussed.

FD11 **11:37 – 11:52**

DISPERSED FLUORESCENCE SPECTROSCOPY OF JET-COOLED METHYLCYCLOHEXOXY RADICALS

JAHANGIR ALAM, MD ASMAUL REZA, AMY MASON, JINJUN LIU, *Department of Chemistry, University of Louisville, Louisville, KY, USA.*

Vibrational structures of the nearly degenerate $\tilde{X}$ and $\tilde{A}$ states of all four positional isomers of the methylcyclohexoxy (MCHO) radicals were studied by jet-cooled dispersed fluorescence (DF) spectroscopy, which unravels the effect of methyl substitution at different positions on the six-membered ring. Experimentally observed vibronic transitions in the DF spectra were assigned based on vibrational frequencies from quantum chemical calculations and predicted Franck-Condon factors that take into account the Duschinsky rotation. DF spectra of 2-, 3-, and 4-MCHO radicals are dominated by CO-stretch progressions or the progressions of CO-stretch modes in combination with the excited vibrational modes. DF spectra of two lowest-energy conformers of the tertiary 1-MCHO radical, chair-axial and chair equatorial, are significantly different from each other and from those of the other three positional isomers. Strong C-CH_3 stretch progressions as well as progressions of its combination bands with the CO stretch modes or the excited modes were observed. Such differences between the isomers and the conformers can be explained by variation of geometry and symmetry of the electronic states of cyclohexoxy upon methyl substitution at different positions. DF study of MCHO provides direct measurement of the energy separation between the $\tilde{A}$ and $\tilde{X}$ states that are subject to the pseudo-Jahn-Teller effect.

FE. Small molecules

Friday, June 26, 2015 – 8:30 AM

Room: 217 Noyes Laboratory

Chair: Robert W Field, MIT, Cambridge, MA, USA

FE01 **8:30 – 8:45**

TOWARDS A GLOBAL FIT OF THE COMBINED MILLIMETER-WAVE AND HIGH RESOLUTION FTIR DATA FOR THE LOWEST EIGHT VIBRATIONAL STATES OF HYDRAZOIC ACID (HN_3)

BRENT K. AMBERGER, R. CLAUDE WOODS, BRIAN J. ESSELMAN, ROBERT J. McMAHON, *Department of Chemistry, University of Wisconsin, Madison, WI, USA.*

Hydrazoic acid (HN_3) is a near-prolate asymmetric top molecule which we have extensively studied in the millimeter-wave region. Having completed an R_e structure determination based on 14 isotopologues of HN_3, we have moved on to analyze the very complex rotational spectra for the first 7 vibrationally excited states, as well as the higher K levels of the ground vibrational state. The excited states include the 4 lowest (out of 6) fundamental modes (ν_5, ν_6, ν_4, and ν_3) and the 3 lowest combination and overtone states ($2\nu_5$, $2\nu_6$ and $\nu_5+\nu_6$). All of these states are totally symmetric (A') except for ν_6 and $\nu_5+\nu_6$, which are antisymmetric (A''). The ro-vibrational states are substantially more intermingled than in most molecules due to unusually wide rotational spacing in HN_3. This intermingling leads to a tangled web of perturbations connecting the various ro-vibrational states: a-type and b-type Coriolis interactions between ν_5 and ν_6, between ν_4 and ν_6, and between $2\nu_6$ or $2\nu_5$ and $\nu_5+\nu_6$, local Fermi resonance between ν_3 and $2\nu_6$, and a strong centrifugal distortion interaction between the ground state and ν_5. Fortunately, we have been able to make extensive use (in both assignment of spectra and fitting of spectroscopic parameters) of previously published high resolution FTIR data for the ν_5, ν_6, ν_4 and ν_3 bands and the pure rotational spectrum of the ground vibrational state.[a,b,c,d] For the ground state, a-type R-branches from $K = 0$ to $K = 9$ and $J = 9$ through $J = 19$ and b-type transitions with $K = 0$ through $K = 2$ and J values up to 56 have been assigned. The datasets for most other states are similarly extensive. Combined millimeter-wave/FTIR multi-state fits have been performed using Pickett's SPFIT program.

[a] J. Bendtsen, F. Hegelund and F. M. Nicolaisen, *J. Mol. Spectrosc.* **118**, 121 (1986)

[b] J. Bendtsen and F. M. Nicolaisen, *J. Mol. Spectrosc.* **119**, 456 (1986)

[c] J. Bendtsen and F. M. Nicolaisen, *J. Mol. Spectrosc.* **124**, 306 (1987)

[d] J. Bendtsen and F. M. Nicolaisen, *J. Mol. Spectrosc.* **152**, 101 (1992)

FE02 **8:47 – 9:02**

MILLIMETER-WAVE SPECTROSCOPY AND GLOBAL ANALYSIS OF THE LOWEST EIGHT VIBRATIONAL STATES OF DEUTERATED HYDRAZOIC ACID (DN_3)

BRENT K. AMBERGER, R. CLAUDE WOODS, BRIAN J. ESSELMAN, ROBERT J. McMAHON, *Department of Chemistry, University of Wisconsin, Madison, WI, USA.*

Hydrazoic acid (HN_3) and DN_3 have qualitatively different rotational spectra, owing in large part to a substantial difference in their *A* rotational constants (345 GHz for DN_3 *vs* 611 GHz for HN_3). Like HN_3, DN_3 has six fundamental vibrational modes, of which four are visible in our millimeter-wave spectra at room temperature. Between 240 and 450 GHz, many pure rotational transitions for the ground vibrational state, ν_5 (496 cm^{-1}), ν_6 (586 cm^{-1}), ν_4 (955 cm^{-1}), ν_3 (1197 cm^{-1}), the first overtones of ν_5 and ν_6, and the combination ν_5+ν_6 have been observed and assigned. Because DN_3 is a light molecule, the rotational energy levels are widely spaced, leading to numerous interactions between rotational states of different vibrational modes. We have drawn on a wealth of previously published ro-vibrational data from high resolution FTIR spectra[a,b,c,d] in our efforts to understand these perturbations. The centrifugal distortion interaction between ν_5 and the ground state of DN_3 is less dramatic than in HN_3 but still significant. DN_3 shows the same set of Coriolis interactions as does HN_3, but again, their magnitude is generally smaller. In DN_3 the ν_4 state is at slightly lower energy than $2\nu_5$, instead of being nearly degenerate with ν_5+ν_6 as is the case for HN_3. Therefore, there are strong local interactions between $2\nu_5$ and ν_4, as well as between ν_3 and $2\nu_6$. A notable advantage in solving the DN_3 problem compared to HN_3 is the substantial increase in the number and diversity of observable *b*-type lines in our frequency region. Furthermore, the smaller *A* value permits higher *K* states to be observed due to a more gradual decrease in state populations. Ground state observations have been extended through K = 11 and through J = 50. Pickett's SPFIT has been employed to carry out multi-state fits using combined datasets of our millimeter-wave data and the published FTIR data.

[a] J. Bendtsen and F. M. Nicolaisen, *J. Mol. Spectrosc.* **125**, 14 (1987)
[b] J. Bendtsen, F. Hegelund and F. M. Nicolaisen, *J. Mol. Spectrosc.* **128**, 309 (1988)
[c] J. Bendtsen and F. M. Nicolaisen, *J. Mol. Spectrosc.* **145**, 123 (1991)
[d] C. S. Hansen, J. Bendtsen and F. M. Nicolaisen, *J. Mol. Spectrosc.* **175**, 239 (1996)

FE03 **9:04 – 9:19**

SIMPLIFIED CARTESIAN BASIS MODEL FOR INTRAPOLYAD EMISSION INTENSITIES IN THE $\tilde{A} \rightarrow \tilde{X}$ BENT-TO-LINEAR TRANSITION OF ACETYLENE

BARRATT PARK, *Department of Chemistry, MIT, Cambridge, MA, USA*; ADAM H. STEEVES, *Chemistry, Ithaca College, Ithaca, NY, USA*; JOSHUA H BARABAN, *Department of Chemistry, University of Colorado, Boulder, CO, USA*; ROBERT W FIELD, *Department of Chemistry, MIT, Cambridge, MA, USA.*

The acetylene emission spectrum from the *trans*-bent electronically excited $\tilde{A}$ state to the linear ground electronic $\tilde{X}$ state has attracted considerable attention because it grants Franck-Condon access to local bending vibrational levels of the $\tilde{X}$ state with large-amplitude motion along the acetylene $\rightleftharpoons$ vinylidene isomerization coordinate. For emission from the ground vibrational level of the $\tilde{A}$ state, there is a simplifying set of Franck-Condon propensity rules that gives rise to *only one* zero-order bright state per conserved vibrational polyad of the $\tilde{X}$ state. Unfortunately, when the upper level involves excitation in the highly admixed *ungerade* bending modes, ν_4' and ν_6', the simplifying Franck-Condon propensity rule breaks down—so long as the usual polar basis (with v and l quantum numbers) is used to describe the degenerate bending vibrations of the $\tilde{X}$ state—and the intrapolyad intensities result from complicated interference patterns between many zero-order bright states. We show that when the degenerate bending levels are instead treated in the Cartesian two-dimensional harmonic oscillator basis (with v_x and v_y quantum numbers), the propensity for *only one* zero-order bright state (in the Cartesian basis) is *restored*, and the intrapolyad intensities are simple to model, so long as corrections are made for anharmonic interactions. As a result of $trans \rightleftharpoons cis$ isomerization in the $\tilde{A}$ state, intrapolyad emission patterns from overtones of ν_4' and ν_6' evolve as quanta of *trans* bend (ν_3') are added, so the emission intensities are not only relevant to the ground-state acetylene $\rightleftharpoons$ vinylidene isomerization—they are also a direct reporter of isomerization in the electronically-excited state.

FE04 **9:21 – 9:36**

OBSERVATION OF LEVEL-SPECIFIC PREDISSOCIATION RATES IN S_1 ACETYLENE

CATHERINE A. SALADRIGAS, JUN JIANG, ROBERT W FIELD, *Department of Chemistry, MIT, Cambridge, MA, USA.*

A new spectroscopic scheme was used to gain insight into the predissociation mechanisms of the S_1 electronic state of acetylene in the 47000-47300 cm^{-1} region. To study this mechanism, H-atom action spectra of predissociative S_1 were recorded. Instead of detecting H-atom via REMPI, an H-atom fluorescence scheme was developed, in which the H-atom was excited to 3s and 3d levels and the fluorescence was detected. The signal-to-noise ratio of H-atom fluorescence-detected action spectra is superior to REMPI detected H-atom spectra. By comparing the LIF and H-atom spectra, there is direct evidence of level-dependent predissociation rates. Some of the line-widths observed in the H-atom spectra are broader than in the LIF spectra, confirming the triplet-mediated nature of S_1 acetylene.

FE05 **9:38 – 9:53**

FULL DIMENSIONAL ROVIBRATIONAL VARIATIONAL CALCULATIONS OF THE S_1 STATE OF C_2H_2

BRYAN CHANGALA, *JILA, National Institute of Standards and Technology and Univ. of Colorado Department of Physics, University of Colorado, Boulder, CO, USA*; JOSHUA H BARABAN, *Department of Chemistry, University of Colorado, Boulder, CO, USA*; JOHN F. STANTON, *Department of Chemistry, The University of Texas, Austin, TX, USA.*

Rovibrational variational calculations on global potential energy surfaces are often essential for investigating large amplitude vibrational motion and isomerization between multiple stable conformers, as well as for understanding the spectroscopic signatures of such dynamics. The efficient and accurate representation of high dimensional potential energy surfaces and the diagonalization of large rovibrational Hamiltonians make these calculations a technically non-trivial task.

The first excited singlet electronic state of acetylene (C_2H_2) is an ideal model isomerizing system. The S_1 state supports both a *trans* conformer and a higher energy *cis* conformer ($T_e^{cis} - T_e^{trans} \approx 2700$ cm^{-1}), separated by a planar near-half-linear transition state ($T_e^{TS} - T_e^{trans} \approx 5000$ cm^{-1}). The low-energy structure of the *trans* well is complicated by strong Coriolis and Darling-Dennison interactions between the near-resonant torsion and asymmetric bending modes. The resulting polyad patterns are eventually broken as the internal vibrational energy approaches that of the barrier to isomerization. In this region, qualitatively new spectroscopic patterns emerge, such as rotational K-staggering and vibrational effective frequency dips.

We examine these effects with an efficient *ab initio* variational treatment. Our global potential energy surface is constructed as a hybrid of a high-level reduced dimension surface, which excludes the two r_{CH} bond lengths, and a lower-level full dimensional surface incorporating the effects of r_{CH} displacement. Diagonalization of the large, sparse Hamiltonian, which contains an exact internal coordinate rovibrational kinetic energy operator, is achieved with an efficient restarted Lanczos algorithm that generates variational energies and wavefunctions. We discuss how our results elucidate the S_1 state's rich variety of spectroscopic features and the insights they provide into the isomerization process.

Intermission

FE06 **10:12 – 10:27**

MILLIMETER-WAVE SPECTROSCOPY OF FORMYL AZIDE ($HC(O)N_3$)

NICHOLAS A. WALTERS, BRENT K. AMBERGER, BRIAN J. ESSELMAN, R. CLAUDE WOODS, ROBERT J. McMAHON, *Department of Chemistry, University of Wisconsin, Madison, WI, USA.*

Formyl azide ($HC(O)N_3$) is a highly unstable molecule ($t_{1/2}\sim2$ hours at room temperature as a gas) that has only recently been studied spectroscopically by UV, IR, Raman and NMR methods.[ab] We have synthesized formyl azide and obtained its absorption spectrum at room temperature over the range 250-360 GHz. As in the case of carbonyl diazide,[c] two conformers are expected for $HC(O)N_3$, with the *syn*-isomer 2.8 kcal/mol lower in energy than the *anti*-isomer (CCSD(T)/ANO1). Calculations at the same level of theory and the same basis set predict the dipole moments for the *syn*-isomer (μ = 1.56 D) and *anti*-isomer (μ = 2.56 D). These calculations also indicate that *b*-type transitions should dominate the *syn*-isomer spectrum, while *a*-type transitions become more significant in the case of the *anti*-isomer. Despite the *anti*-isomer having a larger dipole moment, the *syn*-isomer still gives rise to all the dominant features of the spectrum. Thus far, five vibrational states (ν_9, ν_{12}, $2\nu_9$, $\nu_9 + \nu_{12}$, ν_{11}) have been studied for the *syn*-isomer, with the highest energy state ν_{11} = 582.6 cm^{-1}. Searches for the spectra of the *anti*-isomer are ongoing.

[a]Banert, K. et al. *Angew. Chem. Int. Ed.* **2012**, 51, 4718-4721
[b]Zeng, X. et al. *Angew. Chem. Int. Ed.* **2013**, 52, 3503-3506
[c]Amberger, B.K. et al. *J. Mol. Spectrosc.* **259**, (2014) 15-20

FE07 **10:29 – 10:44**

MILLIMETER-WAVE ROTATIONAL SPECTRUM OF DEUTERATED NITRIC ACID

REBECCA A.H. BUTLER, CAMREN COPLAN, *Department of Physics, Pittsburg State University, Pittsburg, KS, USA*; DOUG PETKIE, IVAN MEDVEDEV, *Department of Physics, Wright State University, Dayton, OH, USA*; FRANK C. DE LUCIA, *Department of Physics, The Ohio State University, Columbus, OH, USA.*

Previous studies of the pure rotational spectrum of deuterated nitric acid, DNO_3, have focused on the ground and first excited state, ν_9. This paper focuses on the next lowest energy vibrational states, covering the spectral range from 128-360 GHz. Two of them are unperturbed,ν_7 and ν_8, and two of them, ν_6 and $2\nu_9$ are highly perturbed. The unperturbed states are fit separately, while the two perturbed states are fit together using both Coriolis and Fermi interaction terms. Each state is fit to within experimental accuracy. We also extend the assignments and update the rotational constants for ν_9.

FE08 **10:46 – 11:01**

THEORETICAL ANALYSIS OF THE RESONANCE FOUR-WAVE MIXING AMPLITUDES: A FULLY NON-DEGENERATE CASE.

ALEXANDER KOUZOV, *Department of Physics, Saint-Petersburg State University, St. Petersburg, Russia.*

Degenerate (one-color) and two-color variants of the resonant four-wave mixing (RFWM) have developed into a sensitive and nonintrusive spectroscopic tool to study molecules in different gaseous environments. Yet, the fully non-degenerate (four-color, 4C) RFWM was scrutinized and implemented only for the Coherent AntiStokes Raman Scattering (CARS) excitation scheme[a],[b] . Here, by using the line-space approach[c], we analyze other 4C-RFWM schemes potentially interesting for the efficient up- and down-frequency conversion as well as for studies of molecular states. Decoupled expressions of the 4C-RFWM amplitudes are derived which allows to predict their polarization dependence.

[a]B. Attal-Trétout, P. Berlemont, and J. P. Taran, Mol. Phys. **70**, 1 (1990).
[b]J.P. Kuehner, S.V. Naik, W.D. Kulatilaka, N. Chai, N.M. Laurendeau, R.P. Lucht, M.O. Scully, S. Roy, A.K. Patnaik, and J.R. Gord, J. Chem. Phys. **128**, 174308 (2008).
[c]A. Kouzov and P. Radi, J. Chem. Phys. **140**, 194302 (2014).

FE09 **11:03 – 11:18**

SAND IN THE LABORATORY. PRODUCTION AND INTERROGATION OF GAS PHASE SILICATES[a].

DAMIAN L KOKKIN, TIMOTHY STEIMLE, *Department of Chemistry and Biochemistry, Arizona State University, Tempe, AZ, USA.*

Given its technological importance, the literature abounds with models for plasma enhanced chemical vapor deposition of the $SiH_4/O_2/Ar$ system. In a continuing effort to identify and characterize the optical spectra of Si_3 generated in a SiH_4/Ar pulsed discharge source[b], we detected, via two dimensional (2D) LIF, a relatively strong electronic transition in the 570-600 nm region that is strongly enhanced by the addition of a small amount of O_2. The excitation spectrum shows resolved band structure at the pulsed laser resolution of $0.5\,cm^{-1}$ and exhibits a radiative lifetime of $1.97\,\mu s$. The dispersed fluorescence exhibits three vibrational progressions and an unusually small splitting of approximately $50\,cm^{-1}$. Here we report on efforts to identify the molecular carrier of these bands, with particular interest paid to species resulting from oxygen impurities in the silane discharge.

[a]NSF CHE-1265885

[b]The electronic spectrum of Si_3 I: the triplet D_{3h} system" Reilly, N. J.; Kokkin, D. L.; Zhuang, X.; Gupta, V.; Nagarajan, R.; Fortenberry, R. C.; Maier, J. P.; Steimle, T. C.; Stanton, J. F.; McCarthy, M. C., J. Chem. Phys. 136(19), 194307, 2012.

FE10 *Post-Deadline Abstract* **11:20 – 11:35**

IMPACT OF COMPLEX-VALUED ENERGY FUNCTION SINGULARITIES ON THE BEHAVIOUR OF RAYLEIGH-SCHRÖDINGER PERTURBATION SERIES. H_2CO MOLECULE VIBRATIONAL ENERGY SPECTRUM.

ANDREY DUCHKO[a], ALEXANDR BYKOV, *Molecular Spectroscopy, V.E. Zuev Institute of Atmospheric Optics, Tomsk, Russia.*

Nowadays the task of spectra processing is as relevant as ever in molecular spectroscopy. Nevertheless, existing techniques of vibrational energy levels and wave functions computation often come to a dead-lock. Application of standard quantum-mechanical approaches often faces inextricable difficulties. Variational method requires unimaginable computational performance. On the other hand perturbational approaches beat against divergent series. That's why this problem faces an urgent need in application of specific resummation techniques. In this research Rayleigh–Schrödinger perturbation theory is applied to vibrational energy levels calculation of excited vibrational states of H_2CO. It is known that perturbation series diverge in the case of anharmonic resonance coupling between vibrational states [1]. Nevertheless, application of advanced divergent series summation techniques makes it possible to calculate the value of energy with high precision (more than 10 true digits) even for highly excited states of the molecule [2]. For this purposes we have applied several summation techniques based on high-order Pade-Hermite approximations. Our research shows that series behaviour completely depends on the singularities of complex energy function inside unit circle. That's why choosing an approximation function modelling this singularities allows to calculate the sum of divergent series. Our calculations for formaldehyde molecule show that the efficiency of each summation technique depends on the resonant type.

REFERENCES

1. J. Cizek, V. Spirko, and O. Bludsky, ON THE USE OF DIVERGENT SERIES IN VIBRATIONAL SPECTROSCOPY. TWO- AND THREE-DIMENSIONAL OSCILLATORS, J. Chem. Phys. 99, 7331 (1993).

2. A. V. Sergeev and D. Z. Goodson, SINGULARITY ANALYSIS OF FOURTH-ORDER MÖLLER-PLESSET PERTURBATION THEORY, J. Chem. Phys. 124, 4111 (2006).

[a]National Research Tomsk Polytechnik University, Physics and Technology Institute, Tomsk, Russia

AUTHOR INDEX

A

Abe, Masashi – TF02, TF03
Abeysekera, Chamara – MA04, FD10
Abramenkov, Alexander – RG08
Adam, Allan G. – TA03, TA07
Agarwal, Ankur – TE09
Agarwal, Jay – TH06
Ahmed, Musahid – WG05, RB05
Ahn, Ahreum – TD06, TD07
Aiba, Shohei – FD08
Akmeemana, Anuradha – MG07
Al-Refaie, Ahmed Faris – MH04, MH13
Alam, Jahangir – FD09, FD11
Aldridge, Leland M. – RJ03
Alekseev, E. A. – WF12
Allison, Thomas K – TF07
Allodi, Marco A. – TF10, TF11, TI09, TI10, RH09, RI03, RI14
Alonso, José L. – TG06, TG08
Amano, Takayoshi – MI04, RD05, RD06, RD07
Amberger, Brent K. – MG09, MG10, FE01, FE02, FE06
Amicangelo, Jay – RJ14
Amiot, Claude S. – MJ07
AMYAY, Badr – TB06
ANDERS, CATHERINE B – RC09
Anderson, David T. – RJ05
Annesley, Christopher – RG10, FB08
Antonov, Ivan – WG12
Appadoo, Dominique – TB07, WG09
Araki, Mitsunori – RI09
Arce, Héctor – WI14
Archer, Kaye A – WH14
Ard, Shaun – RA06
Ariyasingha, Nuwandi M – MA04, FD10
Armacost, Michael D. – TE09
Armentrout, Peter – RA06
Arnold, Sean – RC07
Arsenault, E. A. – MG05, RH13
Arunan, Elangannan – RB10
Asselin, Pierre – RB10
Asvany, Oskar – MI01, RB03, RJ01
Aviles Moreno, Juan-Ramon – RF14, RF15

B

Bagdonaite, Julija – MF01
Bailey, Jeremy – MH09
Bailey, Josiah R – TG14
Bailleux, Stephane – RD04, FD08
Bakker, Daniël – WG03, RB02
Balabanoff, Morgan E. – RJ05
Banerjee, Pujarini – WJ10, WJ11, WJ12
Bao, Xun – MI12, MI14
Baraban, Joshua H – TH11, TJ09, RD12, RG14, FE03, FE05
Bardakci, Tayyibe – FC06
Barnum, Timothy J – TJ06, TJ07, TJ09
Bartlett, Joshua – RA06, FA05
Basterretxea, Francisco J. – WA02, RF11
Baudhuin, Melissa A. – TA10, RA10
Baumann, Esther – WF07
Beale, Christopher A. – MH10
Bell, Aimee – WG10
Belloche, Arnaud – RI05, RI06
Belov, Sergey – WF12
Benidar, Abdessamad – MH11
Benke, Kristin – RE01
Benner, D. Chris – MH05, MJ11, RF02, RF08
Bensiradj, Nour el Houda – RH14
Berden, Giel – MI13, WG02, RB04, FC09
Bergin, Edwin – WI08
Bermúdez, Celina – TG06
Bernath, Peter F. – MH09, MH10, TB07, RD10
Betz, Thomas – MG11, TD02
Bevan, John W. – WJ04, WJ08
Bhatta, Ram – RG06
Białkowska-Jaworska, Ewa – RF11
Billinghurst, Brant E – TB03, TB07, RB06
Binns, Marshall – RA01
Bird, Ryan G – TG14
Bittner, Dror M. – RA04, RH04, RH05
Bjork, Bryce J – WF02
Blake, Geoffrey – TF10, TF11, TI09, TI10, WI13, RC01, RH08, RH09, RI03, RI14
Bocklitz, Sebastian – TD11
Bocquet, Robin – FB01
Boopalachandran, Praveenkumar – RA10
Booth, Ryan S – RG10, FB08
Booth, S. Tom – WI04
Böttcher, Fabian – TJ12
Boudon, Vincent – TB06, RE06
Bowen, K P – WG06
Bowen, Kit – MI15
Bowman, Joel – FD03
Brageot, Emily – TE03
Brathwaite, Antonio David – RA09
Bray, Caitlin – RF01, RF02, RH13
Brice, Joseph T. – TH05, WA01
Brinker, Katelyn R. – RC02
Brister, Matthew M – TI05
Broderick, Bernadette M. – TH06, RJ07, RJ09
Brooke, James S.A. – RD10
Brown, Gordon G – TE08, RC07, FD01
Brown, James – RG02
Brown, Linda – MH05, MJ11, MJ14, TG01, TG02, FB04
Bruneau, Yoann – TJ05
Brünken, Sandra – MI01, RB03, RJ01
Bucchino, Matthew – RA01
Buckingham, Grant – WG05
Bui, Thinh Quoc – RE05, RF01, RF02
Buma, Wybren Jan – RI04
Bürger, Hans – FC04
Burkart, Johannes – TF06
Burkhardt, Andrew – WI04
Burrmann, Nicola J. – RJ13
Butler, Rebecca A.H. – FE07
Bykov, Alexandr – FE10

C

Caballero-Mancebo, Elena – TD05
Cairncross, William – MF11
Caminati, Walther – RH06
Campbell, Ian – RJ14
Candian, Alessandra – RI04
Cao, Wenjin – MI10, MI11, RA11
Carlotti, Massimo – RF10
Carlson, Michaela – RE01
Carollo, Ryan – TJ05
Carrat, Vincent – TJ03
Carrington, Tucker – RG02, RG03
Carroll, Brandon – TF10, TF11, WI04, WI12, WI13, RC01, RH08, RH09, RI03, RI14
Carter, Jason P – WG09
Case, Amanda – TI12
Casey, Sean M. – TA10
Cernicharo, Jose – RD08, RI07
Chakraborty, Tapas – WJ10, WJ11, WJ12
Chan, MAN-CHOR – TA05
Chang, M.-C. Frank – TE03
Changala, Bryan – TH11, WF02, RD12, FE05
Chen, Li-Wei – FD02

Chen, Ming-Wei – TH09
Chen, Wang – MJ01
Chen, Wenting Wendy – MJ05
Chen, Yuning – TF07
Cheng, Cunfeng – WF05
Cheng, Lan – TH01, RA08, FD07
Cheng, Wang-Yau – TF12
Cheung, Allan S.C. – TA05
Chew, Kathryn – TI11
Cho, Young-Sang – MF08
Choi, Myong Yong – TD06, TD07
Choi, Yoon Jeong – MG05, RC04
Chow, C S – MI12
Christensen, Lance – RE05
Christopher, Casey – TA09
Cich, Matthew – WF08, WF09
Cocinero, Emilio J. – TD05, WA02, RF11
Codd, Terrance Joseph – TH09
Coddington, Ian – WF07
Cohen, Michael – WH08
Colvin, Sean – TD09
Cooke, S. A. – MG05, RC04, RH12
Coplan, Camren – FE07
Corby, Joanna F. – WI11
Cornell, Eric – MF11, MF12
Corral, Inés – TI04
Cossel, Kevin – MF11, MF12
Coudert, L. H. – WF11, WJ04, RF04
Cox, Richard M – RA06
Coy, Stephen – TJ08, TJ09
Crabtree, Kyle N – MG01, TE08
Craig, Norman C. – MG04, TC09, RC05
Craig, Stephanie – MI09
Craver, Barry – TE09
Crawford, Timothy J – MH05, MJ14, RD07
Crespo-Hernández, Carlos E. – TI04, TI05, TI06
Crim, Fleming – TI12
Crockett, Nathan – WI10, WI13
Crozet, Patrick – TF09
Cubel Liebisch, Tara – TJ12
Cueto, Maite – RD08
Cuisset, Arnaud – TB05, WG11, FB01
Cummins, Christopher – MJ06

D

Dagdigian, Paul – MF07
Dahlstrom, Julie – RI08
Dai, HAI-LUNG – TI14, RE08
Daily, John W – WG05, RB05
Daly, Adam M – TE03, TG01, TG02, TG03, TG06, RD07
Daly, Ryan W – TE01
Dantus, Marcos – TI02
Dattani, Nikesh S. – MF08, MF09, MF10, MJ07
Davis, Benjamin G. – WA02
Dawadi, Mahesh B. – RG06
Dawes, Richard – WA03, FC08, FD05, FD06
De Lucia, Frank C. – TE01, TE09, TG08, RI01, FB02, FB03, FE07
de Oliveira, Nelson – WG07, WG08, RD09
de Vries, Xander – RI14
Dean, Jacob C. – RG01
DeBlase, Andrew F – WH14
Delahaye, Thibault – RF05, FA08
Delcamp, Jared – WJ13
Delvin, Jack – TA04
Demaison, Jean – MG04, TG04, FC04
DePalma, Joseph W – MI08
Deprince, Bridget Alligood – WA04
Devasher, Rebecca – RC06
Devi, V. Malathy – MH05, MJ11, RF02, RF08
Dewberry, Chris – TE02, RF12, RF13, RH01, RH02
Diddams, Scott – WF07
DiScipio, Regina – TI06
Dixneuf, Sophie – RE09
Dlott, Dana – FB05
Dobrev, Georgi – TF09
Dollhopf, Niklaus M – WI12, WI13
Dolson, David A – TC06, RC09
Domenech, Jose Luis – RD08
Doney, Kirstin D – RG09
Dorris, Rachel E. – MG08
Douberly, Gary E. – TH04, TH05, TH06, WA01, WA06, RJ07, RJ08, RJ09
Doyle, John M. – WF02, FA06
Drouin, Brian – TE03, TG01, TG02, TG06, RD06, RD07, RF02, FB04
Dubowsky, Scott E. – FB07
Duchko, Andrey – FE10
DUCOURNEAU, Gaël – TB05
Duffy, Erin M. – MI06
Duflot, Denis – FD08
Dulick, Michael – MH09, MH10
Dumont, Elise – TI08
Duncan, Michael A – RA09, RE04, RG12, RG13
Dupré, Patrick – MJ10, WF10
Dzugan, Laura C. – FC03

E

Ecija, Patricia – TD05, WA02, RF11
Eden, J. Gary – MJ05, MJ08
Edsell, Anastasia – FC07
Eibl, Konrad – TG12
Eidelsberg, Michele – RD09
Eliet, Sophie – FB01
Elliott, Ben – RD07
Ellison, Barney – WG05, RB05
Emerson, N J – WI01
Endo, Yasuki – RH07
Endres, Christian – RB07, RB08
Erickson, Trevor J. – MJ06, RF09
Ernst, Wolfgang E. – FA02, FA07
Escribano, Rafael – RI12
Esma, Gunes – FC05
Esselman, Brian J. – MG09, MG10, FE01, FE02, FE06
Evans, Corey – MF13, WG09
Everitt, Henry O. – FB03
Ewing, Paul R. – TE09
Eyler, Edward E. – TJ05, RJ03

F

Fan, Lin – MI13
Faucheaux, Jacob A – RG05
Faye, Mbaye – TB06
Federman, Steven – WI05, RD09
Feifel, Raimund – WG03
Fermann, Martin – TF05, WF06
Fernández, José A. – WA02
Fernandez-Lopez, Manuel – WI14
Field, Robert W – MA04, MJ06, MJ13, TJ06, TJ07, TJ08, TJ09, RF09, FD01, FE03, FE04
Finneran, Ian A – TF10, TF11, TI09, TI10, RC01, RH08, RH09
Fischer, Theo – RJ10
Flagey, Nicolas – WI05
Flaud, Jean-Marie – TG04, RF10
Fleisher, Adam J. – WF01, WF04, RF03
Folkers, T W – WI01
Foltynowicz, Aleksandra – TF05, TF08
Forbes, D – WI01
Foreman, Hannah – FB09
Forthomme, Damien – MF07, WF09
Fortman, Sarah – TG08, RI01
Fournier, Gilles – WG11
Fournier, Joseph – MI08, MI09, TD09, WJ09
Frank, Derek S. – RH11, RH12
Freund, R W – WI01
Friedel, Douglas – WI11, WI15
Fu, Dejian – RF07
Fujita, Chiho – MH14
Fujiwara, Takashige – TI03

Fukushima, Masaru – MJ12, TH12

G

GABARD, Tony – TB06
Gaigeot, Marie-Pierre – RB02
Galica, Scott E. – RJ03
Gallagher, Tom – TJ02, TJ03, TJ04
Galvin, Thomas C. – MJ05
Gamache, Robert R. – MH06, MJ11
Gans, Berenger – WG07
Gao, Jiao – TG11
Gao, Juehan – MI12, MI14, RB04
Garand, Etienne – MI06, MI07, TD10
Garavelli, Marco – TI08
Garcia Rey, Natalia – FB05
Gardner, Adrian – FA04
Garrod, Robin T. – RI05
Gaster, Sydney A – TE08, RC07
Gato-Rivera, Beatriz – MF03
Gauss, Jürgen – TD01, RA08
Gavilan, Lisseth – RD09
Geballe, Thomas R. – WI07
Gellman, Samuel H. – TD08
Georges, Robert – MH11, TB01, RB10, RE06
Ghosh, Supriya – TD04
Gibson, Bradley M. – RE07
Giorgetta, Fabrizio – WF07
Gipson, Courtney N – MJ04
Giuliano, Barbara Michela – TD02, RH06
Giussani, Angelo – TI08
Gnanasekar, Sharon Priya – RB10
Goldshlag, William – MJ08
Goldsmith, Paul F – WI05
GOLUBIATNIKOV, G Yu – WF12
Gong, Justin Z – TI15
González, Leticia – TI04
Good, Jacob T – TF10, TF11
Gord, Joseph R. – MG13
Gordon, Iouli E – MH01, MH02, MH03, RD10
Gorlova, Olga – TD09
Goto, Miwa – WI07
Gottlieb, Carl A – RD12
Gou, Qian – RH06
Goubet, Manuel – RB10
Goudreau, E. S. – TA07, TB04
Gould, Phillip – TJ05
Grabow, Jens-Uwe – MA02, WF11
Grames, Ethan M – MJ04
Grant, Mikayla L. – MG07
Grau, Matt – MF11
Gray, Jeffrey A. – RC03
Green, Susan – TE02
Greene, C. H. – TJ12
Gresh, Dan – MF11, MF12
Grimes, David – TJ06, TJ07, TJ08, TJ09
Groenenboom, Gerrit – MJ09
Groner, Peter – MG04, TG01, TG02, TG04
Grubbs II, G. S. – RC02, RC10, RH12
Gruet, Sébastien – WG11, RB07, RB08
Gu, Qun Jane – TE03
Guillemin, J.-C. – RI02, RI06, RI07
Guillemin, R – WG06
Gupta, Harshal – WI09, WI10
Gurusinghe, Ranil M. – TG09, TG10
Gutle, C. – WF11

H

Habgood, Matthew – RH05
Haenni, Benjamin C. – MG10, TI12, RJ13
Halfen, DeWayne T – WI01, RA02
Hall, Gregory – MF07, TI13, WF08, WF09
Hall, Taylor M – TE08, RC07
Hamlow, Lucas – MI13, MI14, FC09
Hammer, Nathan I – MI15, WJ13
Harada, Kensuke – RH07
Hargreaves, Robert J. – MH09, MH10, TB07
Harms, Jack C – MJ04
Harris, Robert J – WI14
Hauser, Andreas W. – FA02, FA07
Havenith, Martina – MA03, RJ10
Hays, Brian – TE04, TE05, TH03, WA04
He, Chenchen – MI12, MI13, MI14, FC09
Heaven, Michael – TA06, RA06, FA03, FA04, FA05
Heays, Alan – WG08, RD09
Heckl, Oliver H – WF02
Heikal, Ahmed A – TI01
Heim, Zachary N. – MG09
Helal, Yaser H. – TE09
Hemberger, Patrick – RB05
Hemmers, O – WG06
Hepburn, John – TC02
Hermes, Matthew – RG04
Hernandez, Federico J – TH05, WA01
Herrero, Victor Jose – RD08, RI10, RI12
Hewett, Daniel M. – MG13, TH08, RG11
Hill, Christian – MH01, MH02, MH03
Hindle, Francis – TB05, FB01
Hirata, So – RG04, RG05
Hirota, Eizi – TH13, RH03
Hobbs, L. M. – RI08
Hodges, James N. – MF05, MF06, FB06
Hodges, Joseph – WF01, WF03, WF04, RF02, RF03, RF05
Hofferberth, Sebastian – TJ12
Holland, Daniel – TF10, TF11
Holt, Jennifer – TE01
Hoo, Jiajun – RF02
Hopkins, Scott – TC02, RB01
Horrocks, Benjamin R – RC08
Hougen, Jon T. – TG07, TG13, WF12
Houlahan, Jr., Thomas J. – MJ05
Hsu, Yen-Chu – TF12, WH02
Hu, Shui-Ming – WF05
Huang, Dao-Ling – RG15
Huang, Meng – TG05, FD04, FD05, FD06
Huang, Wenyuan – WJ02
Huang, Xinchuan – MH06
Huang, Yu-Hsuan – FD02, FD03
Huet, T. R. – WF11, RB10, RF14, RF15
Huff, Anna – RH02
Hull, Alexander W. – MJ06, RF09
Hunt, Katharine – MJ09

I

Ichino, Takatoshi – TH01
Ilyushin, V. – WF12
Indriolo, Nick – WI08
Inomata, Risa – WH05
Ioppolo, Sergio – RI03, RI14
Ishikawa, Haruki – WH05, WH06, WH12, WH13
Ishiwata, Takashi – MJ12, TH13
Ito, Kenji – WG07
Ivanov, Maxim – WJ07
Iwakuni, Kana – TF02

J

Jäger, Wolfgang – TC08, TD04, TG11, WJ02, WJ03
Jaiswal, Vishal K. – TI08
Jang, Heesu – MG03
Janik, Ireneusz – TH14
Jansen, Paul – TJ01
Jawad, Khadija M. – TH08
Jenkins II, Paul A. – MF05, MF06
Jensen, Per – MI02
Jian, Tian – RH15
Jiang, Jie – WF06
Jiang, Jun – MJ06, MJ13, RF09, FE04

Jiménez-Redondo, Miguel – RI10
Joalland, Baptiste – MA04, FD10
Johanssson, Alexandra C – TF05, TF08
Johnson, Britta – TH02
Johnson, Christopher J – MI08
Johnson, Mark – MI08, MI09, TD09, WH14, WJ09
Jordan, Kenneth D. – WH14
Joyeux, Denis – WG07, RD09
Jusko, Pavol – RB03

K

Ka, Soohyun – MG03
Kanamori, Hideto – RJ06
Kang, Hyuk – MI05
Kang, Justin M. – MG07
Kaniecki, Marie – TI02
Kannengießer, Raphaela – TG12
Karman, Tijs – MJ09
Kasahara, Shunji – TH13, FA09, FA10
Kasahara, Yasutoshi – WH06, WH12, WH13
Kassi, Samir – MH11, TF06, RE06
Kaufmann, Matin – RJ09, RJ10
Kawaguchi, Kentarou – MJ01
Kawasaki, Hiroyuki – RJ06
Kawasaki, Takayuki – WH05
Kawashima, Yoshiyuki – RH03
Keel, S C – WI01
Kelleher, Patrick J – MI08
Kelly, John T. – MI15
Kern, Jeffrey S. – WI15
Khan, Nazir D. – WJ05
Khodabakhsh, Amir – TF05, TF08
Kidwell, Nathanael – RG01
Kim, Jihyun – MG03
Kim, Jongjin B. – TH01
Kim, JungSoo – RA06
Kim, Rod M. – TE03
Kimutai, Bett – MI12
King, Adrian – RH05
Kingsley, J – WI01
Kisiel, Zbigniew – TB02, TG08, RF11
Kleinbach, Kathrin Sophie – TJ12
Kleiner, Isabelle – TG13, FB04
Klemperer, William – RH11
Kline, Neal – FD04, FD05, FD06
Klose, Andrew – WF07
Kluge, Lars – MI01
Knee, Joseph L. – RH13
Knezz, Stephanie N. – RJ13
Kobayashi, Kaori – MH07, MH14
Kocademir, Mustafa – FC06
Kochanov, Roman V – MH01, MH02, MH03
Kocheril, G. Stephen – MF05, MF06
Koeppen, Nicole – WI06, RE07
Kokkin, Damian L – TA01, TA02, TA04, RA07, FA06, FE09
Kolesniková, Lucie – TG08
Kolomenskii, Alexander – RF06
Konar, Arkaprabha – TI02
Konder, Ricarda M. – TA03
Kondo, Makoto – WH06
Korn, Joseph A. – TH08, RG11
Kouzov, Alexander – FE08
Kowzan, Grzegorz – TF05, WF06
Kozyryev, Ivan – FA06
Kregel, Steven J. – TD10
Krim, Lahouari – RI11
Krin, Anna – TE06, TE07
Krishnan, Mangala Sunder – TC10
Kubasik, Matthew A. – MG13
Kumari, Sudesh – MI11
Kumru, Mustafa – FC06
Kuo, Jer-Lai – FC01, FC02
Kurusu, Itaru – WH13
Kusaka, Ryoji – WH10, WH11
Kwabia Tchana, F. – TB06, TG04, RB09

L

Laane, Jaan – MG05
Laas, Jacob – WA04
Lach, Grzegorz – MF09
Lackner, Florian – FA02
Lafferty, Walter – TG04
Lambrides, E. – WI07
LAMPIN, Jean François – TB05
Langer, William D – WI05
Lapinov, Alexander – WF12
Le, Anh T. – TI13
Le Roy, Robert J. – MF08, MF09, MF10, MF13, MJ02, MJ03, MJ07, TC02, RJ04
Leavitt, Christopher M. – TH05, WA01
Lee, Chien-Chung – WF06
Lee, Chun-Woo – TJ10
Lee, Jeonghun – TJ04
Lee, Katherine – WI14
Lee, Kelvin – MG02
Lee, Kevin – TF05, WF06
Lee, Sang – TH07
Lee, Timothy – MH06
Lee, Yuan-Pern – FD02, FD03
Lees, Ronald M. – TB03
Lefebvre-Brion, H. – TJ13, TJ14
Legon, Anthony – WJ01, RA03, RA04, RA05, RH04
Lehmann, Kevin – MH12
Leicht, Daniel – RJ10
Lemaire, Jean Louis – RD09
Leonov, Igor I – WJ04
Leopold, Doreen – TA10, RA10
Leopold, Ken – TE02, RF12, RF13, RH01, RH02
Lesarri, Alberto – TD05, WA02
Lesslie, Michael – WG02
Leung, Helen O. – TC03, WJ05, WJ06
Leung, Tong – TC02
Lewen, Frank – RB07, RB08, RD02, RD03, FB04
Lewis, Brenton R – WG08
Li, Biu Wa – TA05
Li, Gang – RD10
Li, Jheng-Wei – FC01
Li, Ying-Cheng – FC02
Li Chun Fong, Lena C. M. – MF09
Liebermann, Hans P. – TJ13, TJ14
Lin, Hui-Yu – FD03
Lin, Ming-Fu – RE01
Lin, Wei – RH10, RH12
Lin, Yen-Wei – RJ02
Lin, Zhou – MI03
Lindle, D W – WG06
Linnartz, Harold – RG09
Linton, Colan – TA03, TA07
Liu, An-Wen – WF05
Liu, Jinjun – TI07, FD07, FD09, FD11
Liu, Qingnan – WF04
Liu, Tze-Wei – TF12
Liu, Xunchen – WJ02
Liyanage, Nalaka – WJ13
Lodi, Lorenzo – TA08
Lolur, Phalgun – FC08, FD05, FD06
Long, B. E. – MG05
Long, David A. – WF01, WF04, RF02
Looney, Leslie – WI11, WI14, WI15
Lopez, Gary – RH15
Louviot, Maud – RE06
Löw, Robert – TJ12
Lozovoy, Vadim V. – TI02
Lu, Xin – RH10
Lu, Yan – WF05
Lucchese, Robert R. – WJ04, WJ08
Lucht, Robert P. – RE02, RE03
Lunny, Elizabeth M – RF01, RF02
Lyons, James R – RD09

M

Ma, Jianqiang – TI14
MacDonald, Michael A – WG04
Mackenzie, Becca – TE02, RF12, RF13, RH01, RH02
Magnuson, Eric – TJ03
Mahé, Jérôme – RB02

Mai, Sebastian – TI04
Maltseva, Elena – RI04
Manceron, Laurent – TB06, TG04, RB09
Maner, Jon – RE04, RG13
Mani, Devendra – RJ10
Mantz, Arlan – MH05, MJ11, MJ14, TG01, TG02
Margulès, L. – RI02, RI06, RI07
Markus, Charles R. – MF05, MF06
Marlett, Melanie L. – MI03
Marquetand, Philipp – TI04
Marsh, Brett – MI06, MI07, TD10
Marshall, Frank E – RC02
Marshall, Mark D. – TC03, WJ05, WJ06
Martin-Drumel, Marie-Aline – MG01, TB02, TE08, RB07, RB08, RD12, FD01
Martínez-Fernández, Lara – TI04
Mascaritolo, Kyle – FA03, FA04
Mase, Takayuki – RH03
Maslowski, Piotr – TF05, WF06
Mason, Amy – FD09, FD11
Maté, Belén – RI10, RI12
Matsuba, Ayumi – TF04
Matsushima, Fusakazu – MH07, MI04
MATSUSHITA, Yuki – RI09
Matthews, Devin A. – TI15, RA08
Mauney, Daniel – RE04, RG13
Maxwell, Stephen E – RF05
McBurney, Carl – TD08
McCall, Benjamin J. – MF04, MF05, MF06, WI06, RE07, FB06, FB07
McCarthy, Michael C – MG01, MG02, TD01, TE08, RB07, RB08, RD12, FD01
McCaslin, Laura – RJ07
McColl, M – WI01
McCoy, Anne B – MI03, TG05, WH07, FC03
McCunn, Laura R. – FB09
McDonald II, David C – RG13
McElmurry, Blake A. – WJ04
McEwen, Bridget C. – RD11
McGuire, Brett A. – WI12, WI13, RI03, RI14
McKellar, Bob – WH01, RB06
McMahon, Robert J. – MG09, MG10, TI12, RJ13, FE01, FE02, FE06
McMahon, Terry – RB01
McMahon, Timothy J – TG14
McMillan, James P. – RI01
McNamara, Louis E. – WJ13
McNaughton, Don – WG09
McRaven, C. – WF08
Medcraft, Chris – MG11, TE07
Medvedev, Ivan – TG08, FB03, FE07
Melko, Joshua J. – RA06
Melnik, Dmitry G. – TH09
Menges, Fabian – MI09
Menten, Karl M. – WI10, RI05
Merer, Anthony – WH02
Merkt, Frederic – TJ01
Merrill, W G – TI12
Mescheryakov, A. A. – WF12
Mesko, AJ – RI13
Messinger, Joseph P. – WJ06
Milam, Stefanie N – RI13
Miliordos, Evangelos – MJ09, WJ09, RG12
Miller, Scott – TD09
Miller, Terry A. – TG05, TH09, TH10, FD04, FD05, FD06, FD07
Min, Ahreum – TD06, TD07
Minei, A. J. – RC04
Mironov, Andrey E. – MJ08
Misono, Masatoshi – TF04
Mizoguchi, Asao – RJ06
Moazzen-Ahmadi, Nasser – WH01
Mohr, Christian – WF06
Moon, Cheol Joo – TD06, TD07
Moradi, Christopher P. – TH04, TH05, TH06, RJ07, RJ08
Moriwaki, Yoshiki – MH07, MI04
Mormile, Melanie R. – RC02
Morris, Patrick – WI09, WI10
Morville, Jérôme – TF09
Motiyenko, R. A. – RF14, RF15, RI02, RI06, RI07
Mouret, Gaël – TB05, WG11, FB01
Muenter, John – TC04
Mukamel, Shaul – TI08
Mullaney, John Connor – WJ01, RH05
Müller, Holger S. P. – RD03, RI05, RI06, FB04
Mundy, Lee – WI14, WI15
Muscarella, Seth – TA01
Mutunga, Fredrick M. – RJ05

N

Nagarajan, Satyakumar – FB02
Nagy, Zsofia – WI09, WI10
Nahon, Laurent – WG07
Nairat, Muath – TI02
Nakajima, Masakazu – RH07
Nakano, Takumi – FA09
Nakashima, Kazuki – TF04
Nava, Matthew – MJ06
Neese, Christopher F. – TE01, TE09, RI01, FB02, FB03
Nei, Y-W – MI12, FC09
Nelson, Rebecca D. – MG07
Nemchick, Deacon – WH08
Nenov, Artur – TI08
Neumark, Daniel – TH01
Newbury, Nathan R. – WF07
Nguyen, Ha Vinh Lam – MG06, TD03, TG07, TG12
Nguyen, Trung – FA06
Nickerson, Nicole M. – TA03
Nikitin, Andrei V. – MJ14, FA08
Nishimiya, Nobuo – MJ02, MJ03
Nishimura, Yoshifumi – FD03
Nishiyama, Akiko – TF04
Niu, Ming Li – MF02, WG08
Nourry, Sendres – RI11
Novick, Stewart E. – RH10, RH11, RH12, RH13
Nunkaew, Jirakan – TJ02
Nyambo, Silver – FC08

O

O'Brien, James J – MJ04
O'Brien, Leah C – MJ04
Obenchain, Daniel A. – RH10, RH11, RH12, RH13
Ocola, Esther J – MG05
Odom, Brian C. – RJ02
Oh, JUNG JIN – MG03
Oishi, Ryo – MI04
Oka, Takeshi – WI07, RI08
Okumura, Mitchio – MA01, RE05, RF01, RF02
Olinger, Aaron C – TE10
Omar, Abdelaziz – FB01
Ono, Shuhei – RF09
Oomens, J. – MI12, MI13, MI14, WG02, RB03, RB04, RI04, FC09
Ooms, Kristopher J – TC05
Opoku-Agyeman, Bernice – WH07
Ordu, Matthias H. – RD03, FB04
Ormond, Thomas – RB05
Orphal, Johannes – RE09
Osburn, Sandra – WG02
Ossenkopf, Volker – WI09, WI10
Ouamerali, Ourida – RH14
Ozeki, Hiroyuki – MH14, RD04, FD08

P

Panchenko, Yurii – RG08
Pandit, Bill – TI07
Park, Barratt – MA04, MJ13, RF09, FE03
Parks, Deondre L – TE08

Parlak, Cemal – FC05
Pate, Brooks – MG07
Patterson, David – TE06, WF02
Pearlman, Bradley W – TA02
Pearson, John – MH08, TG01, TG02, TG03, TG06, WI09, WI10, RB09, RD05, RD06, RD07, FB04
Peebles, Rebecca A. – MG07, MG08
Peebles, Sean A. – MG07, MG08
Perez, Cristobal – TD05, TE07
Pérez-Ríos, Jesús – TJ12
Perrin, Agnes – TG04, RF10, FC04
Perry, Adam J. – MF05, MF06
Perry, David S. – RG06
Petkie, Doug – FE07
Petrignani, Annemieke – RI04
Pfau, Tilman – TJ12
Phillips, Dane – FB03
Piancastelli, M N – WG06
Piau, Gérard Pascal – WG11
Pihlström, Ylva M. – RD11
Pineda, Jorge L – WI05
Pino, Gustavo A – TH05, WA01
Pirali, Olivier – TB02, TB05, WG11, RB07, RB08, RB09, RB10
Plusquellic, David F. – WF01
Pollum, Marvin – TI04
Porambo, Michael – MF04
Potapov, Alexey – RJ01
Pototschnig, Johann V. – FA02, FA07
Pound, Marc W. – WI15
Power, William P. – TC02
Pratt, David – TG14
Pride, Michael A – RC02
Pringle, Wallace C. – MG05
Pszczółkowski, Lech – RF11

R

Radi, Peter – FA01
Rathore, Rajendra – WJ07
Rauch, Kevin P. – WI15
Rauer, Clemens – TI04
Raymond, Kyle T – MJ08
Reber, Melanie Roberts – TF07
Redlich, Britta – RB03
Reed, Amanda – FA03, FA04
Reed, Zachary – WF03, RF05
Reichardt, Christian – TI04
Reid, Scott – WJ07, FC08
Reiland, G P – WI01
Reilly, Neil J – WJ07, RD12, FD07, FD09
Remijan, Anthony – WI04, WI11, WI12, WI13
Rey, Michael – MJ14, RE06, FA08
Reza, Md Asmaul – FD09, FD11
Rice, Johnathan S – WI05
Ridolfi, Marco – RF10
Rieker, Greg B – WF07
Rijs, Anouk – WG03, RB02, RD01
Rivalta, Ivan – TI08
Rivera, Cara Rae – RH13
Rivera-Rivera, Luis A. – WJ04, WJ08
Rizzo, Maxime – WI14
Robichaud, David – RF02
Rodgers, M T – MI12, MI13, MI14, FC09
Rojo, Michellle – RC02
Ross, Amanda J. – MJ07
Ross, Stephen Cary – TB04
Rothman, Laurence S. – MH01, MH02, MH03
Roudjane, Mourad – TH09
Roy, Althea A. M. – WA04
Roy, P. – TB05
Roy, Pierre-Nicholas – RJ04
Rudolph, Heinz Dieter – MG04
Rupasinghe, Priyanka – RF01, RF02
Ruth, Albert A – RE09
Rutkowski, Lucile – TF05, TF08, TF09
Ryland, Elizabeth S – RE01
Ryzhov, Victor – WG02

S

Saladrigas, Catherine A. – FE04
Salomon, T. – RB07, RB08
Salumbides, Edcel John – MF01, MF02, MF03, WG08
Sams, Robert – FB04
Sardar, Rajesh – TI07
Sasada, Hiroyuki – TF01, TF02, TF03
Satija, Aman – RE02, RE03
Sauve, Genevieve – TI06
Schaefer III., Henry F. – TH06
Schäfer, Mirko – RD03
Scheidegger, Simon – TJ01
Schellekens, Bert – MF03
Schibli, Thomas R – WF06
Schlagmüller, Michael – TJ12
Schlecht, Erich T – TE03
Schlegelmilch, B. – WI07
Schlemmer, Stephan – MI01, MI02, RB03, RB07, RB08, RD02, RD03, RJ01
Schmidt, Deborah – WI02, WI03
Schmiedt, Hanno – MI02
Schmitt, Thomas – RD02
Schmitz, David – TD02, TE06, TE07
Schmitz, Joel R – TC06
Schnell, Melanie – MG11, TD02, TE06, TE07, RD01
Schnepper, D. Alex – TA10
Schnitzler, Elijah G – WJ03
Schuessler, Hans – RF06
Schwaab, Gerhard – RJ10
Schwan, Raffael – RJ10
Schwartz, George – MH12
Schwarz, Cara E. – MG10
Schwenke, David – MH06
Sears, Trevor – MF07, TI13, WF08, WF09
Sedo, Galen – MG12
Segarra-Martí, Javier – TI08
Segura-Cox, Dominique M. – WI14
Seifert, Nathan A – MG07
Semeria, Luca – TJ01
Senyel, Mustafa – FC05
Sera, Hideyuki – TF02, TF03
Serrato III, Agapito – RH12
Shaffer, James P – TJ11
Shen, Linhan – RE05
Sheps, Leonid – WG12
Sheridan, Phillip M. – RA01
Shipman, Steven – TC07, TE04, TE05, TE10, FC07
Shu, Ran – TE03
Shubert, V. Alvin – TD02, TE06, TE07
Shuman, Nicholas S. – RA06
Sibert, Edwin – TH02, WH09, WH10, WH11, RG01
Siller, Brian – FB06
Sims, Ian – MA04
Sinclair, Laura – WF07
Singer, James – WG10
Singh, Vipin Bahadur – WJ14, RG07
Sironneau, Vincent – RF03
Sjouwerman, Loránt O. – RD11
Slaughter, Kai – MJ07
Smith, Houston Hartwell – RI13
Smith, Jonathan M. – TI14, RE08
Smith, Mary Ann H. – MH05, MJ11, MJ14, TG01, TG02, RF08
Smith, Peter L – RD09
Sneden, Chris – RD10
Soulard, Pascale – RB10, RJ11
Spada, Lorenzo – RH06
Spaun, Ben – WF02
Springer, Sean D. – WJ04
Srikantaiah, Sree – FB03
Srivastava, Santosh Kumar – RG07
Stahl, Wolfgang – MG06, TD03, TG07, TG12
Stanton, John F. – MG02, TH01, TH09, TH11, TI15, RA08, RB05, RD12, RG14, RJ07, FD07, FE05

Stark, Glenn – RD09
Stearns, Jaime A. – RG10, FB08
Steber, Amanda – RD01
Steeves, Adam H. – FE03
Steimle, Timothy – TA01, TA02, TA04, RA07, FA06, FE09
Stephens, Susanna L. – RA04, RA05, RH04, RH05
Stewart, Jacob – TA06
Stoffels, Alexander – MI01, RB03
Stollenwerk, Patrick R – RJ02
Stolte, W C – WG06
Storm, Shaye – WI14
Storrie-Lombardi, Michael – RC02
Stout, Phillip J. – TE09
Strobehn, Stephen – MI12
Stwalley, W.C. – TJ05
Suas-David, Nicolas – MH11, RE06
Suhm, Martin A. – TD11
Suits, Arthur – MA04, FD10
Sullivan, Michael – TA06
Sumiyoshi, Yoshihiro – RH07
Sun, Yu Robert – WF05
Sung, Keeyoon – MH05, MJ11, MJ14, TG01, TG02, RB09, RF02, FB04
Suzuki, Mari – MI04
Suzuki, Masao – MJ02, MJ03
Swann, William C – WF07
Szalay, Peter – FA08

T

Tabor, Daniel P. – WH09, WH10, WH11, RG01
Tada, Kohei – TH13
Takahashi, Kaito – FC01
Talicska, Courtney – MF04
Talipov, Marat R – WJ07
Tammaro, Stefano – TB05
Tan, Yan – WF05
Tanaka, Keiichi – RH07
Tanarro, Isabel – RD08, RI10, RI12
Tang, Adrian – TE03
Tang, Jian – MJ01
Tarbutt, Michael – TA04
Tarczay, Gyorgy – RJ12
Tashkun, Sergey – MJ14
Tatamitani, Yoshio – RH03
Tennyson, Jonathan – MH04, MH13, TA08
Teuben, Peter J. – WI15
Teunis, Meghan B – TI07
Tew, David Peter – RA03, RA05
Thapaliya, Bishnu P – RG06
Thomas, Javix – TD04, TG11, WJ02
Thomas, Jessica A. – MG14
Thomas, Phillip – RG03
Thompson, Michael C – WH03, WH04, RG14
Thorwirth, Sven – MG01, RB03, RB07, RB08, RD02, RD12
Tielens, Xander – RI04
Timerghazin, Qadir – WJ07
Tokaryk, D. W. – TA03, TA07, TB04
Tötsch, Niklas – MI09
Tran, Henry – TH09, TH10
Tran, Thi Ngoc Ha – RF05
Tremblay, Benoît – RJ11
Tripathi, G. N. R. – TH14
Truong, Gar-Wing – WF07
Tschumper, Gregory S. – MI15
Tsukiyama, Koichi – RI09
Tubergen, Michael – TG09, TG10
Twagirayezu, Sylvestre – WF08, WF09
Tyuterev, Vladimir – MJ14, FA08

U

Ubachs, Wim – MF01, MF02, MF03, WG08
Uchida, Kanako – MH07
Uhler, Brandon – WJ07
Uriarte, Iciar – TD05, WA02, RF11
Urness, Kimberly N. – FB09
Usabiaga, Imanol – WA02

V

Vaccaro, Patrick – TI11, WH08
Van, Vinh – MG06
van der Avoird, Ad – MJ09
van Dishoeck, Ewine – WG08
van Wijngaarden, Jennifer – MG12, WG10, RB07, RB08
Vander Auwera, Jean – TB06, RF10
VanGundy, Robert A. – RA06, FA05
Varberg, Thomas D. – TA02, RA07
Vazquez, Gabriel J. – TJ13, TJ14
Vealey, Zachary – TI11
Verkamp, Max A – RE01
Viggiano, Albert – RA06
Villanueva, Geronimo L. – MJ11
Vogt, Natalja – MG04, TG04, FC04
von Helden, Gert – WG01
Voros, Tamas – RJ12
Voss, Jonathan – MI06, MI07
Vura-Weis, Josh – RE01

W

Wagner, Ian C – RI13
Walker, Nick – MF13, WJ01, RA03, RA04, RA05, RC08, RH04, RH05
Walker, Zachary – RC02
Walsh, Patrick S. – TD08, WH09, WH10, WH11
Walters, Adam – MH08, RD03
Walters, Nicholas A. – FE06
Walters, Richard S. – RA09
Walton, Jay R. – WJ08
Waltz, Terese A – RJ13
Wang, Jin – WF05
Wang, Lai-Sheng – RG15, RH15
Wang, Lecheng – RJ04
Wang, Xiaohong – FD03
Wang, Yi-Jen – WH02
WARD, TIMOTHY B – RA09, RG12
Warner, Brian – FB09
Warner, S H – WI01
Watanabe, Kyohei – MH07
Wcislo, Piotr – MH01, MH02, MH03
Weber, J. Mathias – TA09, WH03, WH04, RG14
Weddle, Gary H – WH14, WJ09
Wehres, Nadine – WA04
Weichman, Marissa L. – TH01
Welty, Daniel E. – RI08
West, Channing – MG12
Western, Colin – TC01, RD10
Westphal, Karl Magnus – TJ12
Westrick, C W – WI07
White, Allen – RC06
Whitehill, Andrew Richard – RF09
Widicus Weaver, Susanna L. – TE04, TE05, TH03, WA04, RI13
Wilhelm, Michael J. – TI14, RE08
Wilkins, Joshua – RJ14
Wilzewski, Jonas – MH01, MH02, MH03
Winnewisser, Manfred – TG08
Witek, Henry A – FD03
Witt, Adolf N. – RI08
Wolk, Arron – MI09
Wolke, Conrad T. – TD09, WJ09
Womack, Carrie – MG01, MJ06, MJ13, RF09, FD01
Wong, Andy – WG09
Wong, Bryan M. – TJ08, TJ09
Woods, R. Claude – MG09, MG10, FE01, FE02, FE06
Woon, David E. – RI15
Wright, Emily – FB09
Wu, Anan – RH10
Wu, Ranran – MI13, FC09
Wu, Weixin – RC04
Wyse, Ian A – TA02

X

Xantheas, Sotiris – WJ09, RG12
Xie, Yizhou – TI07
Xu, Hong – MF07
Xu, Li-Hong – TB03, WF12
Xu, Lisa – WI15
Xu, Shuang – TA09
Xu, Yunjie – TD04, TG11, WJ02

Y

Yachmenev, Andrey – MH13
Yagi, Reona – WH12, WH13
Yamada, Koichi MT – RJ01
Yamamoto, Ryo – FA09, FA10
Yang, Bo – MI13
Yang, Dong-Sheng – MI10, MI11, RA11
Yang, Shaoyue – MH12
Yatsyna, Vasyl – WG03
Ycas, Gabriel – WF07
Ye, Jun – MF11, MF12, WF02
Ye, Yu – TE03
Yeh, S. C. C. – WI07
Yoo, Ji Ho (Chris) – MF13
Yoon, Young – TH07
York, Donald G. – RI08
Young, Justin – RA01
Young, Justin W. – RG10, FB08
Yu, Shanshan – MH08, TG06, WA05, RB09, RD05, RD06, RD07, RF02
Yukiya, Tokio – MJ02, MJ03
Yurchenko, Sergei N. – MH04, MH13, TA08

Z

Zack, Lindsay N. – MA04, RA01, FD10
Zakharenko, Olena – RF14, RF15
Zaleski, Daniel P. – WJ01, RA03, RA04, RC08, RH04, RH05
Zenchyzen, Brandi L M – WJ03
Zgierski, Marek Z. – TI03
Zhang, Kaili – RE01
Zhang, Ruohan – TA02, RA07
Zhang, Yuchen – MI11, RA11
Zhao, Dongfeng – RG09
Zhao, YueYue – TG07
Zhaunerchyk, Vitali – WG03
Zhou, Jia – MI07, TD10
Zhou, Yan – MF11, TJ06, TJ07, TJ08
Zhu, Feng – RF06
Zhu, Guo-Zhu – RG15
Zhu, Yanlong – MI12, MI14
Zingsheim, Oliver – RB07, RB08, RD02
Zinn, Sabrina – MG11, RD01
Ziurys, Lucy – WI01, WI02, WI03, WI04, RA01, RA02
Zou, Luyao – TH03
Zwier, Timothy S. – MG13, TD08, TH08, WH09, WH10, WH11, RB05, RG01, RG11

A Very Special Thanks to Our Sponsors!

Journal of Molecular Spectroscopy

The *Journal of Molecular Spectroscopy* presents experimental and theoretical articles on all subjects relevant to molecular spectroscopy and its modern applications. An international medium for the publication of some of the most significant research in the field, the Journal of Molecular Spectroscopy is an invaluable resource for astrophysicists, chemists, physicists, engineers, and others involved in molecular spectroscopy research and practice.

- Fast editorial times (total time from submission until final decision < 5 weeks)
- High refereeing standards
- Fast journal publication times (final corrected article online < 8 weeks)
- Wide dissemination: over 7,000 institutes worldwide have access to the journal
- High exposure of your article online: >130,000 article downloads per year on www.sciencedirect.com
- High archival value; cited half-life >10.0 years

Feature Articles

The *Journal of Molecular Spectroscopy* publishes peer reviewed Feature articles - these articles will give an overview of areas of particular significance in molecular spectroscopy. They may review and consolidate an area of theoretical development or a collection of experimental data, in each case offering some new insights. The articles may also summarize the present status of a rapidly developing and/or evolving field. All the articles should serve as introductions to areas of spectroscopy other than one's specialty and should be particularly valuable to students entering the field.
Feature articles will be solicited by invitation of the Editor. However, the Editor invites you to suggest, with a reasonable level of detail, a topics that could be of interest. Self-nominations by potential authors are particularly encouraged.

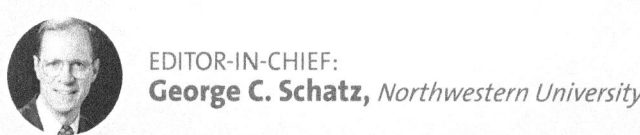

Newport.
The Brands of Innovation.

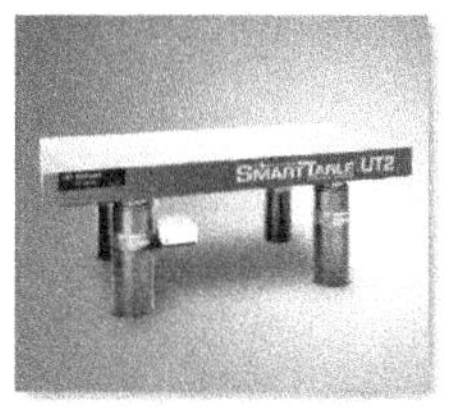

Newport
Optical Tables

Newport
XMS Linear Motor Stages

New Focus
Velocity™ Widely Tunable Laser

Spectra-Physics
Sirah Pulsed Dye Lasers

Spectra-Physics
Quanta-Ray Pulsed Nd:YAG Lasers

At Newport our mission is to continuously evolve our knowledge and experience in order to deliver innovative products and solutions that advance our customer's technologies. To deliver upon our mission, we feel it is vital to possess expertise and experience across a broad spectrum of technologies and interconnected products.

For over 40 years, Newport has continued to grow and today is built upon world-class brands such as Corion®, New Focus™, Oriel® Instruments, Richardson Gratings™ and Spectra-Physics®. Alone, each of these brands has a rich history of product innovation and expertise. Together, we provide a synergy of knowledge across a broad spectrum of products along with the ability to deliver unsurpassed solutions and integration.

Today, Newport is one company comprised of industry leading brands, stronger and more responsive to the full breadth of your product needs. We thank you for over 40 years of support, and we look forward to many more years of partnership.

Visit **www.Newport.com/brands-5** or call **800-222-6440** for more information.

Enjoy FREE donuts, sponsored by Newport

Newport. Family of Brands – ILX Lightwave® • New Focus™ • Ophir® • Oriel® Instruments • Richardson Gratings™ • Spectra-Physics® • Spiricon®

Transmit and Receive Systems Covering the 70GHz-3THz Spectrum

VDI offers a wide variety of transmit and receive systems covering the 70GHz-3THz spectrum. These systems incorporate VDI's frequency extension and mixer components coupled with commercially available microwave oscillators and amplifiers.

For transmit systems, VDI can configure them with or without a drive oscillator. A VDI Amplifier / Multiplier Chain (AMC) requires a customer supplied low frequency source (typically <20GHz, 10dBm nominal). A VDI Transmitter (Tx) integrates a source (oscillator or synthesizer) with the VDI AMC. A VDI Mixer / Amplifier / Multiplier Chain (MixAMC) requires a customer low frequency local oscillator. A VDI Receiver (Rx) integrates the LO drive oscillator with the Mixer and LO Chain for turn-key operation.

Standard AMCs and MixAMCs have been developed to provide high performance RF drive multiplication and downconversion for full waveguide band coverage. These systems can be used to extend traditional spectrum analyzers and signal generators into the THz and mm-wave ranges. VDI's standard AMC and MixAMCs offer various modes of operation. VDI AMCs can be operated in standard frequency mode (<20GHz, 10dBm nominal) or high frequency RF drive mode (<45GHz, 0dBm nominal). VDI MixAMCs can also operate in standard and high frequency LO drive modes. Customers also have the option to operate MixAMCs for block-downconversion (<20GHz IF) or as a spectrum analyzer extender. Standard AMCs and MixAMCs are available from WR15 (50-75GHz) to WR1.0 (750-1,100GHz).

VDI offers both narrow-band high-power and broadband low-power systems. High power systems use VDI's D-series X2 multipliers to achieve maximum multiplier efficiency and power handling. VDI has developed many high power systems for special customer applications, such as a novel multiplier based source with output power of 160mW at 200GHz.

Reconfigurable / modular AMCs are also available upon request.

The Coblentz Society – fostering understanding and application of vibrational spectroscopy

THE COBLENTZ SOCIETY

Call for Award Nominations

Visit www.coblentz.org for more information

THE COBLENTZ SOCIETY

ABB Bomem-Michelson Award: ABB sponsors the Bomem-Michelson Award to honor scientists whom have advanced the technique(s) of vibrational, molecular, Raman, or electronic spectroscopy. Contributions may be theoretical, experimental, or both. The recipient must be actively working and at least 37 years of age. The nomination should include a resume of the candidate's career as well as a synopsis of the special research achievements that make the candidate an eligible nominee for the ABB sponsored Bomem-Michelson Award. Nominations for the award are open between February 1st and **May 1st** each year. Further information regarding the ABB Bomem-Michelson Award can be found at www.coblentz.org/awards/the-bomem-michelson-award.

Coblentz Award: The Coblentz Award is presented annually to an outstanding young molecular spectroscopist under the age of 40. The candidate must be under the age of 40 on January 1st of the year of the award. Nominations should include a detailed description of the nominee's accomplishments, a curriculum vitae and as many supporting letters as possible. Annual updates of files of nominated candidates are encouraged. Nominations for the Coblentz Award are open between January 3rd and **July 15th** each year. Further information regarding the Coblentz Award is available at www.coblentz.org/awards/the-coblentz-award.

Craver Award: The Craver Award is presented annually to an outstanding young molecular spectroscopist whose efforts are in the area of applied analytical vibrational spectroscopy. The candidate must be under the age of 45 on January 1st of the year of the award. The work may include any aspect of (near-, mid-, or far-infrared) IR, THz, or Raman spectroscopy in applied analytical vibrational spectroscopy. Nominees are welcome from academic, government, or industrial research. Nominations must include a detailed description of the nominee's accomplishments, curriculum vitae or resume, and a minimum of three supporting letters. Nominations for the Craver Award are open between March 30th and **August 30th** each year. Further information about the Craver Award is available at www.coblentz.org/awards/the-craver-award.

Ellis R. Lippincott Award: The Ellis R. Lippincott Award is presented annually in recognition of significant contributions and notable achievements in the field of vibrational spectroscopy. The medal is jointly sponsored by the Coblentz Society, the Optical Society of America and the Society for Applied Spectroscopy. Recipients must have made significant contributions to vibrational spectroscopy as judged by their influence on other scientists. Because innovation was a hallmark of the work of Ellis R. Lippincott, this quality in the contributions of candidates will be carefully appraised. Nominations for the award are open between January 1st and **October 1st** each year. Nominations should be submitted to: Lippincott Award Chairperson, awards@osa.org. Further information regarding the Ellis R. Lippincott Award is available at www.coblentz.org/awards/the-lippincott-award.

Honorary Membership: The Coblentz Society awards honorary memberships in the Society to people who have made outstanding contributions to the field of vibrational spectroscopy or any other field related to the purposes of the Society. Nominations close on **February 1st** each year, with awards announced at the Annual Members Meeting at Pittcon and presented at FACSS. Send your nomination for 2015 to Dr. Mark Druy, Coblentz Society President at madruy@gmail.com.

ISMS MEETING VENUE INFORMATION

All contributed talks will be held in the Chemistry complex (and immediately adjoining buildings). The plenary talks will be held across the quad (about 600') in Foellinger Auditorium.

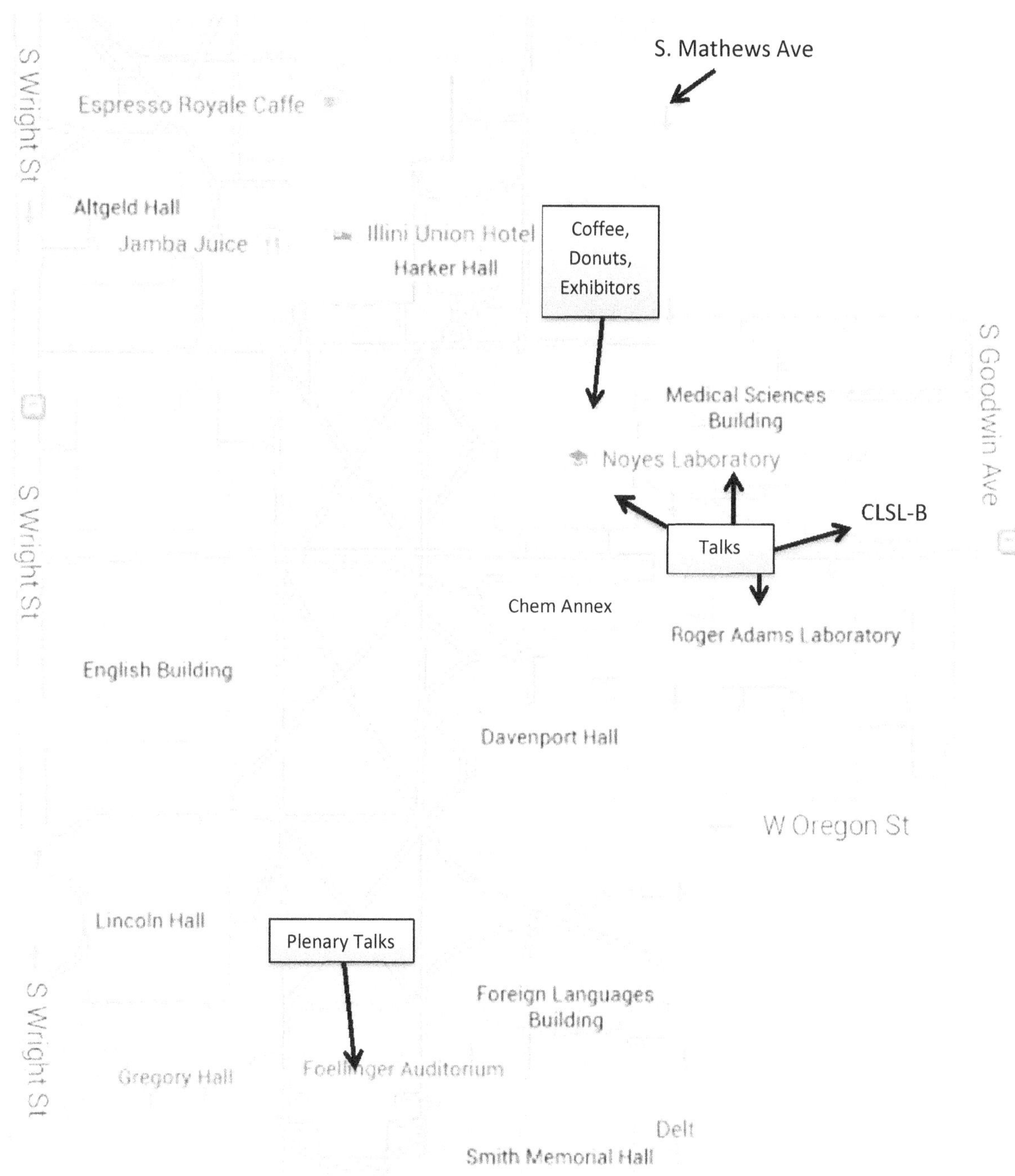

ACCESSIBLE ENTRANCES

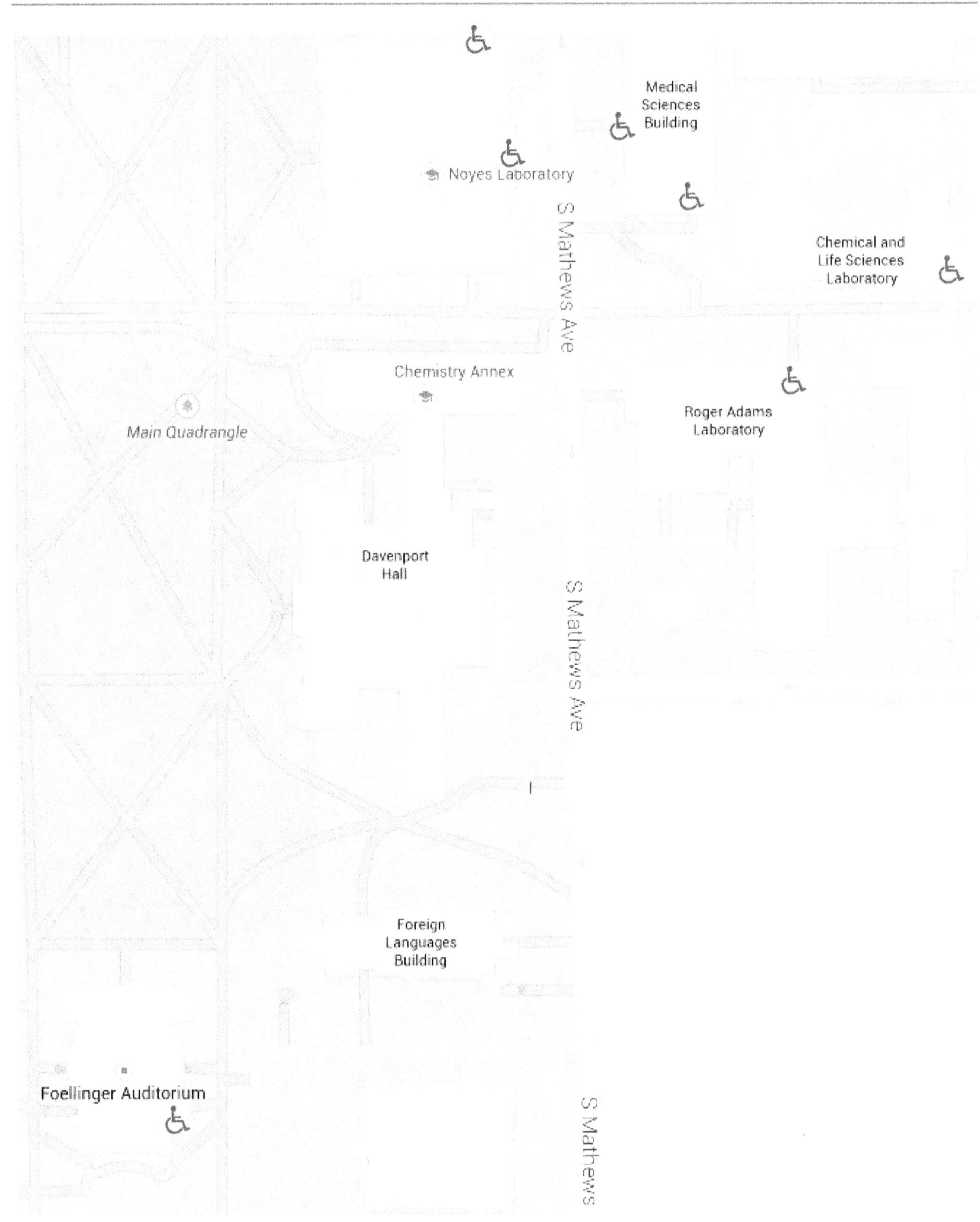

NOYES LABORATORY (NL)

Noyes Laboratory houses our Registration and Exhibitor/Refreshment Rooms (163/165), the Computer Lab (151), two lecture halls (NL 100 and NL 217), and the Chemistry Library.

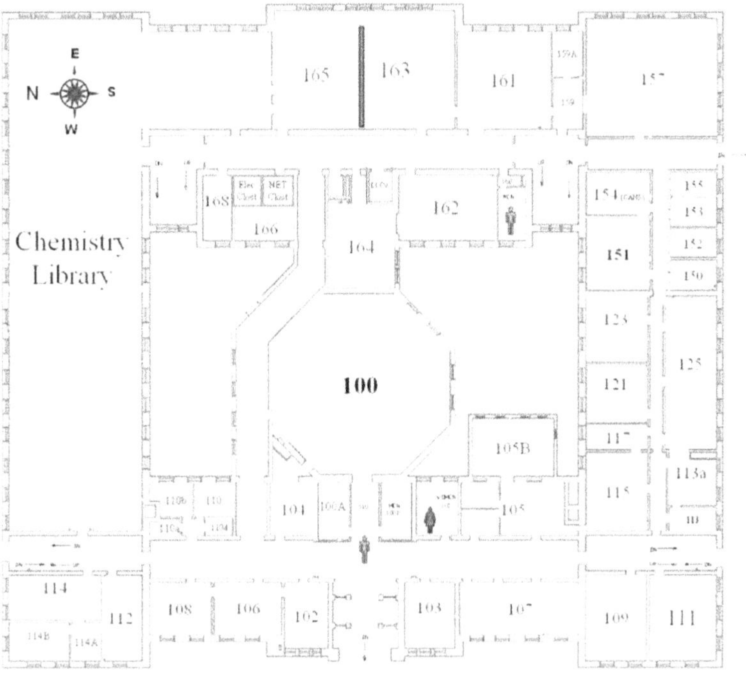

Noyes Laboratory - 1st Floor

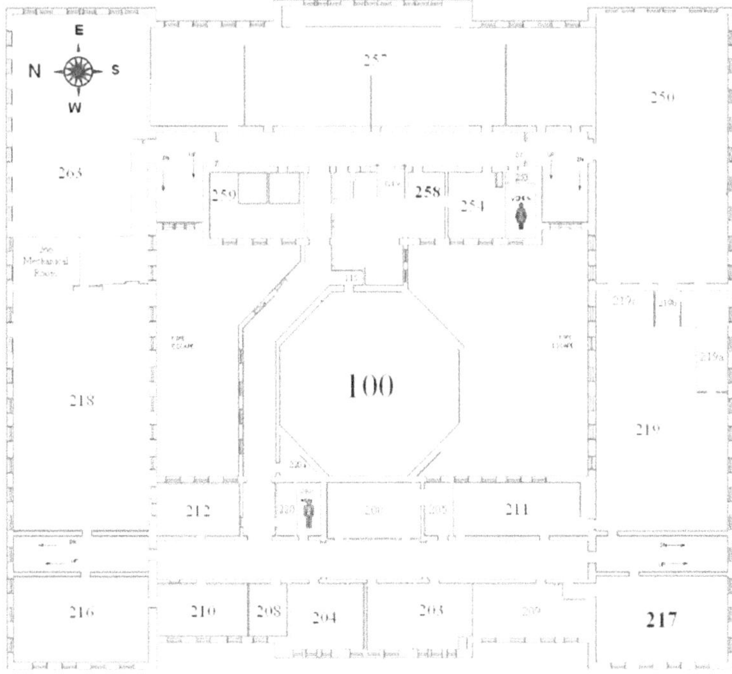

Noyes Laboratory - 2nd Floor

ROGER ADAMS LABORATORY (RAL)

Roger Adams Laboratory is across the street to the east of Chemistry Annex. It has one lecture hall (RAL 116). Please note that in Roger Adams Lab, the ground level is called "Ground" and the First Floor is equivalent to the Second Floor in the other buildings.

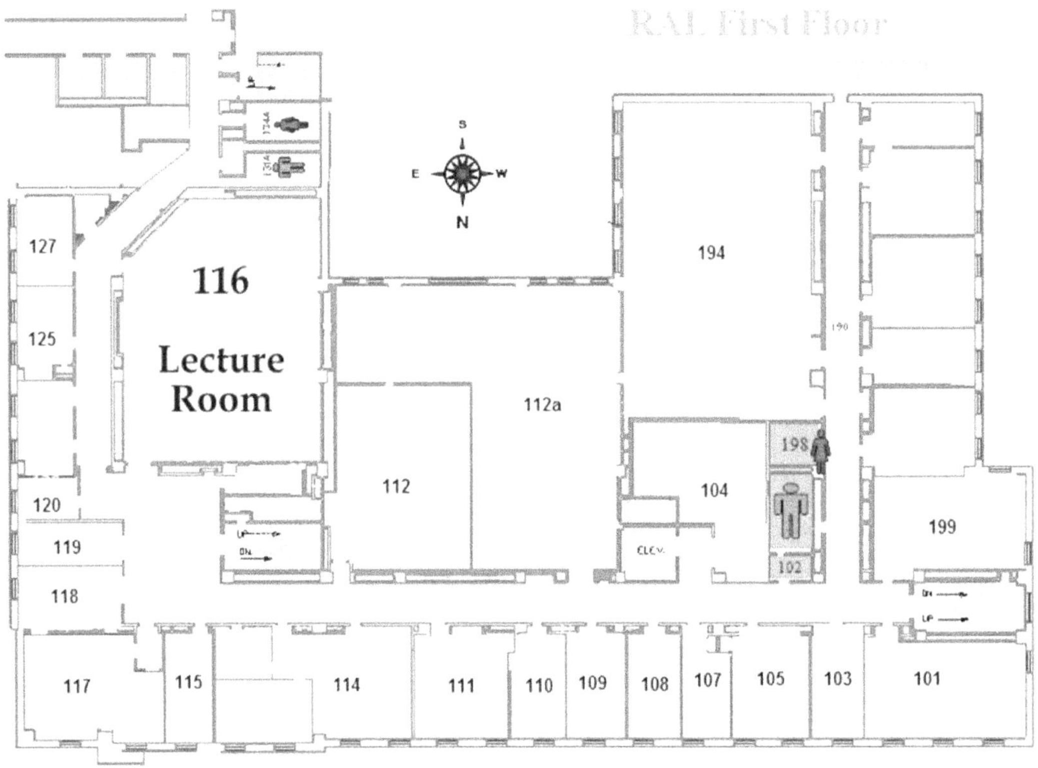

CHEMICAL AND LIFE SCIENCES (CLSL)

CLSL is a multi-wing building located across the street to the east of Noyes Laboratory. The lecture hall (CLSL B102) is in the B wing across the pedestrian walkway to the northeast of Roger Adams.

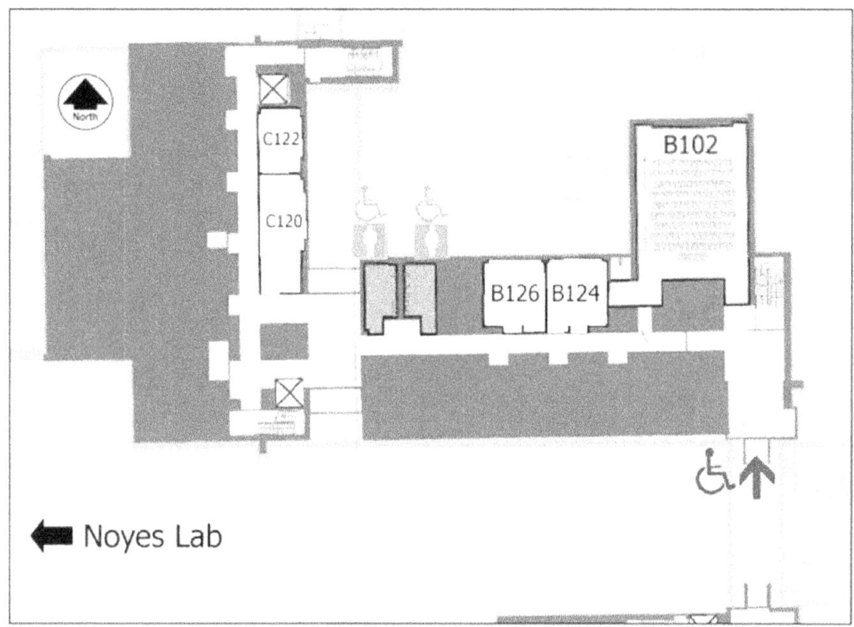

MEDICAL SCIENCES BUILDING (MSB)

Medical Sciences is across the pedestrian walkway to the north of Roger Adams. It has one lecture hall (MSB 274).

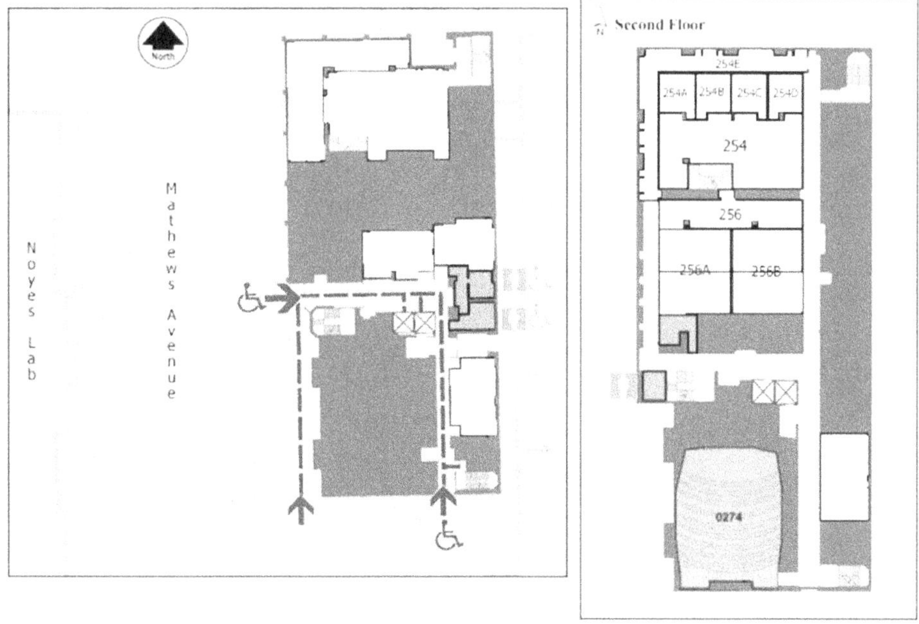

Foellinger Auditorium (Plenary and Intermission)

Foellinger Auditorium is located at the South end of Quad. The main doors on the North (quad) side will open at 8:10 AM (the side ADA/wheelchair door will be open around 8:00 AM). There is seating on the main level and the upper balcony. There is no elevator in the building.

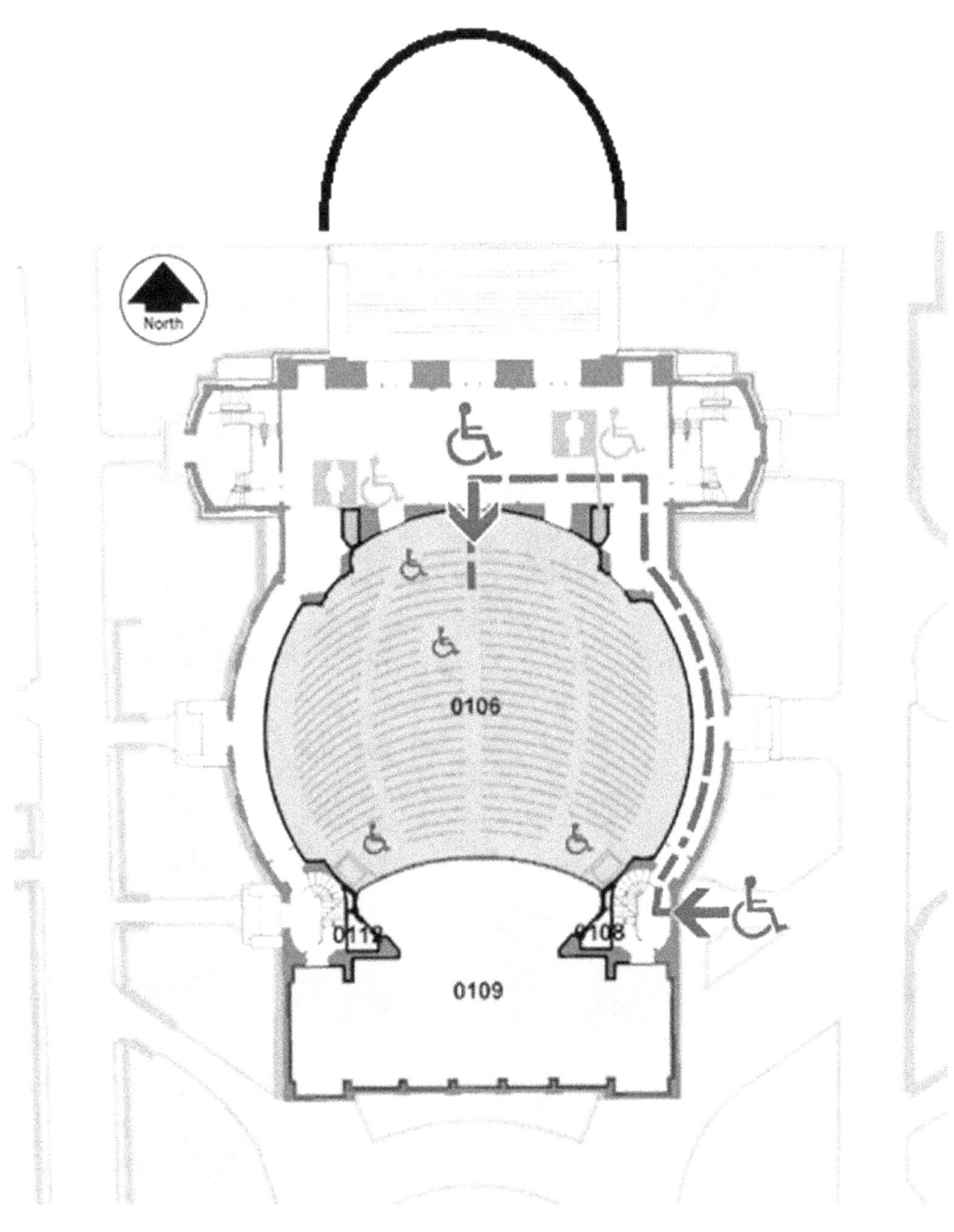

PARKING (E14) TO BOUSFIELD DORM

If you purchase a parking permit and are staying at the dorm, you will park in lot E14 (any spot). E14 is nearly due south of Bousfield Hall Dorm.

Parking enforcement begins at 6:00 AM on Monday, so you will need to have your car in lot E14 with your permit displayed before then. There are many parking meters on E. Peabody Drive (and in the lot across from Bousfield) if you wish to park closer for short periods (25 cents/15 minutes – generally between 6 AM and 6 PM, but check the meter because some go until 9 PM).

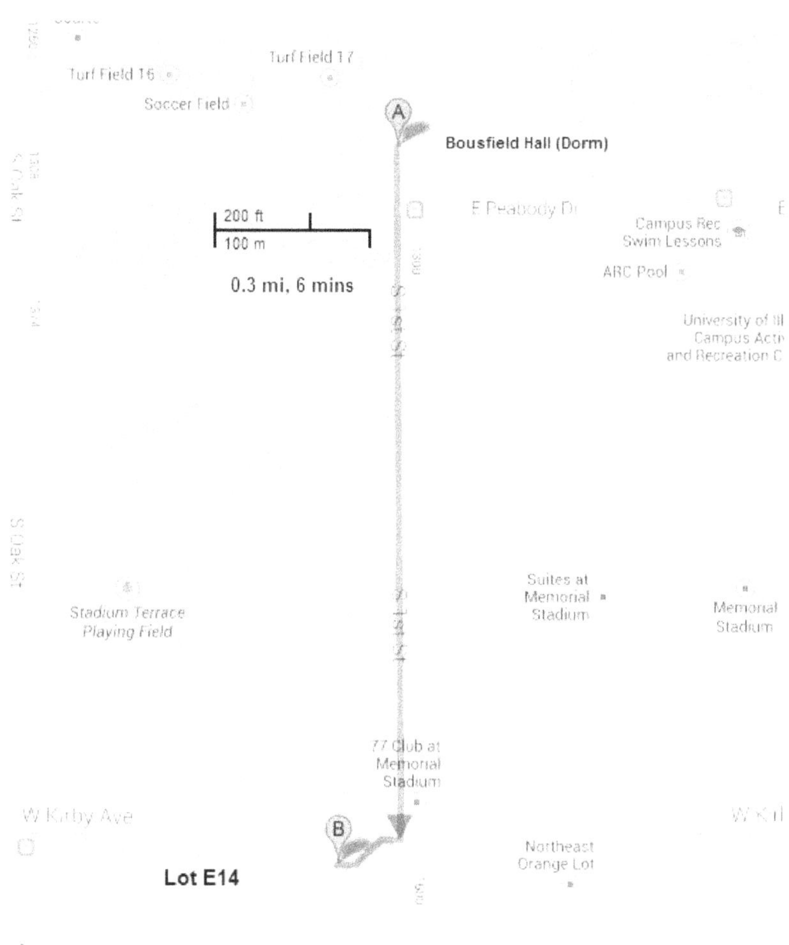

BOUSFIELD/NUGENT DORM to MEETING VENUE (walking)

Bousfield & Nugent Halls are just under a mile (15-20 minute walk) from the main symposium buildings

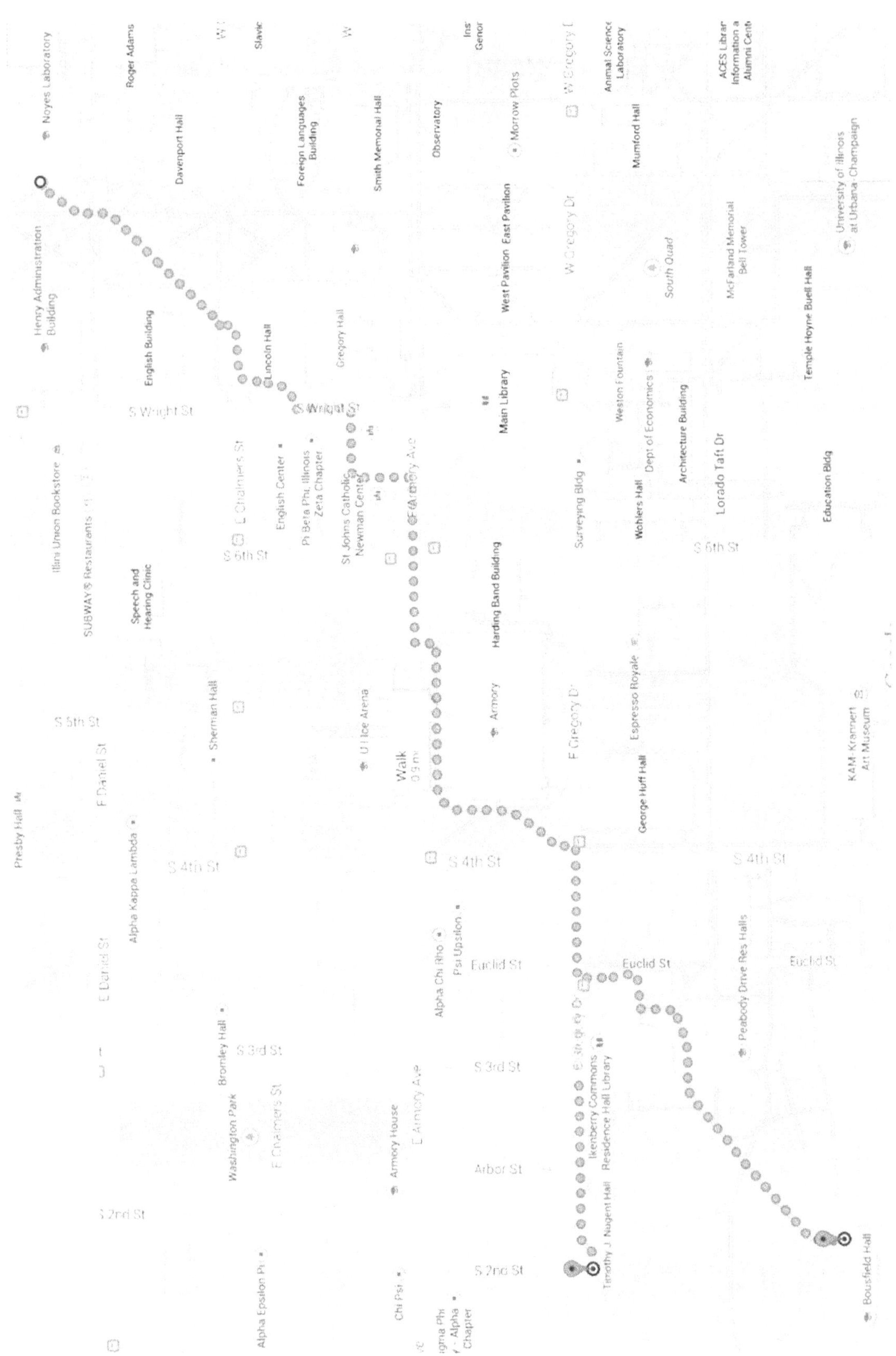

BOUSFIELD/NUGENT DORM to MEETING VENUE (bus)

There is convenient and free bus service between Bousfield/Nugent Dorms and 1 block from the meeting venue. The Yellow Line picks up on the corner of First and Peabody (Bousfield), and also on Gregory Drive (Nugent) in front of Ikenberry Commons, and drops off at the Wright Street Terminal (just outside of the Henry Administration Building). Return locations are the same but across the street. The Yellow Line will also take you to downtown Champaign, but you will need to pay for your return (only iStops are free). Approximately every 10 minutes during the day.

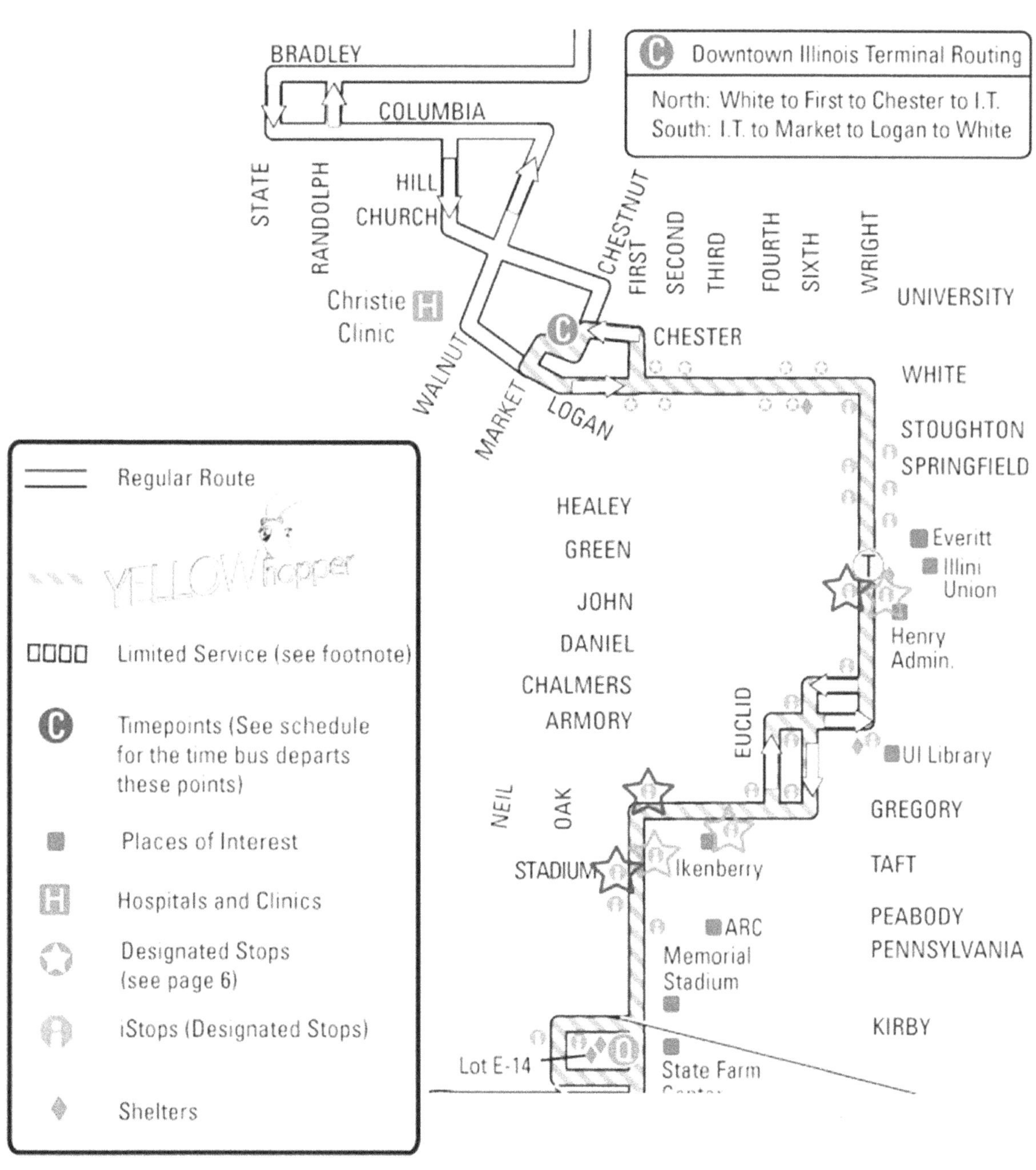

The Gold Line picks up on the corner of First and Peabody, and also on Gregory Drive in front of Ikenberry Commons and drops off at the Krannert Center (across the street from CLSL-B). Return locations are across the street. Runs every ~10 minutes during the day (offset from the Yellow Line by 5 minutes).

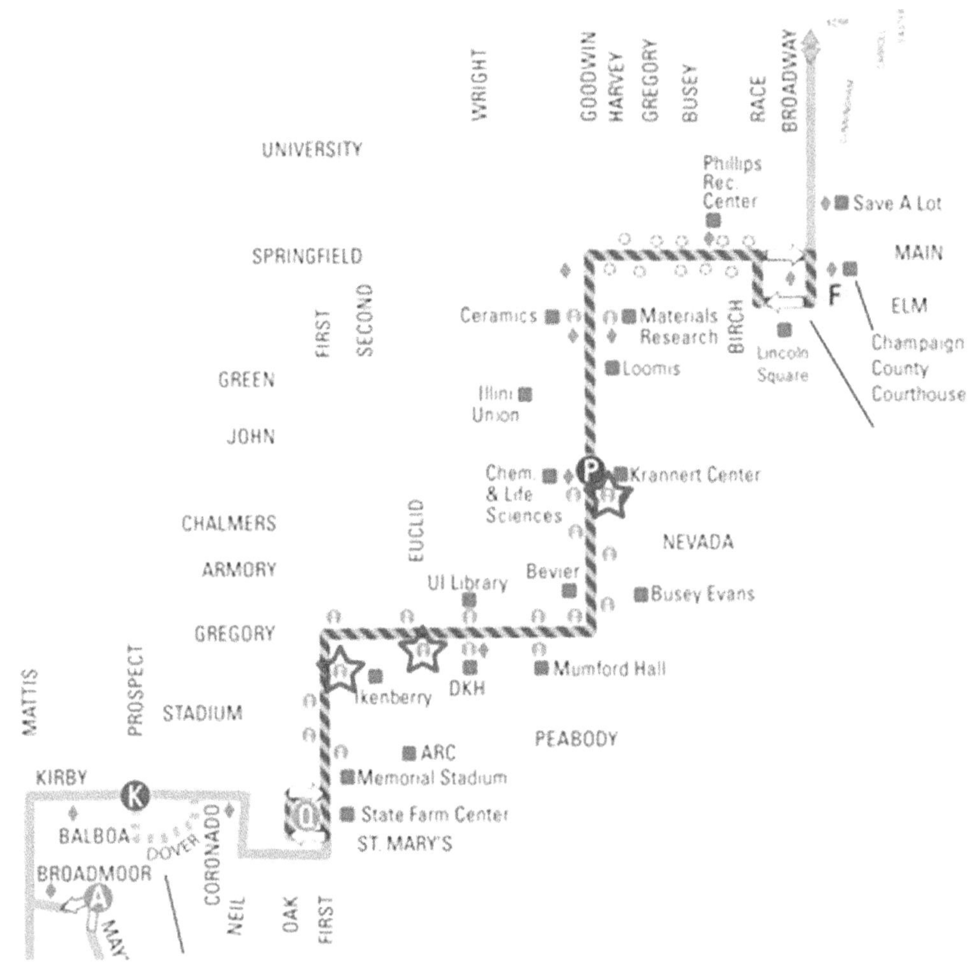

Bus Stops (Yellow Line = Left Arrow, Gold Line = Right Arrow, Foellinger Auditorium (Plenary) and Noyes Lab = Stars)

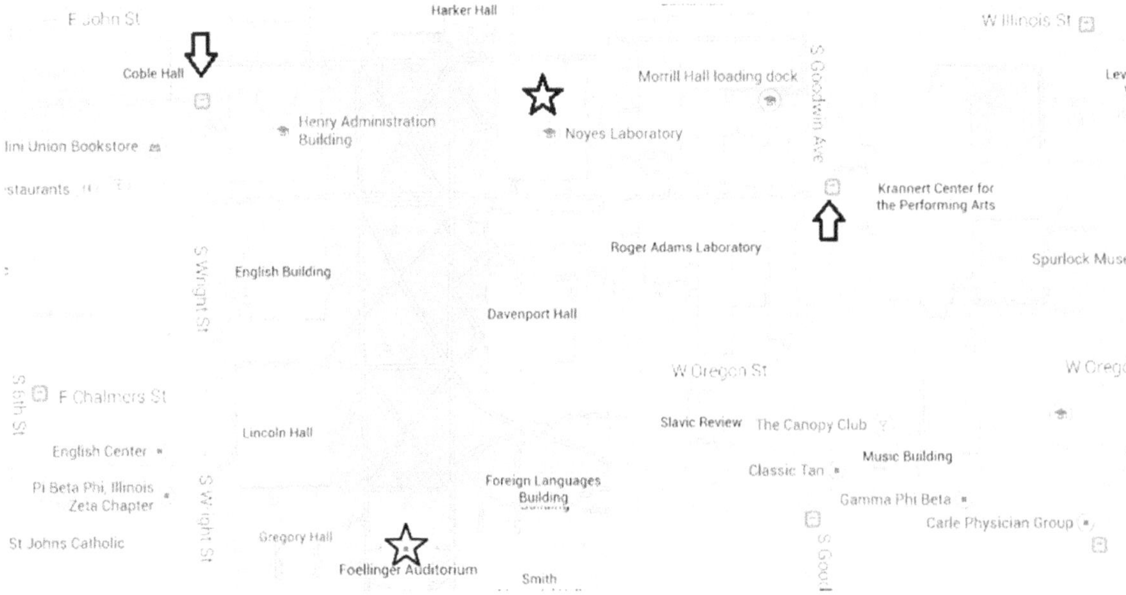

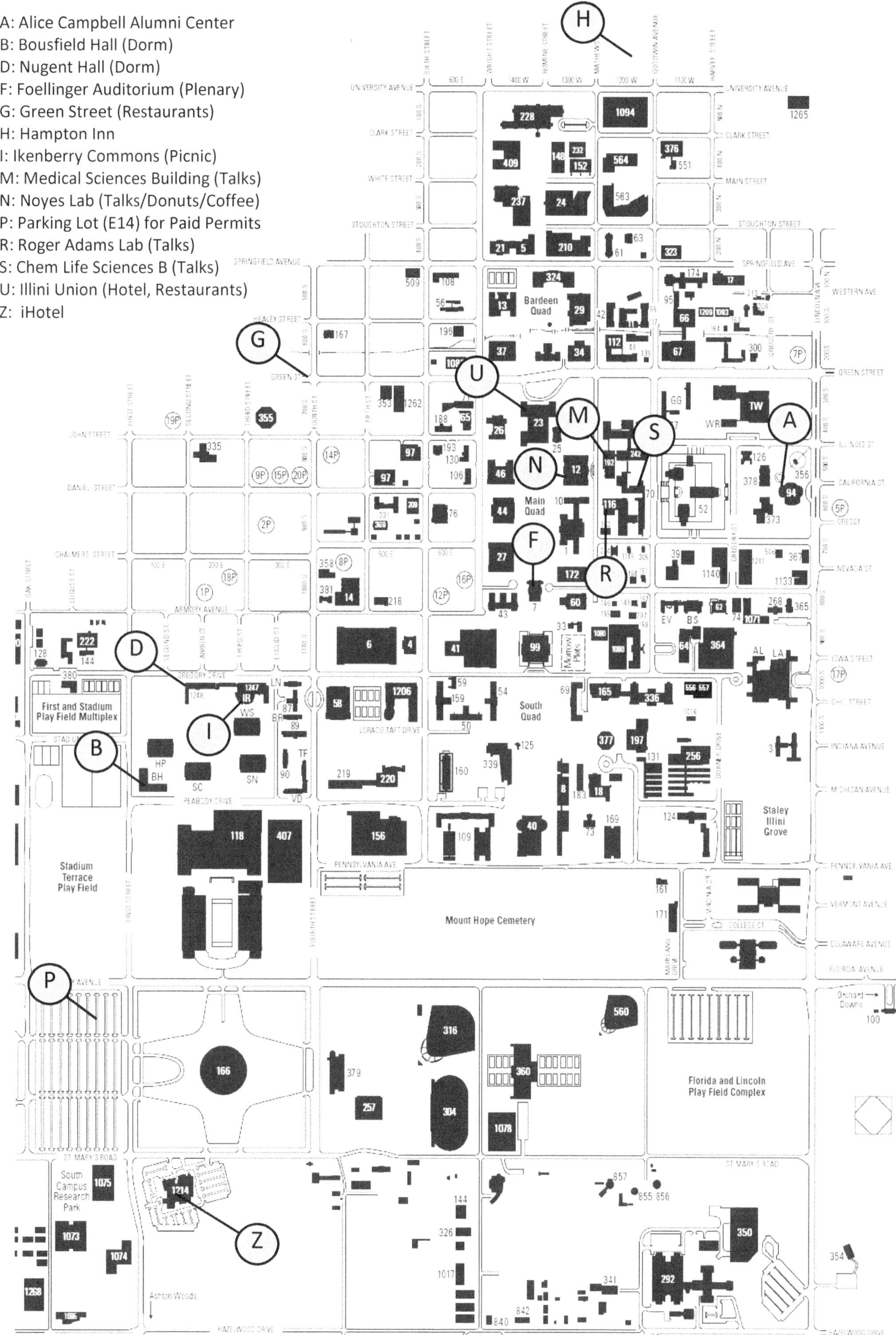

A: Alice Campbell Alumni Center
B: Bousfield Hall (Dorm)
D: Nugent Hall (Dorm)
F: Foellinger Auditorium (Plenary)
G: Green Street (Restaurants)
H: Hampton Inn
I: Ikenberry Commons (Picnic)
M: Medical Sciences Building (Talks)
N: Noyes Lab (Talks/Donuts/Coffee)
P: Parking Lot (E14) for Paid Permits
R: Roger Adams Lab (Talks)
S: Chem Life Sciences B (Talks)
U: Illini Union (Hotel, Restaurants)
Z: iHotel
Bardeen Quad
Main Quad
South Quad
First and Stadium Play Field Multiplex
Stadium Terrace Play Field
Mount Hope Cemetery
Staley Illini Grove
Florida and Lincoln Play Field Complex
South Campus Research Park

NOTES

NOTES

www.ingramcontent.com/pod-product-compliance
Lightning Source LLC
Chambersburg PA
CBHW080801180526
45168CB00006B/2290

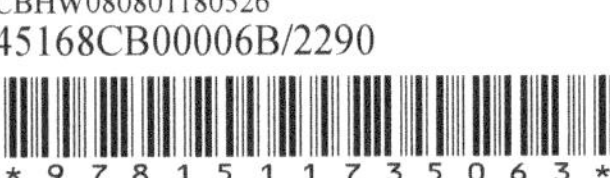